D1233430

Science in Victorian Manchester

Science in Victorian Manchester

Enterprise and Expertise

Robert H. Kargon

The
Johns Hopkins University Press
BALTIMORE AND LONDON

The Johns Hopkins University Press, Baltimore, Maryland 21218
The Johns Hopkins Press Ltd., London

Library of Congress Cataloging in Publication Data

Kargon, Robert Hugh.
  Science in Victorian Manchester.
  Bibliography: p. 267
  Includes index.
  1. Science—England—Manchester—History. I. Title.
Q127.G4K37   509'.427'33   77-4556
ISBN 0-8018-1969-5

*To Dina, Jeremy,
and Marcia*

Contents

Contents

Preface

The city is typical of Gesellschaft in general. It is essentially a commercial town and, in so far as commerce dominates its productive labor, a factory town. Its wealth is capital wealth which, in the form of trade, usury, or industrial capital, is used and multiplies. Capital is the means for the appropriation of products of labor or for the exploitation of workers. The city is also the center of science and culture, which always go hand in hand with commerce and industry. Here the arts must make a living; they are exploited in a capitalistic way. Thoughts spread and change with astonishing rapidity. —Ferdinand Tönnies

The twentieth century, for good and for ill, is a scientific century. Science as professional, specialized activity and as a language with which to interpret the natural and social world has assumed a position of fundamental importance. On the one hand, through science-based industry it has had a critical impact upon the economic base of "advanced" industrial nations. Through scientific medicine and agriculture the character of life throughout the world has been touched by it. In another sense, too, science as method of interpreting the world has permeated all areas of human endeavor; it has become the standard by which all else is measured. This hegemony of science is a legacy of the nineteenth century, and historians are duty-bound to explore the issues surrounding its evolution.

The nineteenth century demonstrated that, rather than remaining gifts of God, scientific practitioners can be *produced,* and science itself can be *invented.* In short, science can result from enterprise. Moreover, rather than remaining merely the contemplation of God's laws, science came to be viewed as a method to be applied to all areas of human endeavor; science became expertise. Great Britain, the pioneer of industrialization in the West, most clearly displays these important changes, and I have chosen its most interesting and exciting industrial city as the locus for my study.

It requires no triumph of historical insight to notice that science in Great Britain at the beginning of the nineteenth century was marked by pockets of brilliant individual achievement and of institutional weakness.

One may without serious distortion characterize the period as the regime of the gentleman-amateur. By the end of the first decade of the twentieth century, science was emerging in a recognizably modern form, with several areas of considerable institutional strength, its practitioners notably "professional" in the sense in which historians and sociologists have chosen to discuss it. This transition, critical for the birth of the modern world, is still imperfectly understood, and the book which follows is intended as a modest contribution toward the goal of reaching a clearer view of the processes which effected the transformation.

I have chosen to examine the scientific community of Victorian Manchester to exemplify these processes, for in 1840, Manchester already possessed an interesting and complex scientific culture which in the succeeding six or seven decades evolved into a world-scientific center of great vitality. As our discipline approaches an awareness of the possibility of a social history of science, the Victorian city will increasingly provide the terrain for fruitful exploration. During the course of the nineteenth century, the city created vital new scientific institutions to meet the changing needs of its society. Without an analysis of the structure of the scientific community as defined through these organizations—without an exploration of the number, type, and quality of these institutions—our understanding of the scientific enterprise will remain sadly incomplete. The term "community," like "professional," is, as Reinhold Bendix has pointed out, "as awkward as it is convenient."[1] For the historian, an awkward, implicit definition often yields flexibility and thereby a kind of strength of which I hope to take advantage. In describing how the Manchester scientific "community" evolved, the development of science-society relations in these crucial years may be illuminated for us. Manchester was remarkable, and singular in the complexity and strength of its institutions. However, its evolution contained elements which clearly reappear in other urban centers, both in Britain and elsewhere.

A word on method. Lytton Strachey held that "the history of the Victorian Age will never be written: we know too much about it." He urged historians to "row out over that great ocean of material, and lower down into it, here and there, a little bucket."[2] We must try to do a little better than that. I have chosen, as an heuristic device, to suggest a *typology* organized around changing definitions of "science," and of its relationship to the larger society, reflected in the types of practitioners and institutions which are important to an understanding of Manchester science at any given time. Consequently, I shall attempt to describe these successive stages of development, fully aware that they overlap both in time and in kind. Claude Levi-Strauss has noted that "Any classification is superior to chaos."[3] From his words I take heart. Typology is particularly useful to the historian, for it admits more readily to change in time than many other forms of classification, as the work of Ferdinand Tönnies and Max Weber demonstrate.

One caveat is in order. The reader should not expect to find in what follows full histories of any of the important scientific institutions of Manchester or intellectual biographies of leading figures. These efforts, while wholly laudable, have not been my aim. We still shall require fuller studies of Roscoe, Schuster, Reynolds, Smith et alii; we shall want histories of the Manchester Literary and Philosophical Society, the Owens College, and of others as well. My intent, rather, is to draw a portrait of the evolution of Manchester science in adaptation to its social context and in response to pressures internal to the scientific disciplines. Whether and how much this portrait contributes to our understanding of Victorian science and society I must of course leave to others to assess.

Acknowledgments

I should like to thank Jeanne Pingree, Imperial College of Science and Technology, for permission to quote from a Liebig letter in the Playfair papers; E.D.G. Robinson, University of Manchester Institute of Science and Technology, for permission to quote from the Joule manuscripts; R. Griffin, British Chemical Society, for permission to quote from a letter of Balfour Stewart; G. Bell and Sons for permission to quote from J. J. Thomson's *Recollections and Reflections;* A. L. Smyth, Manchester Literary and Philosophical Society, for permission to quote from the Joule-Playfair correspondence; Ernest Benn, Ltd., for permission to quote from E. E. Fournier-D'Albe's *Life of Sir William Crookes.*

I am very grateful for the courtesies extended to me by the above institutions and their librarians, the Royal Society, London, Mr. I. Kaye, the University of Manchester Library, the John Rylands Library, the University College (London) Library, and the University Library, Cambridge, England.

During the course of my research for this book, I received travel grants from the National Endowment for the Humanities, and from the American Philosophical Society. Without their generous support I should not have been able to proceed. I am very much in their debt.

Science in Victorian
Manchester

1
The Structure of Science in Manchester in 1840

To some the city of Manchester in the early twentieth century seemed an unlikely place for the pursuit of science. Upon his arrival, Henry Moseley—one of physics' brightest lights—was resigned to "being in the desert." His Oxford tutor called it "a ghastly place and I feel sorry for anyone who is condemned to profess and teach in its murky atmosphere."[1] Smoky, dirty, aggressively *bürgerlich* and obviously lacking the genteel virtues of Oxford or Cambridge, Manchester was nonetheless a splendid, shining world-center for scientific research and teaching. Its university and its urban scientific institutions, enjoying unmatched community support, provided a magnet for the best and brightest young investigators. In Ernest Rutherford's physics laboratory Manchester possessed the nursery of modern physical science, and it could boast of many of the glittering names of contemporary physics—Bohr, Chadwick, Moseley, Geiger, Darwin, and numerous others. It can be argued with considerable justice that the city had, in the first two decades of the twentieth century, one of the most exciting and productive communities in the history of recent science. Yet merely seventy years earlier the city's scientific community was considered a provincial backwater with little claim to the world's attention. The evolution of Manchester from its modest state at the beginning of the Victorian period into one of scientific eminence, and the conditions of this passage, is the subject of the undertaking which follows.

The Revolutionary City and the Scientific Community

Lying at the foot of hills which stretch northward from it, the city of Manchester rises on the left bank of the river Irwell, between that river and two smallish ones, the Irk and Medlock. On the right bank of the

Irwell lies its sister city Salford. In the mid-eighteenth century, the countryside to the east of Manchester was thinly populated, marshy land. By 1840, the beginning of our story, this land was the most densely populated in England. The Manchester region had been, during this period, the scene of an industrial revolution which transformed the face of the county and the character of its people.

The simple population statistics are dramatic enough. In 1773 the population was estimated at about 24,000. By 1801, the first official census determined its numbers at well over 70,000. By 1841, Manchester and Salford had grown to the stupendous size of over 300,000. Economic opportunity accounted in the main for this demographic marvel. It has been estimated that in the first decade of the nineteenth century immigrants to the city accounted for a third of its increases and by the third decade of the century for over three-fifths.[2] What in the 1780s had been a market town, already attracting attention to its bustle and activity, was by 1840 an extraordinary agglomeration which excited and terrified contemporary observers, who viewed Manchester as both monstrous and awe-inspiring. "Amid the fogs which exhale from this marshy district," a French critic-admirer wrote, "and the clouds of smoke vomited forth from the numberless chimneys, Labour presents a mysterious activity, somewhat akin to the subterraneous action of a volcano."[3]

The immediate cause of this growth was the industrial revolution, and the industrial revolution in turn was shaped by *cotton*. David Landes has succinctly summed up this complex upheaval in three simple principles: the substitution of machines for human effort and skill; the substitution of inanimate (especially steam) for animate sources of power; and use of new and plentiful raw materials. Of this last, cotton was by far the most important. In 1760 Britain imported 2.5 million pounds for its domestic industry; by 1787 consumption had increased to 22 million pounds, and by the 1830s to 366 million pounds.[4] The locus of these changes was Lancashire, and Manchester was its most significant center.

Whereas in the last half of the eighteenth century Manchester merchants were still largely independent producers peddling rough fustians on horseback, and the town's Tuesday Market still bore traces of its medieval roots, by the mid-nineteenth century, Manchester was Britain's second city, a place where great fortunes were made and, simultaneously, dismal industrial conditions wrecked numerous lives. In the 1780s the elite of Manchester society was still composed of gentlemen, physicians, lawyers, clergymen, and small "putting-out" manufacturers. The elite groups possessed all the aspirations of their peers in other regions. The institutions of the town—the clubs, churches, libraries, theaters, and so forth—supplied them with the cultural satisfactions appropriate to their stations. By the 1840s, however, the social situation had altered, and the market town's institutions were either outgrown or outmoded. The city was by then rife with class division, dominated by a powerful *bourgeoisie,* and marked by a powerless proletariat.[5] Manchester was both pioneer and victim.

The wealth and prestige of the eighteenth-century town lay with its clergy, professional men, and merchant-manufacturers. From the 1770s on, however, the economic scene changed rapidly. Inventions, especially in spinning, stimulated a cotton boom. While some of the wholesale merchants of the town invested in the new equipment, many of the new factory industrialists emerged not from the putting-out industries nor from merchandizing, but rather from the manufacture of textile machinery, as, for example, did John Kennedy, Robert Owen, and James McConnel. With the advent of the power loom, weaving sheds were added to the spinning mills. By 1827 half the country's power looms were in Manchester and nearby Stockport.[6] The mechanization of the cotton industry stimulated a proliferation of engineering and machinery-making firms. Manchester and Salford directories of the latter part of the eighteenth century trace the phenomenal growth of engineering activities, coordinated with the steep rise in cotton manufacture.[7] In 1844 a foreign writer observed what to him was a remarkable scene:

The manufactories and machine shops form as it were, a girdle around the town. . . . Factories seven stories in height, rear their lofty fronts along the banks of the Irwell and along the borders of the canals. . . . The waters of the Irk, black and fetid as they are, supply numerous tanneries and dye-works; those of the Medlock supply calico-printing establishments, machine shops and foundries.[8]

Within the relatively short span of the Biblical three score and ten, Manchester was entirely transformed. The suddenness of the change was universally noted: "At Manchester," Léon Faucher wrote in 1844, "industry has found no previous occupant, and knows nothing but itself. Everything is alike and everything is new; there is nothing but masters and operatives."[9]

Linked with this transformation in the minds of many was Manchester's interest in science and technology. Faucher continued: "Science, which is so often developed by the progress of industry has fixed itself in Lancashire." Faucher's Mancunian translator added: "With respect to science, the whole phenomenon of Manchester society is but a continual series of investigations into, and practical application of, scientific knowledge."[10]

Owing to technical success, Manchester was in fact widely regarded not only as the center of Britain's most vital industry but the center of invention and therein of progress. The *People's Journal* summarizes a view which many Mancunians would have heartily endorsed: "Invention, physical progress, discovery are the war-cries of today. Of this great movement Manchester is the centre. In that lies its especial importance. That work which it seems the destiny of the nineteenth century to accomplish is there being done." Manchester was a mark of the tendencies of its time and, as the *People's Journal* continued, "a clue, more or less perfect, to the social condition of the future."[11]

An unbiased observer would have to conclude, however, that in 1840 the scientific community of the city of Manchester was relatively undeveloped—that the city was, compared with the great metropolitan centers and even with many university towns, a scientific backwater. To be sure, the city boasted the presence of the internationally renowned Dr. John Dalton and ceremoniously trotted him (or his reputation) out when the occasion seemed to warrant it. But apart from this slender claim Manchester seemed to possess little to merit the admiration or incur the envy of London or even of Glasgow.

Still, backwater though it might have appeared, Manchester, even in 1840, actually possessed a complex structure of scientific institutions and organizations. Despite its provinciality, Manchester provided the setting for a scientific community which served all segments of society and which performed several functions relating to fruitful diversity among conceptions of "science." The premier scientific organization, both in age and reputation, was the famous Literary and Philosophical Society of Manchester, established in 1781 by prominent members of the city elite to encourage discovery and to facilitate the dissemination of literary and scientific culture among those gentlemen best prepared to make use of it.

The "Lit. & Phil." was joined in 1821 by the Natural History Society, whose major goal, it appears, was the preservation of interesting collections and the convening of like-minded amateurs of the fields and streams of south Lancashire which were beginning to be heavily scarred with the effluences of industrialism. Soon the Royal Manchester Institution, providing both an art gallery and public lectures in the sciences, joined them in their effort to enrich the private lives of the *haute-bourgeoisie* and the public life of the city. The Manchester Mechanics' Institution (established 1825), the New Mechanics' Institution (established 1829), and the Salford Mechanics' Institution (established 1839) brought the wave of interest in science and technology to the working class or, at least, to that advanced segment of it able to profit from science courses, public lectures, and reading rooms.

These institutions, performing a variety of functions and aiming at several levels of society, were joined in the 1830s by two new types of scientific organizations, heralds of things to come both socially and intellectually in the city of Manchester. The first was the Statistical Society (1834), dedicated to collecting and analyzing data concerning the major problems beginning to affect the burgeoning industrial complex of Manchester and its environs. Its aim was decidedly reformist, attempting to give some solid base to its members' wide-ranging criticisms of conditions and to provide some impetus for reform.

Ultimately activist in its aims also was the Manchester Geological Society (1838), which brought together for the presentation of papers and public lectures those concerned with the economic development of the region and those natural history "buffs" in love with the vanishing beauties of the surrounding countryside. The appearance on the scene in the 1830s of these two new "concerned" groups signaled the coming end of an era: an end to the eighteenth-century conception of science as the recreation of leisured gentleman and of the busy "professionalist." The theme of the need for scientific knowledge for the health of the city and the success of its industries was the slender thread which bound together all segments of the Manchester scientific community both in the new institutions of the 1830s and in the older ones as they were continuously transformed in the decades to follow. It was a theme which was not recognized at once and not by all sections of the Manchester elite, but its binding powers grew in the years ahead and helped to produce, in fact as well as in name, a Manchester scientific community.

The assessment of Manchester science in 1840 will have to include a detailed examination of the aims, practices, and composition of the institutions which provided a setting for the practice and dissemination of science; it will have to go beyond the individual societies and attempt to grapple with the more complex relationships that obtained among them. Any serious examination will require some picture of the interlocking directorates, the special functions within the community that each played, the special conceptions of "science" that each formulated, and the share of the available talent that was apportioned among them. In short, nothing less than a portrait of the "ecology" of scientific institutions in the city will suffice. This chapter will be devoted to supplying the elements of this portrait and to the attempt to develop some picture of the dynamic relationships among the various organizations and men who populated them. Succeeding chapters will attempt to assess the internal and external forces tending to disrupt the system and to describing the new situations resulting from such changes.

The Manchester Literary and Philosophical Society and the Manchester Scientific-Cultural Network

The most prestigious scientific institution in the city in 1840 was by all odds the Literary and Philosophical Society, founded in 1781. Most accounts credit Dr. Thomas Percival (1740–1804) with being the *vis motrix* of the society. Educated at the Warrington Academy and Edinburgh and Leyden Universities, Percival settled in Manchester in 1767. He had already become a fellow of the Royal Society of London two years previously and, upon setting up his medical practice in Manchester, continued his philosophical interests. Around him grew an informal weekly intellectual club; as the membership increased, the original format became cumbersome and the original plan was extended.

The preface to volume 1 of the society's *Memoirs* (1785) gives an official account and may perhaps supply some sense of motivation. The literary and philosophical societies formed in the different parts of Europe are not only a means of diffusing knowledge but also provide a means to encourage "a greater number of important discoveries." The societies offer the "respectable sanction of societies of men of the first eminence and learning" and also an "easy mode of publishing" the products of their intellectual interaction. "Science, like fire," the preface continues, "is put in motion by collision."[12]

The origins of the society were impeccably genteel:

Many years since, a few Gentlemen, inhabitants of the town who were inspired with a taste for Literature and Philosophy, formed themselves into a kind of weekly club for the purpose of conversing on subjects of that nature. These meetings were continued, with some interruption for several years; and many respectable persons being desirous of becoming members the numbers were increased so far as to induce the founders of the Society to think of extending their original design. Presidents and other officers were elected, a code of laws and a regular Society constituted and denominated THE LITERARY AND PHILOSOPHICAL SOCIETY OF MANCHESTER.[13]

The bylaws limited the membership, at first, to fifty ordinary members, supplemented by an undetermined honorary membership, eligibility for the latter based on the member's "residing at a distance from Manchester" and by his having "distinguished himself by his literary or philosophical publications; or favoured the Society with some paper."[14]

The bylaws indicated that the subjects for discussion and for the presentation of papers included natural philosophy, theoretical and experimental chemistry, literature, civil law, commerce, and the arts. Excluded, pointedly, were "Religion, the Practical Branches of Physic, and British Politics."[15] The exclusion of religion and politics as subjects for debate and discussion had ample precedent and equally sufficient and obvious practical advantages. That the practical parts of medicine were not to be discussed prevented the society, on the one hand, from becoming a narrowly focused professional society (more than half the founding members were physicians, surgeons, or apothecaries) and, on the other, from opening a Pandora's box of potential disputation.

An historian of the society, Francis Nicholson, has preserved a list from the original minutes of the founding members. Providing an index to the cultural elite of the town, the list includes the Reverend Thomas Barnes (1747-1810), minister of Cross Street Chapel; George Bell, M.D. (1755-1804), honorary physician to the Manchester Infirmary; Thomas Henry (1734-1816), apothecary to the infirmary and, until the arrival of Dalton, the town's leading chemist; Peter Mainwaring, M.D. (1696?-1785), first president of the society; James Massey (1713-1796), first president of the infirmary and one of the early presidents of the society; Thomas Percival, mentioned above; and Charles White (1728-1813), honorary surgeon to the infirmary and one of the four fellows of the Royal Society of London among the original members.

During the first year of operation many new members were added, first to a limit of forty, and afterwards, in October 1781, to fifty. After May 1781 candidates for membership had to be proposed by three members and elected by ballot. By the time of the publication of the first volume of the *Memoirs* in 1785 there were forty-three ordinary members. Of these a subgroup of fourteen comprised the Committee of Papers, whose duties included approbation of papers read to the society to be included in the *Memoirs*. They were authorized to select parts of a paper for publication if the whole be deemed unsuitable. They were charged with the awarding of prizes and the setting of premiums. This committee was to be elected by the membership and generally included its most active portion. It is to be supposed that the membership was in fact not overly zealous. It is reported by Nicholson that admission of new members required the presence of at least thirteen members for the ballot. Very often, however, the quorum was not met.[16]

The committee included all the officers of the society *ex officio* as well as six additional members. The 1785 committee included George Duckworth, an attorney; the Reverend Ralph Harrison, a classicist; Thomas Kershaw, a calico printer or handkerchief manufacturer; and John Leigh Phillips, a cotton manufacturer and collector of natural history specimens. Of fifteen papers contributed to the first volume by ordinary members of the society, twelve — or eighty percent — were authored by those serving on the committee. The other three were given by a single man, Alexander Eason, M.D., and even he had been an officer during 1781-1785, the years in which the papers were written. In the second volume of the *Memoirs,* all of the fourteen papers contributed by members were by members of the committee, the rest of the volume being contributions of the honorary membership. It seems clear that, in the early years of the society at least, the Manchester Literary and Philosophical Society was overwhelmingly the creature of its officers, and in particular of Percival, Barnes, Henry, and

White. These men were joined in the closing years of the eighteenth century by Dr. John Ferriar (member 1786, officer 1787 onward), by Dr. Edward Holme (member 1794, officer 1794 onward), and ultimately by John Dalton (member 1794, officer since 1800, but a member of the committee of papers almost immediately).

It is perhaps no distortion to single out Thomas Henry as representative of the society's elite.[17] He served as an officer of the society virtually without interruption from the date of its inception until his death in 1816. From 1807 until 1816 (except for 1809) he was the group's president. Henry, a fellow of the Royal Society of London, was an apothecary with an important medical practice in the town. He founded Henry and Company, chemical manufacturers, and invented a new way of preparing magnesia alba ("Henry's Magnesia") which formed the mainstay of the firm.

It was Henry who in 1781 delivered before the society an address which delineates clearly the tone and character of the early years of the society, a character which carried over well into the nineteenth century. In his paper entitled "On the Advantages of Literature and Philosophy in General, and especially on the Consistency of Literary and Philosophical with Commercial Pursuits," Henry entered upon an apologia for the existence of the society and a defense of its concerns. Henry's address may be read in retrospect as a birth cry of new class divisions. He is first of all attempting to demonstrate the fitness of the group for polite literature and science, and the appropriateness of the subjects for the Manchester gentleman. "A taste for polite literature and the works of nature and of art," he wrote, "is essentially necessary to form the Gentleman. . . . Affluent circumstances and abundant leisure give the Gentleman great advantages over his inferiors. . . . The cold and heavy hand of poverty chills and represses the efforts of genius; wealth cherishes and, if I may be allowed the metaphor, *manures* and pushes it to maturity."[18] Henry's rather ripe metaphor might have brought smiles to a metropolitan reader, but the message is clear enough. The congruence of talent and concern are important elements in the success of the society. Unfortunately, he continued, he regrets but must admit that some merchants and manufacturers still fail to see the advantages of such studies to the "Gentleman and the professionalist."

The commercial man, say they, should confine his knowledge to trade. His compting-house should be his study; his ledger his hourly amusement. Gold and silver are the only metals with which he ought to be acquainted; and of these to know no more than the different coins into which they are formed, and the current price of bullion. For poetry, painting and music he must have no attachment, no taste for engravings but those of bank bills, and if a single philosophical idea should enter his head, these inveighers against knowledge would expect to see his name immediately in the list of bankrupts.[19]

That some imprudent young man might be carried away and devote too many hours to philosophical amusements and too few to business cannot be denied. But why, Henry continues, blame these studies for the mischief arising "from the improper arrangement of time"?[20]

"Will not the time he can spare from business be more usefully employed in the study of history . . . or at an air pump, an electrical machine or a microscope than . . . at the tavern, the gaming table or the brothel?[21] What is more, the gentleman or professionalist, manufacturer or merchant can realistically expect such intellectual pleasures: "It is one thing to be a *professed* scholar or philosopher and another to possess such a degree of information on a subject as is compatible with our other avocations. To be a *complete* astronomer would almost monopolise the business of a man's life. To procure a general but satisfactory idea of the motions . . . only a moderate degree of application is necessary."[22] It is clear that for Henry the Lit. & Phil. was not primarily for the "professed" scholar or the "complete" scientist, but rather for the amateur of these studies, a man who can benefit from science as an avocation. Moreover, there are indeed commercial benefits and industrial utility which may accrue to members and to the community. Chemistry, for example, "is the cornerstone of the arts."[23] In a very interesting and valuable five page discourse Henry recounts the utility and bearing of the science of chemistry on the arts and manufactures. For example, in dyeing and coloring, an area of concern to many Mancunians, the basic principles are chemical. But, "the misfortune is, that few dyers are chemists and few chemists dyers. Practical knowledge should be united to theory in order to produce the most beneficial discoveries."[24] In brief compass Henry's early paper affords a major insight into motivation behind the society's inception, into the character of the organization, and into the conception of science of the amateurs who controlled it. It might be argued that Henry's apologia was a carefully constructed piece of propaganda designed to lure into the society's operations those members of the Manchester industrial elite who were in fact most reluctant, and whose support would make a crucial difference in the society's fortunes. There is no clear evidence, however, that Henry was anything less than sincere in his professions. His entire career and the early history of the society under his influence would seem to argue against that interpretation.

What of the amateur character of science at the Lit. & Phil.? Henry's assurances that one need not be a professed or "complete" scientist to participate in and benefit from the scientific endeavor give us a clear hint of the direction of his thought. Science is valuable in several regards: first, it is a study which befits the town elite — the professional men and the industrial-commercial — and, secondly, it is an endeavor which may in fact yield practical advantages, for the principles behind much manufacture are scientific. Surely, clear knowledge of such principles will benefit trade. In short, the subjects of study of the Lit. & Phil. are worthy of the marriage of polite learning and utility, and therefore they are worthy of the time and efforts — within clear limits — of the industrious gentleman.

The *eloge* of Thomas Henry written by his son William, a famous chemist in his own right, gives particular insight into the scientific character of the man who may with much justice be termed the father of Manchester science:

Without claiming for Mr. Henry the praise of great original genius, we may safely assert for him a very considerable share of that inventive talent which is commonly distinguished by the term *ingenuity.* This was especially displayed in the neatness and success, with which he adapted, to the purposes of experiment, the simple implements that chance threw in his way; for it may be proper to observe that, at no period of his life, was he in possession of a well furnished laboratory, or of nice and delicate instruments of analysis or research. . . . And when it is considered that his investigations were carried on not with the advantages of leisure, ease and retirement but amidst constant interruptions. . . it will be granted that he accomplished much more than might have been expected.[25]

Among the most accomplished of the Manchester amateurs, Henry was largely self-taught: "His reading was . . . entirely self-directed; and, by means of such books as chance threw into his way, he acquired a share of knowledge, creditable both to his abilities and his industry."[26] After an apprenticeship and some years of experience, Henry moved to Manchester from Knutsford. In his new town Henry quickly formed friendships with like-minded young men, such as Thomas Percival, who, along with Henry, provided much of the zeal which was required for the success of the Lit. & Phil.

This zeal was, moreover, not the ardor of the specialist; Henry's interests spanned the range of the society's: "It was the habit of his mind [his son William continues] when wearied by one occupation, to seek relief not in indolent repose but in a change of objects. . . . He had a share of general information and a flow of animal spirits that rendered him an instructive and agreeable companion."[27]

The second volume of the *Memoirs* contains a proposal by the Rev. Barnes which adds substance and detail to the sentiments of Mr. Henry. Barnes's paper, "A Plan for the Improvement and Extension of Liberal Education in Manchester" (read April 9, 1783), recommends the establishment of an institution which would connect together "LIBERAL SCIENCE AND COMMERCIAL INDUSTRY!" The aim, of course, would be the "application of school learning to superior objects . . . Natural philosophy, the Belles Lettres, and Mathematics; together with some attention to History, Law, Commerce and Ethics." Of these, the lectures in natural philosophy would pay special attention to chemistry and mechanics "because of their intimate connection with our manufactures." The course of study would be particularly geared to those headed for the "learned professions" and to "those Gentlemen designed for the COMMERCIAL line," for it would "give that general insight into science which might answer the noblest purposes of mental cultivation."[28]

In fact the College of Arts and Sciences had already come into being by the time of the publication of the *Memoirs,* under the patronage of the Earl of Derby, the presidency of Thomas Percival, and a board of governors largely drawn from the leading figures of the Lit. & Phil. Lectures were delivered on chemistry by Thomas Henry, on fine arts by George Bew, on the progress of the arts and manufactures by Barnes, and on moral philosophy by Barnes as well. The college was forced to close within two years for lack of support, but Barnes proceeded in 1786 to found a second institution, the Manchester Academy (later New College), a successor to the Warrington Academy. It was to New College that John Dalton was named to teach mathematics, natural philosophy, and chemistry.

The coming of John Dalton to Manchester and his election to the Literary and Philosophical Society mark a new era in the history of the society (this comment has often been made) and signal the emergence of a new scientific type in the city. In his essay "Science and Technology in the Industrial Revolution," Arnold Thackray has called attention once again to Dalton as the exemplar of a new class of men in science — lower middle class, provincial, Dissenting — who in their devotion to science contrasted sharply with the more dilettante amateur tradition in the existing scientific institutions.[29] Indeed, Dalton provides an interesting comparison with his friend, Thomas Henry, who, after all, represents the best of eighteenth-century Manchester amateurism. Musson and Robinson have written extensively on the problems of science and industry in the late eighteenth century and have documented the tremendous interest in science on the part of some members of the industrious classes and among professional men in Manchester,[30] particularly those who found their way into the Literary and Philosophical Society. The resources available for self-education in the Lancashire area for these men are impressive: public lectures, libraries, reading rooms, Dissenting academies, local scientific societies, lyceums, and so forth. Yet these resources served mainly to enrich an essentially "amateur" interest. Philosophy was necessarily secondary to trade, and though the faith was there that the two could and would reinforce each other, as the career of Thomas Henry shows, still very few could with justice be termed devotees of any particular discipline or branch of scientific endeavor. This was not, of course, true in the case of Dalton; hence the term "new man" is applicable.

Wherein lay Dalton's "newness"? His early training, under the patronage and tutelage first of Elihu Robinson and later John Gough and his access to the Quaker network of men concerned with science and industry do not alter the basic contention that Dalton, especially in chemistry and natural philosophy, was essentially self-taught.[31] His arrival in Manchester was occasioned by his accepting a post at Barnes's New College, and quite naturally Dalton at once was swept up into the scientific and cultural life of the city. Dalton's public lectures in Kendal, and the publication of his *Meteorological Observations and Essays* shortly after his arrival promoted Dalton almost at once to a position of prominence. He was elected to the society in 1794, attended his first meeting as a member in October of that year, and immediately was selected a member of the powerful Committee of Papers, the group entrusted with the quality of the *Memoirs*. Dalton's position as tutor, his energy in publishing, and above all his devotion to his subjects of inquiry place him apart from most of the other members of the society. It is more than hindsight which leads one to assert that it was inevitable that Dalton would soon rise within the society to a position of leadership.

In 1800 Dalton resigned from New College to begin a career of private teaching and research. He gave hourly lessons in mathematics, calligraphy, grammar, physics, and chemistry for a fee of (at first) one shilling (and later two).[32] His earnings were supplemented by fees from public lecturing and by making commercial chemical analyses. Aided by a decision of the Literary and Philosophical Society to offer him a room for teaching and experimenting, Dalton made out quite well indeed. He was able to equip his rent-free laboratory relatively well from his earnings. This was the base from which he launched the career which was to win him international acclaim and to afford Manchester virtually its only major cultural ornament for some years.

In May of 1800 Dalton was elected secretary of the society, a task which entailed a fair amount of correspondence and society business. He held the post until 1809 when he was elected vice-president, a position to which he was re-elected each year until his succession to the presidency in 1817. From April 1817 until his death in 1844 Dalton presided over, and to a large extent remade the image of, the society. The claim — often made — that owing to Dalton's preferences the society increasingly emphasized science at the expense of other studies is difficult to assess. To be sure, under Dalton about sixty-five to seventy-five percent of the papers published in the Lit. & Phil. *Memoirs* were scientific, but this reflects a trend apparent already in the 1790s, and one sees no sharp breaks with tradition either with the coming of Dalton into the society or with his accession to the presidency. The predominance of scientific subjects may reflect the composition of the Committee of Papers, for the officers of the society increasingly were drawn from its medico-scientific members. It may, after all, be an inaccurate reflection of the substance of the meetings. Not all papers presented before the society were published; the Committee of Papers, it will be recalled, carefully screened them. Finally, the upsurge in scientific papers may be merely the measure of Dalton's relative vitality vis à vis the rest of the membership — a reflection of his devotion against their dilettantism. It was always difficult to get the membership to contribute papers for discussion; during Dalton's years in the society, he himself contributed an extraordinarily large percentage of papers to the published *Memoirs*.

Whether the quality of papers increased under Dalton (aside from his own contributions) is likewise difficult to state with any certainty. Certainly the pace of publication of the *Memoirs* was nothing to boast of. In the years 1800–1844, when Dalton was an officer of the society until his death, only six volumes of *Memoirs* appeared; at times Dalton contributed as much as forty percent of the papers. But one theme is perfectly clear: after the passing of great eighteenth-century Manchester science leaders (Thomas Percival, Thomas Henry, etc.), the name Dalton was virtually synonymous with science in Manchester. There were, of course, others — men like Thomas Henry's son William, who wrote an important chemistry text — but it was Dalton who was the doyen and overseer of science in Manchester, a position which no doubt brought him much pleasure — but little else for many years.

Yet in the 1820s the honors were to come thick and (relatively) fast. In 1822 he became at long last a fellow of the Royal Society, and in 1826 it was announced that he was to receive its Royal Medal. Perhaps more significantly, the French Academy of Sciences elected him a foreign associate in 1830, one of eight such positions in that organization. In 1833 he received a civil pension, and in 1832 a doctorate in civil laws from Oxford. A subscription was begun in his own city for a statue to be erected in his honor "to wipe away," as a local newspaper put it, "their unaccountable neglect of an individual who has raised the intellectual character of Manchester higher than that of Birmingham."[33] Outside the confines of the Lit. & Phil., in the years of his greatest intellectual vigor and scientific success, the city had taken little pride in Dalton. When given the hint by the metropolises of the world, however, Manchester polished its greatest intellectual jewel.

In 1840 then, at the beginning of our story, Dalton was a world-renowned philosopher and was now publicly acclaimed in Manchester as well. He presided, however, as a pitiful image of his former self. In April 1837 he suffered the first of a number of damaging strokes. By the end of the following year he was able to read a paper before the society "On Arseniates and Phosphates," which was subsequently rejected by the Royal

Society for publication. His last years in the presidency were marked by a rapid decline in powers both mental and physical, a decline documented by many sad anecdotes which it would be pointless to reproduce here.[34]

The Lit. & Phil. was, then, limping along with a stricken leader in 1840; the vice-presidents elected in that year were not able to give the society the vigorous scientific leadership it required. Among them were Dalton's old friend Dr. Edward Holme (to whom we shall return shortly) and John Moore, described by Angus Smith as "not a man of great force of character, but . . . amiable and intelligent," who took "pleasure in going among scientific men."[35] The secretary was Peter Clare, Dalton's friend and now constant companion; the Council (the new name for the Committee of Papers since 1822) of the society included only one man with any scientific reputation, John Davies, James Joule's tutor in chemistry. These were not, it appears, exciting years for the Lit. & Phil. No volume of *Memoirs* had appeared since 1831. The total income of the society for the year 1840 was only about £208, a small figure compared with the Bath Society's £444 or the Lit. & Phil. of Bristol's £650 or even the Royal Manchester Institution's £800.[36]

The Manchester Natural History Society

The second-ranking scientific institution of Manchester, in longevity, in 1840 was the Manchester Society for the Promotion of Natural History, formed in 1821 for the specific purpose of saving for the use of gentlemen of the city the excellent entomological and ornithological collection left by the late J. Leigh Phillips, a prominent member of the Lit. & Phil., and at the time owned by T. H. Robinson. On June 30, 1821, a distinguished group met in a room in St. Ann's Place for the purpose of forming a society of natural history. Those attending included prominent members of the Literary and Philosophical Society: John Moore, James Ainsworth, Dr. Edward Holme, Dr. Hardie, and T. H. Robinson himself. It was resolved that a society be formed under the name of the Manchester Society of Natural History, that each gentleman subscribe ten pounds for the purchase of the collection, that for the continuation and maintenance of the society an annual subscription of a guinea be assessed, and that when the number of subscribers reached thirty all future admissions be by ballot only. At once thirty names, including the most distinguished in the city, were enlisted and a general meeting called. The society took up quarters in St. Ann's Place and chose their officers. Gifts and donations began to be received, larger promises secured, and a library opened. In the following years, despite periodic economic fluctuations severely affecting the area, the society prospered. In 1835 a brand new building was opened and a young man, Mr. W. C. Williamson (later to be the first professor of zoology, botany, and geology at the Owens College), was hired as the society's first curator and general manager of the museum.[37]

The main business of the society was the maintenance of the collection as a private museum for the subscribers or governors. Only members, their relatives, and their guests were permitted to utilize and enjoy the collections. In 1837 Richard Cobden accepted the presidency of the society; by 1838 there were 433 "hereditary governors," nine honorary members, six corresponding members, and seventy-seven annual subscribers. Once in the new building, the Natural History Society began efforts to widen its appeal and expand its role as a cultural asset of the city. Up until 1838 the society was solely a private club; in that year the museum was opened to visitors at a shilling for each admission, and the council of the society was authorized to open the exhibitions to other non-subscribers such as school children and working men. The receipts in 1839 show the admission of 1286 "strangers" (middle-class non-subscribers), 97 residents of Manchester, 177 "mechanics," 25 school children admitted at six pence, and 14 "Sunday scholars" admitted at 3 pence. Within two years the admissions revenue doubled. However, the elite character of the society having been compromised, it began to have difficulty for the first time in replacing deceased governors; the total receipts of the society in the late 1830s began to fall off. The downward trend continued throughout the 1840s, so that the annual report of 1846 reported only 298 hereditary governors and 47 annual subscribers. As the public role of the society and the quality of its collection increased, the number of subscribers—the mainstay of the society—declined.

The Natural History Society is inextricably bound up with the life of its president, Dr. Edward Holme. Holme was the last of the important eighteenth-century virtuosi who dominated the cultural scene of Manchester for so long. Holme was a native of Kendal who studied at the Manchester Academy and afterwards at Edinburgh, Göttingen, and Leyden, where he received his M.D. in 1793. He settled in Manchester and took up medical practice there, after serving as amanuensis and companion to Thomas Percival. Almost immediately he was swept up in the intellectual-cultural currents around the Infirmary and Cross Street Chapel group led by Percival, Thomas Henry, Charles White, and the Reverend Barnes. Holme, the "walking dictionary" as Percival called him, formed close friendships with the elite of the Lit. & Phil., of which he became a member and officer in 1794. His intimate friends included Ferriar, William Henry, P. M. Roget, who then practiced medicine in Manchester, John Moore, John Leigh Phillips, who first interested Holme in natural history, and, not least, John Dalton.

Holme's interests spanned those of his intimates: botany, antiquities, *belles lettres,* chemistry, entomology, etc. His obituary in the *Manchester Guardian* vouchsafed him the success his group sought: "He contributed

... to advance the taste of the more influential ranks in the town and ... to awaken [a] love of literary and scientific pursuits." He became the first president of the Natural History Society and served in that position until his death in 1847. With the passing of Dalton he assumed the scientific-cultural leadership of the city, serving as president also for the Literary and Philosophical Society, the Chetham Society, and the Portico Library. Upon his death, his main institutional bequest went, however, to the University College, London, which also received his extensive library. He had, the *Guardian* reports, originally willed it to the Natural History Society, but evidently the democratization of the organization in the late 1830s and 1840s was not to his taste.[38] Holme was the last mainstay of the old guard. With him passed an era.

The Royal Manchester Institution:
Nihil Pulchrum Nisi Utile

One of the more vigorous contributions to Manchester's public scientific life was made by the Royal Manchester Institution, which sponsored lecture series and *conversazione,* thereby raising the level of scientific discourse in the city considerably. It was at the Royal Institution, too, that the first honorary professorships were located; the most vigorous among them was the chemistry professorship held first by Dr. Lyon Playfair. The Royal Institution's scientific role has been virtually forgotten, owing largely to the fact that the institution began as and later actually became primarily an organization designed to promote the fine arts, in short, ultimately an art gallery.

The history of the Royal Institution is generally given as follows. In the summer of 1823 three young men from Manchester, William Brigham, Frank Stone, and David Parry, visited an art exhibition at Leeds. Upon their return, they convened a meeting of artists designed to promote an annual exhibition of works of art. Mr. Dodd, an auctioneer, offered premises in King Street in return for one quarter of the admissions proceeds. In September a group known as the Associated Artists of Manchester convened and the "gentry of Manchester" were invited to discuss the project. By October the plan had escalated and a leading role was assumed by an influential merchant and public figure, Mr. George W. Wood, who already figured largely in the cultural affairs of the city as member and officer of the Literary and Philosophical Society and of the Natural History Society.[39] On the first of October, 1823, a meeting was called at the Exchange, in the name of the "Institution for the Encouragement of the Fine Arts," to formalize events. By this time it is evident that Wood was the prime mover behind the organization; it was he who gave the principal

address at the October 1st meeting. In it he declared that Manchester
would demonstrate to the world that England was not a mere nation of
shopkeepers. Wood's plans far outstripped the original intentions of the
Manchester artists. Towards the end of the proceedings, for example,
Wood suggested

the propriety of adding to the institution an establishment similar to the School of
Arts in Edinburgh for the purpose of affording to working mechanics an opportunity of acquiring the rudiments of chemical and mechanical knowledge. Two
courses of lectures on chemistry and mechanics were given in the year and the
persons who attended them paid an annual subscription of 15 shillings. In addition
to the lectures there was a library of books on scientific subjects which was open to
the subscribers. It was gratifying to learn that no less than 400 persons availed
themselves of the advantages thus held out to them. If such an institution succeeded in Edinburgh no one could doubt of its success here where an acquaintance
with chemistry and mechanics was of so much importance to the labouring
classes.[40]

Wood's aims therefore were threefold: first, to establish a gallery for the
exhibition of works of fine art; secondly, to provide Manchester's first
facility for the regular presentation of public lectures in the arts and
sciences; and, thirdly, to provide technical instruction for the laboring
classes.

Within three weeks a list of contributors to what was by now known as
the "Manchester Institution for the Promotion of Literature, Science, and
the Arts" appeared in the *Guardian,* and the names included such famous
ones as Moseley and Birley, as well as lesser lights on the Manchester
scene. In November trustees were nominated; the chairman was G. W.
Wood. Later that month the *Guardian* bolstered Wood's expanded plans
with an editorial:

With some persons, possibly, it may be a stronger motive for attaching themselves
to the proposed institution that the money expended in its formation is ultimately
certain to be refunded in direct and substantial advantages. The fields of mechanics and scientific ingenuity are still far from being explored. . . . To no place
have greater advantages heretofore accrued from mechanical skill or chemical
investigation than to the town in which we live; by none can greater be expected
in future than those to which Manchester has a right to look forward.[41]

Almost at once the new institution took hold. A relatively large number of
gentlemen of the town followed the lead of the lord of the manor, Sir
Oswald Mosley and became hereditary governors at £42, a lesser number
became life governors at £26, and many became annual governors at two
guineas per annum. The governors and their immediate families were
entitled to admission to the exhibitions and permanent collection and

attendance at the lectures without charge. As in the Natural History Society, the facilities were to be closed to non-subscribers. Wood's plan for a type of mechanics' institute withered away, probably owing largely to the same aversion to the lower orders which many subscribers to the Natural History Society discovered in the late 1830s and 1840s. Wood was not deterred; he transferred his efforts in this area almost at once to the formation of a Mechanics' Institution (which will be discussed in the next section of this chapter). By April 1824 the receipts had grown to a stupendous £23,000. After considering several sites for their proposed building, they chose a plot of land adjoining Dr. Henry's house in Mosley Street, a house which they also purchased. The architect Charles Barry of London was selected to design the building, which was to include exhibition rooms, a lecture theater, natural history rooms, a council room, and a ladies' ante-room.[42]

It is interesting to note that while these proceedings were taking place, a public debate about the scientific requirements of the city was occurring in the pages of the *Manchester Iris: a Literary and Scientific Miscellany*. In a letter dated 16 September 1823, a correspondent who signed himself "Z." wrote a long and interesting letter which deserves examination. The views of Z. coincide with those of G. W. Wood; indeed, it is possible that Z. was in fact Wood himself:

It has often been a matter of surprise and regret to me, that in a town containing so many scientific men as Manchester, there are yet no means of diffusing a general knowledge of Philosophy and the Sciences throughout the various classes of the community, and which might so well be effected through the medium of public lectures. . . . Were there lectures delivered regularly on Chemistry, Hydraulics, Optics, Pneumatics and the various branches of Natural Philosophy, I am sure they would be well attended. . . .

Turning to the "Institution for the Promotion and Encouragement of the Fine Arts," Z. continues:

I trust that the projectors will take a more extensive view of what such an institution is capable of effecting than that which they now seem to contemplate. It may be made to embrace other branches of the Arts and Sciences of general interest. . . . From the enlightened spirit and cultivated taste which are gradually gaining ground amongst my townsmen, I do not despair of very shortly seeing this evil removed, namely, THE WANT OF REGULAR LECTURES IN MANCHESTER ON THE DIFFERENT BRANCHES OF THE ARTS AND SCIENCES.[43]

In the very next issue of the *Iris*, a correspondent signing himself "A Calico Printer" was worried that Z.'s public lectures would be limited to the "higher and middle classes" and suggested that the lectures be "almost gratuitous."[44] The printer further complained that the Lit. & Phil. comprised only a small proportion of "our scientific and literary townsmen" and that "it will not be disputed that much of its time is occupied with futile hypotheses and unprofitable disquisition." In the next issue of the *Iris*, "A Buyer of Calicoes" sarcastically retorted that "it would be quite delightful to see the Manchester Calico Printers, who have hitherto been looked upon as a rather tight-fisted generation, coming forwards and very liberally offering their calicoes 'almost gratuitously.' "[45] In the same issue, a Mr. S. X. supported Z. on the great utility of public lectures, which, he wrote, "supply a sort of Royal Road to Knowledge in which difficulty is smoothed and substantial progress greatly facilitated."

On October 11 the *Iris* itself carried a lead article lauding the "Institution for the Fine Arts," and Z. took up Calico Printer's praise of the Andersonian Institution of Glasgow and was pleased to report to the readers of the *Iris* that the institution so widely talked about and the subject of Z.'s previous letter had in fact assumed the wider scope suggested for it: "A prevailing spirit for improvement seems to pervade every class of the community, and the recently formed Public Institutions and alterations which have been made in the town will ever redound to the credit of the inhabitants. May we not indulge the hope that Manchester will one day rank as high in the literary and scientific world as it now does in a commercial point of view?"[46]

Wood's high hopes (and therefore Z.'s) were only partially realized. The Royal Institution did in fact expand its operations to include promotion of the sciences and literature as well as the fine arts, and lecture series were in fact initiated. However, the benefits of them were not extended to the working class; that is, the Royal Manchester Institution did not become another Andersonian or Edinburgh School of Arts. It was the Manchester Mechanics' Institution which was to attempt to assume this role.

George W. Wood appears to have been a key figure in many of the scientific institutions of the city, as well as in public life in general. He was born in 1781 in Leeds, the son of the Rev. William Wood, the man who succeeded Priestley as minister of Mill Hill Chapel and who himself had scientific interests which he seems to have imparted to his son. George moved to Manchester about 1801 and in the course of time became a leading merchant in the city. His monument in the Upper Brook Street Chapel in Manchester records that "having early in life engaged in commercial pursuits . . . he quitted the pursuits of wealth for the nobler objects of public usefulness." He was elected the first M. P. for Manchester in 1832 and later was returned as Whig M. P. for Kendal from 1837 until his death in 1843. His scientific interests were wide: he was a fellow of the Linnaean and Geological Societies, a member and vice-president of the Manchester Literary and Philosophical Society (a member of its council since 1810), and a moving figure in the founding of the Royal Institution and the Mechanics' Institution.[47]

The Manchester Mechanics' Institution

In his agitation to expand the aims of the Royal Institution, Wood was only partially successful. Although the scope of the Royal Manchester Institution was extended to include public lectures and exhibitions in science as well as literature and art, Wood's aim of extending these benefits to the working man was abortive. As would be the Natural History Society's experience later, the upper middle class in Manchester was unwilling to support financially and by personal attendance any organization or institution which would provide for the mixing of classes. The laboring classes were effectively excluded from the benefits of the Royal Manchester Institution. It was for the most part the paternal desire on the part of enlightened businessmen in Manchester to raise the caliber of the working man that led to the founding of the Manchester Mechanics' Institution.

Before the Royal Institution agitation had died down, a meeting was convened on 7 April 1824 at the Bridgewater Arms to discuss the formation of a new institution designed to spread knowledge about the application of science to manufacture. The leading figures in bringing about the meeting were William Fairbairn, already a famous engineer and large employer of labor, Thomas Hopkins, who as alderman was to play a major municipal role in the mid-nineteenth century, and Richard Roberts, the inventor of the self-acting mule. They were joined by four influential businessmen: George William Wood and his partner George Philips; Joseph Brotherton, like Wood a future member of Parliament and well-known businessman; and Benjamin Heywood, the banker.[48] Most of the group were Dissenters by religion (at least three — Wood, Fairbairn, and Heywood — were Unitarians) and Whig or radical politically. Virtually all were prominent members of the Literary and Philosophical Society. Wood, it may be recalled, had urged the extension of worker education on two counts: economic benefit and moral uplift. At the meeting, with Benjamin Heywood in the chair, Wood's resolution, seconded by Roberts, made these aims perfectly explicit; he resolved that "the Instruction of Artisans in those branches of science and art which are of practical application . . . is of the utmost importance — enabling them more thoroughly to understand their business, giving them a greater degree of skill in the practice of it and leading them to improvement with a greater security of success."[49] The resolution passed unanimously. The assembly also laid down the structure of the proposed new institution. The property and control were to be vested in the honorary members, that is, those gentlemen who invested ten guineas for a life membership or one guinea annually. The honorary members were to elect twenty-one to a board of directors. Artisans and mechanics were to be invited to subscribe to the facilities offered at a pound per year paid quarterly. A committee was appointed to carry out the resolutions which included Wood, Fairbairn, Brotherton, Heywood, Roberts, John Davies, as well as the famous entrepreneurs Greg, Kennedy, and Murray.[50]

Within a month the committee reported back with a prospectus. The aims of the founders were concisely put and are worth quoting at length:

> The Manchester Mechanics' Institution is formed for the purpose of enabling Mechanics and Artisans, of whatever trade they may be, to become acquainted with such branches of science as are of practical application in the exercise of that trade. . . . It is not intended to teach the trade of the Machine-maker, the Dyer, the Carpenter, the Mason or any other particular business, but there is no Art which does not depend, more or less, on scientific principles and to teach what those are, and to point out their practical application will form the chief objects of this institution.[51]

For the mechanic or artisan, such instruction would provide a secure means of advancement and a way to personal, moral improvement: "The Value to the Mechanic of the acquirements which it is thus intended he should be enabled to make he will find in the most likely means of advancing his success and prosperity—in the agreeable and useful employment of his leisure—and in that increased respectability of character which knowledge has always a tendency to confer."[52] The means would be lectures, a well-stocked library, models and instruments for the use and instruction of the subscribers. The prospectus makes clear, however, that the venture is *for* and not *by* the mechanics and artisans. Control is securely in the hands of the honorary members; the laboring classes will have to rest content with the knowledge that all is in their interest: "If the Mechanics come forward earnestly to partake of these benefits now placed within their reach, they may feel confidently assured that the institution will be liberally upheld and conducted with an uniform regard to their interests and advantage."[53]

The chairman (president) Benjamin Heywood further stressed all these themes in the first address to the members. In his opening address before the Mechanics' Institution in 1825 Heywood reiterated that "there is no art which does not depend more or less on scientific principles."[54] He pointed to the work of Hargreaves, Murdoch, and, above all, Watt: "There cannot indeed be a more beautiful and striking exemplification of the union of science and art than is exhibited in the steam engine."[55] In chemistry, Heywood continued, advances in bleaching, dyeing, and calico printing all depend on knowledge of the chemical principles which underlay the arts. Therefore, the goal of the Mechanics' Institution is to enable the mechanic to gain a greater knowledge of his business, to qualify him to make valuable improvements in it, and to demonstrate the way to a profitable use of leisure time. It is essential that Great Britain provide facilities to actualize these aims, for foreign competition, particularly from the vigorous United States, may threaten Britain's industrial superiority.[56]

Two years later, Heywood was concerned with refuting critics of the Institution, particularly those who feared the social consequences of an educated working class. It was charged, for example, that Mechanics' Institutions were dangerous, that they made workers dissatisfied with their station in life. Heywood retorted that he was "at a loss to see how we are disturbing the proper station of the working classes. . . . We do not alter their relative position."[57]

The question of control, however, loomed larger and larger. John Davies, vice-chairman, expressed the views of the management: "If they [the subscribers] came for instruction, they were, of course, incompetent to manage."[58] The artisans and mechanics, however, saw the problem from a different perspective. In 1829 a breakaway group formed themselves as a "Society for the Promotion of Useful Instruction," better known as the New Mechanics' Institution, under the presidency of Rowland Detrosier (ca. 1800–1834). Detrosier was a former fustian cutter who by dint of laborious self-education elevated himself to a position of working class leadership. By the end of his short life he had become a well-known lecturer on science and morals before working class audiences.[59]

On 5 January 1829 Detrosier aided in the founding of the Banksian Society in Manchester, a group of workmen-amateur botanists. The first president of the group was Edward Hobson (died 1830), a servant in a manfacturer's house, whose leisure hours were devoted to botany, mineralogy, and entomology. The purpose of the society was "the acquisition of knowledge in the sciences of botany, entomology, mineralogy, geology, etc." to be effected by conversations, papers, occasional lectures, a modest library, and a collection of specimens. The first meeting was presided over by Detrosier, who presented the first address. He told his audience that "the progress of learning is no longer dreaded save by those who profit by ignorance of the multitude."[60] Knowledge for the worker was the key to his redemption. The society was short-lived; after the death of its first president, successors Peter Barrow and Joseph Eveleigh were unable to sustain the initial momentum. By the beginning of 1836 many members were in arrears and departing, and in August of that year the group sought amalgamation with the Mechanics' Institution.[61] By March 1829 Detrosier had assumed the presidency of the New Mechanics' Institution, which by that time had secured premises in Poole Street and which ran successfully for a number of years before it reunited with the parent organization.[62]

Besides serving as the chief officer, Detrosier lectured before the New Mechanics' Institution in chemistry, natural philosophy, astronomy, and morals, all the while holding a managerial position in a local factory. Bronterre O'Brien complained in 1831 that the working class had no ideology beyond middle class nostrums shouted in a harsher accent; to a large extent this charge was true of Detrosier. Like Wood and Heywood, Detrosier had complete faith in the elevating power of science. Lecturing solely on science and morals, he was convinced that the union of the two disciplines would provide a path for the workers towards improvement: "Science creates wealth," he declaimed before the New Mechanics' Institution, "but it is morality that perfects man."[63] Unlike the founders of the Mechanics' Institution, however, Detrosier believed that insight into the physical world was a means to the end of political power for the workers. "Numbers without union are powerless; union without knowledge is useless."[64] By the end of his life, Detrosier was a convert to political economy and an advocate of working-class and middle-class radical cooperation. In fact, Detrosier was the epitome of all the virtues (save one — docility) that the founders of the Mechanics' Institution expected in an educable worker: "Detrosier was sober; Detrosier was industrious; Detrosier was painstaking, precise; Detrosier respected himself; Detrosier not only respected others, but he loved his species."[65]

The insurgent institution was not without effect upon its parent. By 1830 the Manchester Mechanics' Institution was in dire straits. Heywood himself reported in October of that year: "I am sorry to tell you that we are by no means in a prosperous condition; the Institution languishes. . . . Our error has been that we have not sufficiently prepared the ground before we have sown the seed."[66] The subscriptions of life and annual members reached an all time low in the years 1830-31. It was not until the slow process of democratization had taken root that matters began to improve. For this Detrosier himself had taken some credit. In 1830 the subscribers began to elect members to the Board of Directors; in 1831 the percentage of directors popularly elected reached fifty percent. By the beginning of 1834 all the directors were chosen by the entire membership, including working class subscribers; by 1847 not a single honorary member remained to lead the group.[67]

Since chemistry and mechanics were the two leading subjects, John Davies, vice-chairman and ultimately vice-president under Heywood, was appointed in 1828 to a lectureship in chemistry. Although the class had a fitful history, it was repeated in 1832, and subsequently in 1847 Daniel Stone, who was also managing director, was appointed to the position, and a laboratory was equipped for the use of the class. Davies himself is a rather shadowy figure. He was elected to the Lit. & Phil. in 1816 and served as librarian to the society from 1819 to 1827. Ultimately, in the 1840s, he became one of its two secretaries. During the 1820s Davies advertised himself as a "Private Teacher of Mathematics, Chemistry and Natural Philosophy," and gave a series of lectures in chemistry at the rooms of the Lit. & Phil. at one and a half guineas for men and one guinea for ladies and children.[68] In 1839 James Joule studied chemistry with Davies and attended Davies' lectures for medical students. From the late 1820s to the 1840s, it was Davies who oversaw the day-to-day scientific activities of the Mechanics' Institution. For the most part he well represented the interests of the honorary members and best expressed their continuing regard for the value of the institution:

> It is to our manufactures that we owe our natural superiority. It is by our manufactures that we must maintain it. We have at present got the start of other nations and we must take care that they do not come up to us. Our prosperity has excited an active competition. . . . There is certainly no way of proceeding by which we may keep in advance of our national competitors, more certainly than by enlightening the mechanics.[69]

By the mid-1830s, the Manchester Mechanics' Institution had recuperated from its doldrums; Heywood was able to report that "now we are in strength and vigour; the sun shines brightly upon us, the clouds which hung over us and the mists which shrouded us are gone."[70] In 1840 the Manchester Mechanics' Institution had secured for itself a respectable position in the structure of the scientific community. Although the receipts were once again declining, at the annual meeting on 28 February 1840 over one thousand subscribers were reported, over 5500 books and fifty periodicals were listed in the library, numerous classes and seventy-nine lectures were held in the previous year, and special societies within the institution attested to the flourishing life of the organization. At the 28 February meeting, John Dalton assumed the chair in Heywood's absence, although his illness prevented him from addressing the assembly. The presence of Dalton as chairman and as one of the directors of the institution signified its respectability and success.

In regaining its institutional vigor, however, the Manchester Mechanics' Institution changed its character. After 1840 classes and lectures on practical knowledge and science declined, and there was a sharp rise in interest in nonscientific subjects. After 1840 more than half of the lectures were on such subjects as poetry, travel, elocution, drama, and history. The original momentum of the organization appears to have expended itself.

The Manchester Geological Society

Another organization with an avowedly practical aim was the Manchester Geological Society, established near the end of 1838 in order to bring together men interested in the practical and theoretical aspects of geology. At the first general meeting of the society, held in January 1839, the honorary secretary Mr. H. C. Campbell stated that "the purpose for which the society was formed was to investigate the organic remains and mineral structure of the earth, but more particularly the surrounding district." By collecting fossils and examining strata laid bare by coal pits, mines, and railway excavations, as well as those appearing naturally, information would be collected and collated which would be available to "assist any geologist and miner [to] enable them to predict with tolerable accuracy what is to be found in any locality . . . and of occasionally being the means of saving sums of money which might be expended in a fruitless search."[71] After electing many of the famous geologists of Britain to honorary membership, a leading figure in the society, John Eddowes Bowman, rose to urge the admission, in special status, of working miners to the meetings. Lyell, Buckland, Sedgwick, *et alia* who had just been elected to honorary membership were at the pinnacle of the temple of science. There were others, Bowman maintained, "nearer the base of the building" whose labors would be extremely useful and "whose practical knowledge might often facilitate the labours of the scientific geologist," but from whom it would be unreasonable to expect an annual subscription. Bowman proposed that miners be admitted as associates at no cost to them. "Receiving this admission as a favour they would be stimulated to exert themselves to supply and collect practical information." They themselves would become more efficient and knowledgeable and better able "to consult the interests of the proprietors."[72] The motion was seconded by Peter Clare, Dalton's friend, but was opposed, interestingly enough, by Francis Looney, an artisan with broad scientific interests. Embarrassed by appearing to oppose the admission of working men, Looney explained that he believed that if miners were sufficiently remunerated for their work they would have sufficient means to pay the annual subscription. He knew many, however, who were not well paid. Looney opposed Bowman's paternalist gesture and voted against his motion, which, nevertheless, overwhelmingly passed. An examination of the membership lists and society rules, however, has turned up no such associates or official provision for them.

At the first annual meeting of the society in October, Dr. William Fleming of Salford took the chair. The secretary, Mr. Edward Binney, reported on the state of the society, which had ninety annual subscribers and six life members. The group had also received substantial donations from Lord Francis Egerton and Mr. James Heywood. Binney addressed the group on the subject of the Lancashire coal fields, an address which was published a year later in the first number of the *Transactions of the Manchester Geological Society* (1841).

In that first volume, James Black, M.D., a vice-president of the society, produced a statement "On Some of the Objects and Uses of Geological Researches," which sums up the attitudes and aims of the organization. After a survey of benefits to natural theology and to sciences related to geology, Black turned to subjects of particular interest to the Manchester business and agricultural communities. Mineral deposits are embedded only in certain geological positions, which are characteristic of distinctive eras of formation. Unless the accompanying rocks are found, therefore, mining expeditions will be wasteful and futile. For example, veins of tin and copper were found mostly to lie in schistose and micaceous rocks joined to granite; bituminous coal in beds of sandstone, shale, and limestone. The line of strike and the dip of the strata of the area will point out the best manner of procuring the coal. Acquaintance with the local geology therefore will be of use in scientifically boring for the coal.

In agriculture, geological knowledge in conjunction with chemical analysis—that is, knowledge of the mineralogical composition of the land—is of obvious value to the planter. Moreover, the agriculturist may be aided in drainage, the improving of adhesiveness in the soil, in reforestation, and in the extracting of valuable minerals from the land.

"Considering then the many great and valuable purposes to which a knowledge of geology may be made available," Black continued, "it is to be regretted that until very lately there has no public encouragement been given, nor any institution erected for the teaching of practical and economical geology in this country—in connection with the support of national or provincial museums—containing, exclusively, specimens of the geological and mineral formations and treasures of the kingdom at large, or of a certain district."[73] In order to supply this lack in Lancashire, and to "promote the taste for the general objects of geology," the Manchester Geological Society was formed. Concluding his address, Black insisted that provincial societies such as the Manchester Geological "will lead to many useful and economical results, besides their becoming the converging points on which the cultivators of the different branches of natural science can meet . . . the facts and discoveries from science throwing its light across the dazzling or bedimmed atmosphere of another, till the object of investigation is illuminated on all sides and truth becomes more revealed."[74]

The value of local geological investigations was exemplified by the paper of Mr. E. W. Binney, a local attorney, based on field work done by him and by his friend, John Leigh, a local physician. "As a provincial society," Binney claimed, "the first attention of the members ought to be mainly directed to the examination of the geological structure of their own immediate neighborhood. Indeed, no course but this is likely to attract the attention of the public, and to shew that Geology is not only valuable as an intellectual pursuit, but that it is one of the most useful of the sciences."[75] In his report, Binney surveys the south Lancashire country, describing in detail the strata composing the new red sandstone formation. He concludes that the coal deposits in the Manchester area were elevated from the level at which they were originally deposited; that the present coal fields were islands and peninsulas around which were the waters "that deposited the members of the new red stone"; and that it is likely that coal deposits exist under the bays of these long-departed waters. However, Binney was sad to report, although a great amount of coal doubtless lies buried there, "at the present price of coal, it is extremely doubtful whether, except in a few special instances coal can be worked to profit under this deposit [i.e., the new red sandstone]."[76]

Of the eleven papers which comprised the first volume of *Transactions,* four were presented by Binney, four were authored by J. E. Bowman, and one by William Fairbairn. These men comprised the most important part of the active membership of the society.

The Structure of Manchester Science in 1840

It seems clear from the above review of Manchester scientific organizations in the period 1781-1840 that there existed in the city a complex, structured community with institutions cutting across class lines, each organization serving a separate function, yet sharing the material and intellectual resources of the city with the others. The premier place was still afforded the Literary and Philosophical Society, founded as we have seen as an eighteenth-century literary and philosophical discussion club and extended to a formal society, "many respectable persons being desirous of becoming Members."[77]

The scientific community of Manchester was virtually born with the creation of the society. Control over the Lit. & Phil. meant scientific-cultural predominance in the city. Indeed, the leadership of the society rested in the hands of a relatively small professional (mainly medical) elite, with twin centers at the Manchester Infirmary and the powerful, progressive Dissenting community at the Cross Street and Mosley Street Chapels. The first presidents of the society (Mainwaring, Massey, and Percival) all were intimately connected with the infirmary. The leading lights of the society (Percival, Thomas Henry, and Barnes) were connected with the Unitarian Cross Street Chapel.

The leadership of the society changed very slowly over the first sixty years of its existence. One can easily pick out the "perennials"—those officers of the society annually re-elected (*pro forma,* it seems) and who exerted tremendous influence in the running of the society. From 1790 to 1847 there were only a handful of presidents: Percival (1790-1804), George Walker (1805-1806), John Hall (1806-1807), Thomas Henry (1807-1816), John Dalton (1817-1844), and Edward Holme (1844-1847). All were connected with either the infirmary or the Dissenting leadership through the Manchester Academy. Walker had been a student at Kendal under Caleb Rotherham, a mathematical tutor at Warrington and eventually Barnes' successor as principal of the second Manchester Academy.

Dalton of course came to Manchester as a mathematical tutor at the academy or New College as it was then called; Holme had been a student at the academy.

The vice-presidents, too, had among them an overwhelming share of "perennials": Charles White (1784-1806), surgeon at the infirmary; Holme (1798-1844, the longest tenure as vice-president); Thomas Henry (1788-1806); William Henry (1807-1836), who succeeded his father as vice-president and was in turn succeeded by his son W. C. Henry in that post in 1837; Peter Ewart (1812-1843); and G. W. Wood (1822-1843).

From this list one can with much justice make the claim that the society was dominated in the first sixty years of its existence by men who had entered the society before 1800, who were drawn into the "old network," the elite group which dominated the society, and who largely shared the aims, values, and background of the founding members of the society. Correspondingly, the *conception of science* which dominated the Lit. & Phil. was also slow to change. The "amateur" conception of science's role and practice changed relatively slowly given the great changes in the social and economic setting of the organization. It was made clear in the very first volume of the *Memoirs* that the Lit. & Phil. was a society founded to enable gentlemen of the city to raise the cultural tone of the area, to effect easier scientific communication, and to encourage scientific work—that individual, spontaneous creative effort which "like fire is put into motion by collision."[78] Even if the society aimed at reducing the cultural disparity between provinces and metropolis and at easing the path of genius, science itself served cultural, religious, and utilitarian ends. The disquisition of Henry discussed at length above sheds much light on the prevalent conception of the role of science. The Lit. & Phil. was not then a professional scientific society as we have come to know it; it was, essentially, a league of gentlemen with cultural aspirations who aimed at the promotion of science and, thereby, of industry and commerce as well.

The "old network" was slow to relinquish leadership and slow to change, but it was not stagnant. It selected promising young men, polished and groomed them for eventual leadership. Some, like Holme, had to serve long years in secondary roles; others, like Dalton, rose relatively quickly to the upper levels of the organization.

When Thomas Barnes, the minister of the Cross Street Chapel and principal of New College, sought a mathematical tutor, he turned to a fellow member of the Lit. & Phil., John Gough of Kendal, who recommended Dalton. A recent biographer of Dalton surmises that his name might have originally been proposed by Edward Holme, a Kendal native who knew Dalton and who was already accepted into the influential circles of the city. After coming to Manchester, as we have recounted above, Dalton quickly was drawn into the scientific-cultural life of the city. His

Meteorological Observations and Essays having just been published, Dalton was proposed for membership in the Lit. & Phil. by the society's two leading figures, Percival and Henry, and by Robert Owen, at that time a prosperous local textile manufacturer. Within two weeks he presented his first paper, the first systematic study of what is now known as color blindness. Within a few years, Dalton—the serious student, the devotee among virtuosi—would by force of his scientific achievement be elevated to the first positions of the community.

By the 1820s a new leadership element would be admitted to important positions in the community. These men, generally, were connected by family and religious ties to the "old network" but possessed a somewhat different professional orientation. Increasingly, men from the business community, rather than the medical ranks, were introduced into the positions of prominence in the society. Wood, Ewart, Eaton, Hodgkinson, Heywood, and Fairbairn began to make their views heard in councils of Manchester's scientific-cultural elite. Wood, Heywood, and Fairbairn were Unitarians and successful men of business; presumably they originally were drawn into the old network by social and religious ties. Heywood, for instance, was the grandson of Percival.

These men, all of whom (with the exception of Ewart) entered the society in the nineteenth century, were men of business, drawn to the spirit of reform abroad in Britain after the end of the continental war. They conceived of the role of science in a way subtly different from that of the founders of the Lit. & Phil. While they agreed in principle with the aims of the society, they were attracted more strongly by the arguments of men like Lord Brougham and George Birkbeck who advocated the *diffusion* of useful knowledge among the industrious classes. These Mancunians aimed at supplementing and enlarging the local scientific scene by taking more seriously and reinterpreting the Lit. & Phil.'s stated commitment to the principle of alliance between science, commerce, and industry. The Lit. & Phil. was not structured to accommodate scientific needs seen in this new way; that is, science for large numbers of people—for the upper middle classes or for workers and artisans. The Lit. & Phil. was, by its constitution and governing conception, a society for the cultural elite. What was needed—in addition—were organizations designed to carry out not merely the promotion of scientific achievement and the elevation of scientific taste in the city, but rather the spread and diffusion of scientific knowledge to that part of the populace which could best make use of it: the middle and working classes.

It was this shift of emphasis which led George W. Wood to attempt (in part successfully) to enlarge the original aims of the Royal Manchester Institution to include public lectures, courses, and so forth, on science, as well as art. It was the same motives which led Wood, Benjamin Heywood,

Fairbairn, and others to urge the establishing of the Mechanics' Institution
for the enlightenment and elevation of the laboring classes although not
by them.

At the annual Geological Society dinner in February 1840, Dr. Black,
vice-president of the Geological Society, summed up the situation very
precisely. Many classes of society in Manchester are at present pursuing
science, he pointed out. Workers, for example, pursue the subject in "a
very earnest and diligent manner," something which all ought to applaud.
Science, according to Dr. Black, makes a man a better "husband, father
and subject."[79] He hoped that the well-known and lamentable perversions
of the working men's time would soon decrease. Black was obliquely referring to the sharpening of political divisions among classes in Manchester
following upon the economic depression of 1839, perhaps to collisions
between the Chartists and the Anti-Corn Law Leaguers. It was the hope of
the founders of the Mechanics' Institution to make better and more docile
workers of their subscribers.[80]

Black's toast was accepted on behalf of the Mechanics' Institution by
S. E. Cottam who reiterated that subscribers to the Mechanics' Institution
were "humble labourers in the paths of science, whose object was to follow
in the trails which were being explored [by the Lit. & Phil. and the Geological Society]." The mechanics "did not aim at research; their object
was to attain the knowledge and walk by the principles which such societies
laid before them. They were gaining elements rather than discovering
principles."[81] There was, then, already a clear division of labor in the
system of scientific organizations in Manchester. The Lit. & Phil. was
charged with encouraging research, an enterprise best reserved to such an
elite organization.[82] The Royal Manchester Institution provided public
lecture series on recent developments in and applications of science for its
mainly upper middle class and professional membership. The Mechanics'
Institution provided education in elementary science for the edification of
clerks, artisans, and mechanics. All the institutions drew for sustenance on
the same group of patrons, for local scientific talent on the same small
group of practitioners, and for leadership upon the same small elite which
figured so largely in the cultural life of the city.

The Manchester Geological Society appears to mark another important
change in the make-up of Manchester science. It was founded by men
much younger than existing scientific leadership (Binney, the Ormerods,
J. Heywood) and, although the presidency of the society was at first in the
hands of nonscientific men,[83] the goals of the group were decidedly practical (see Black's summary of the society's aims in the preceding section).

The list of officers for 1840–41 is particularly instructive. The president
of the society was Lord Francis Egerton, fellow of the Geological Society
of London, an area member of Parliament, and later to be Earl of Elles-

mere. Egerton was a local worthy who served also as president of the Manchester Agricultural Society and who was selected president (at least in
part) because of his patronage of the society, his station, and his undoubted interest in geological matters. The vice-presidents included three
other members of the Geological Society of London: the aforementioned
Bowman, William Hulton, and James Heywood, later to become a fellow
of the Royal Society (1859) and future member of Parliament. Heywood,
along with Edward Binney and H. M. and G. W. Ormerod, is generally
considered among the real founders of the society. Born in 1810, he was
the grandson of Percival and the brother of Benjamin Heywood, so often
referred to above. After receiving a considerable bequest from the estate
of his uncle B. A. Heywood, James withdrew from Heywood's Bank and
attended Cambridge. Owing to his nonconformity, he was prevented from
taking his degree. He removed this impediment by promoting the Cambridge Reform Bill in Parliament, and took his degree in 1857. He eventually served as president of the Geological Society as well as of the Statistical
Society.[84] The secretaries were Binney and J. F. Bateman, soon to figure
importantly in the construction of the waterworks.

The members of the council included James Black, M.D., a former vice-
president of the society, William Fairbairn, Eaton Hodgkinson, F.R.S.,
and Francis Looney. The banker was Sir Benjamin Heywood. The general
membership, although small, reflected substantial changes which were to
come about in the Manchester scientific community. Along with a few of
the names from the early Manchester cultural elite like Dalton, Peter
Clare, and John Moore were the names of young men who were to figure
importantly in the two decades to come: Thomas Ashton of Hyde, J. F.
Bateman, J. Baxendell, Edward Binney, Fairbairn, W. R. Greg and R. H.
Greg, John Hawkshaw, and Eaton Hodgkinson. The presence and active
participation of these men signal the beginnings of a new era in Manchester science: the passing of the traditional amateur concern with science as a cultural asset, with an elevation of the taste of the city, with
science as an antidote to parochialism as well as a stimulus towards the
improvement of industry and agriculture — that is, as relief of man's estate.
These leaders — the Henrys, Percival, Barnes, Holme, in large measure
Dalton, too — were in the main self-trained, had wide and often diffuse
scientific interests, and — Dalton of course excepted — being professional
men were usually able to devote only a small proportion of their time and
energies to the pursuit of any special science. Men like Fairbairn, Hodgkinson, Binney, Baxendell, *et alia* emerged from a different matrix. Most
began their adult lives in a working or business capacity rather than a
professional one. In the Manchester Geological Society, despite the organization's obeisances to wealth and station, a scientific type relatively new to
community leadership was taking hold. For lack of a better word, one may

term this type the *devotee,* a man who while still generally self-trained is far more concerned with preparing himself for the scientific frontier, with *contributions* to scientific knowledge, and with serious applications of his work in the practical sphere.

The incipient changes in the Manchester scientific community signaled by the composition and concerns of the Geological Society were made plainer still by the coming of the British Association to Manchester in 1842. The British Association, founded in 1831 largely by provincial and Scottish scientists "to give a stronger impulse and a more systematic direction to scientific inquiry,"[85] met annually in a different city of the British Empire to survey progress in the various branches of science and to present papers of current interest.

The scientific community of the city was exhilarated by the opportunity to play host to so many important scientific dignitaries. The Royal Institution was fitted up to serve as the major meeting center, with sectional meetings to be held there, at the Literary and Philosophical Society, the Athenaeum, and at the Mechanics' Institution. Over two hundred new members were recruited for the British Association from Manchester and the surrounding area. Exhibitions, soirees, and dinners were planned. The engineer J. F. Bateman designed a membership ticket embossed with a medallion sporting a likeness of Dalton.[86] The coming of the British Association served also as the occasion to honor Dalton once more; a memorial to Dalton, presenting him with a medallion, was sponsored by George Faulkner, editor of *Bradshaw's Journal.*

The president for the year was, as customary, selected from local scientific dignitaries. The obvious choice from Manchester was John Dalton, who was, however, too ill to assume these duties. Lord Francis Egerton, M.P., member of the Geological Society of London, president of the Manchester Geological Society and of the Manchester Agricultural Society was chosen to serve in his stead. Dalton agreed to serve as vice-president, along with William Charles Henry and Sir Benjamin Heywood, among others. The local secretaries were Peter Clare, Dr. William Fleming, and James Heywood. It was, however, in the sectional committees, the committees responsible for the business of the convention, that younger, scientifically trained men became important for the first time. Along with those dignitaries necessarily honored by the association, like Dalton, Rev. William Herbert, Dean of Manchester, Edward Holme (president of the medical section), and G. W. Wood (president of the statistics section), were names less prominent on the local scene: Lyon Playfair (chemistry), Binney (geology), Fairbairn, Eaton Hodgkinson, and J. F. Bateman (mechanical science).

It was Playfair's privilege to present an abstract of Justus von Liebig's "Report on Organic Chemistry Applied to Physiology and Pathology," which was a highlight of the meeting. Playfair and Edward Schunck, another newcomer to the Manchester scientific scene, were appointed to important chemical committees receiving grants for special research. Other recipients were Binney, James Heywood, G. W. Wood (statistics), Fairbairn, Hodgkinson (mechanical science).

Egerton's presidential address was an exercise in modesty. Disclaiming any scientific pretensions and acknowledging his selection as resulting from "the accidents of local connexion with the [meeting] place," Egerton insisted that he "would gladly have served as a door-keeper in any house where the father of science in Manchester [Dalton] was enjoying his just pre-eminence."[87] It was in the transactions of the special sections, however, not in Egerton's historical survey, that the important changes for Manchester science were taking place. In the chemical section, for example, two technical papers were delivered by Dr. Lyon Playfair, by the unknown Mr. J. P. Joule, by Dr. John Leigh, by John Davies, and by John Dalton, although under the circumstances Dalton's papers were read by Peter Clare. In the geological section, local honors were upheld by Edward Binney's paper on the Lancashire coal field. Fairbairn and Bateman figured prominently in the mechanical section.

When, therefore, the vigorous British Association for the Advancement of Science met in Manchester in 1842, it exposed a dichotomy in the Manchester community. Institutional control and honors were held for the most part by men who had long been prominent locally (and sometimes, as in the case of Dalton, nationally) but who were far from the frontiers of their disciplines. When major scientists from Britain and abroad came to the city, they came to listen to *new* voices: to Joule, Binney, and Leigh; to Fairbairn, Bateman, Playfair, and Schunck. The impact of the meeting was enormous. It meant a large influx of these new, younger scientists in the existing societies. Their access to the great scientific names and their superior familiarity with current scientific research ensured their rapid rise within these organizations. In short, the stability of the system of Manchester scientific institutions was shaken; when equilibrium was restored, great changes were rendered visible.

2

The Emergence of
the Devotee: The Changing Face
of Amateur Science

Science as a "Calling"

During the 1840s and 1850s, time and social change took their toll on
the ruling circles of Manchester; before the decade was out Dalton, the
doyen of Manchester science, was dead, and the eighteenth-century men,
the remaining members of the "old network" which had begun the organization of science in the city, were gone too. Naturally enough, new groups
gradually emerged to replace them. New conceptions of science and scientific organization infused the community. As we have seen, by the third
and fourth decades of the century, new institutions had been founded in
the town which enlarged and extended those already existing; men like
G. W. Wood and William Fairbairn, caught up in the utilitarian fervor
sweeping the British middle classes, were instrumental in founding the
Royal Manchester Institution and Mechanics' Institution for the diffusion
of useful knowledge to the middle and lower classes. The Geological Society, too, reflected larger changes: the increasing specialization of science
demanded a fractionalization of scientific societies, and the largely undoubted uses of geology for the economy of the region encouraged "practical men" to band together for the study and dissemination of practical
science.

In short, the "gentlemanly" brand of amateur science, the nineteenth-
century variety of virtuoso inquiry, was gradually giving way under an
assortment of pressures. The science of "professionalists" like Percival or
the Henrys was yielding—slowly, perhaps, but surely—to the science of the
utilitarians. Socially, the scientific community dominated by professional
men, especially medical men, altered its constitution, and other segments

of the middle class, more vigorous in their prosecution of new ideas, assumed importance. Business men, like Benjamin Heywood and G. W. Wood, while not scientists themselves, succored and encouraged what they believed to be of utmost importance: the creation of institutions to disseminate practical (i.e., scientific) knowledge to the productive classes of society. Ultimately, however, scientific leadership itself passed to a new group, often a bit impatient with the leisurely pace and parochial character of the science of the gentleman amateurs. This new group, generally from the less prestigious segments of the middle class and sometimes self-made men, were *devotees* of science, who saw science as their "calling." Far more serious about the prosecution of scientific activity than their predecessors, the devotees possessed the inspiration proper to a true sense of vocation and were concerned with research at the frontiers of science, with publication, with "keeping up" with the output of the great practitioners of their specialities, with scientific communication—in short, with many of the things which we associate with the professional pursuit of science.[1] Yet, they were not indeed true professionals. They were, generally, self-taught men, who for a variety of reasons did not care to seek or were not able to obtain careers in science, but earned their livelihoods in business. These businessmen-savants, however, sought their *identity* and, perhaps paradoxically in Mr. Gradgrind's Coketown, even their *status* in the scientific pursuit.[2] At first, the devotees were men accepted by the ruling elite precisely because of their expertise—Dalton represents an early form. Gradually, however, the socially distinguished gave way to the pawnbrokers, the brewers, the engineers, and the solicitors. These men were not truly a part of the old network; they represented the accession to powerful positions in the scientific community of new associations, new groups, new circles. They had less patience with the dilettante and the local dignitary who wished to have his say on natural philosophy, merely because he had expressed an interest in those matters. The devotees were far less tolerant vis-à-vis the cultivation of science than their predecessors. Under their rule, the Literary and Philosophical Society became solely a scientific society, one which was well regarded throughout Britain. The devotee in Manchester science provides, in short, a bridge between the leisurely, gentlemanly science of Percival and the Henrys and the thoroughly professionalized, efficient, consistently up-to-date operations of the academics such as Frankland, Roscoe, Balfour Stewart, and Arthur Schuster, who dominated the community in the last half of the nineteenth century.

Of course, this process was one which was occurring in one form or another all over Great Britain. Men like Lyell, Murchison, de la Beche, Brewster, Grove, and many, many others took part in a great reformation of British science. In part, the founding of the British Association for the Advancement of Science and the establishment of the numerous special-

ized societies give testimony to this point.[3] In 1830, Lyell spoke of the
"scientifics" with whom he could intelligently communicate on important
matters.[4] In 1840, Whewell wrote of the need for a new term to describe
the serious cultivators of science; he called them "scientists."[5] Had not the
term been used in so many ways since Whewell's invention of it, we could
justifiably reserve it for our devotees. This chapter in the development of
the Manchester scientific community will provide a portrait of the community as it changed during the period in which the devotees rose to
prominence. The men whose career profiles will be offered to the reader
are the most influential, the most prominent among the group. By no
means are they the only examples which could have been chosen; they
are, however, among the most important.

William Sturgeon and the Royal Victoria Gallery

Throughout the 1830s, the Mechanics' Institution, in order to survive,
was broadening its goals and extending its services. The science classes, as
we have seen, were only intermittently successful, and although lectures
were mostly on scientific subjects until 1840, nonscientific subjects, such
as poetry and travel, were introduced in 1830. By 1840 about half the
lectures were on such nonscientific subjects as language, education, history, and music.[6] Moreover, the Mechanics' Institution never achieved its
original goal of collecting and preserving scientific and technical models
and apparatuses for teaching and illustrative purposes.[7] It was the alteration and evolution of the Mechanics' Institution which precipitated efforts
to create a new institution to replace or complement the new, democratized Mechanics' Institution.

On 21 March 1839 a "numerous and highly respectable meeting of
gentlemen" was held at the York Hotel to discuss the possibility of establishing an "Institution of Practical Science."[8] H. H. Birley, local Tory man
of civic affairs and infamous in some circles as leader of the charge at
Peterloo two decades previously, was in the chair. Birley outlined the basic
aims of the assembly: to form an institution which would provide a collection of apparatus "combining philosophical instruction and general enter-
tainment"; present demonstrations of elementary principles; exhibit the
progress of those sciences fundamental to industry; stimulate research and
invention by prizes, etc.; and, like its model, the Adelaide Gallery of
London, be attractive to the younger members of the Manchester community. William Fairbairn rose to add his strong support to such an
organization which would have "the very best effects upon the population
at large, particularly upon the younger classes." Eaton Hodgkinson added
that he was particularly concerned to have exhibits of models, machinery,
and instruments. "As we in Manchester possessed such great ingenuity in

our workshops," he maintained,"we should have some institution in which all these machines might be collected and shown and the elements of them explained."

John Davies of the Mechanics' Institution told the group that he was old enough to remember when the only society in town for the diffusion of knowledge was the Literary and Philosophical Society. He could remember the time when that group was virtually unknown outside its membership; when Dr. Dalton would go about the city unrecognized. Now things had changed and there were many other organizations. Still, the proposed new institution was, in his words, a "desideratum." "Our other institutions," Davies said, "had tended much to circulate speculative knowledge; but in a manufacturing district like this we wanted an institution which would enable the artisan . . . to see all the modifications and arrangements of different machines so as to enable him to perfect any that might be submitted to his notice."[9] The Mechanics' Institution had aimed at presenting the elements of the sciences to artisans and mechanics; it had faltered, Davies implicitly seemed to say, in this latter goal. Certainly, Fairbairn and Hodgkinson appeared to have been disaffected; they no longer took as active a role in the Mechanics' Institution's affairs as they had previously.[10]

A committee was appointed to proceed, and within a fortnight a prospectus appeared in the *Guardian* for a joint-stock company to be called the "Manchester Institution for the Illustration and Encouragement of Practical Science." Twelve thousand pounds was to be raised by selling shares of £10 each. Subscribers were to be admitted at two-guinea family memberships, one-guinea individual memberships, and one-shilling entrance fees. The potential shareholders were assured that the institution provided "an eligible opportunity for safe investment" and could be expected to "yield a very liberal interest."[11] A directing committee was formed with Birley as chairman, and which included Davies, Fairbairn, Hodgkinson, Richard Roberts, J. A. Ransome, John Hawkshaw, George Peel, and Joseph Peel. It is interesting to note that the committee included three founders of the Mechanics' Institution, two of whom no longer participated in the direction of that organization (Fairbairn and Roberts), and the other of whom was known to be out of sympathy with the new democratized direction (Davies).[12] It is clear from the prospectus that the laboring classes would have little to say about the operation of the new organization. In the first place, the new institution was planned as a profit-making company, controlled by the shareholders and the directors. The proposed entrance fee (about a shilling) was considered expensive for all but the relatively well-to-do.[13] Interestingly, the York Hotel meeting was attended by some who objected to the elitist direction proposed for the new institution, and they met on the fourth of April to suggest counter-resolutions on the questions at hand. The second assembly described themselves as "gentlemen who have been some time engaged in making arrangements for establishing a Gallery of Art and Science on popular principles."[14] The session was chaired by George Jackson of Salford; those attending charged the York Hotel group with establishing a private profit organization under the guise of a public interest institution, and urged that it be publicly controlled, open to all classes, and run at lower fees than previously announced.

Jackson's constituency was small, and the original plans proceeded. In February the Practical Science Institution was renamed the Manchester Royal Victoria Gallery for the Encouragement of Practical Science; the board of directors announced that Mr. William Sturgeon was hired as the first superintendent of the gallery and would commence with a course of six lectures on electricity.[15] On May 30, the official announcement of the opening appeared in the *Guardian*. Queen Victoria had consented to appear as patroness; Birley remained the chairman and among the directors were Fairbairn, Hodgkinson, Joseph Whitworth, the manufacturer, and John Hawkshaw. John Davies was no longer on the board. The newly published objectives of the gallery differed little from those stated at the York Hotel meeting of the previous year.[16]

The Royal Victoria was modeled after the Adelaide Gallery of the Strand, London, opened in 1830 to stimulate popular interest in science and engineering. Sir George Cayley was instrumental in its founding; Telford and Wheatstone worked there. A visitor to the Adelaide Gallery described it in the following manner: "Clever Professors were there, teaching elaborate science in lectures of twenty minutes each. Fearful engines revolved and hissed and quivered. Mice led gasping sub-aqueous lives in diving bells. Clockwork steamers ticked round and round a basin perpetually to prove the efficacy of invisible paddle wheels. There were artful snares laid for giving galvanic shocks to the unwary."[17] The Adelaide Gallery was operated by the Society for the Illustration and Encouragement of Practical Science. In their catalog for 1835 the gallery is described as "for the exhibition of objects blending instruction with amusement," containing practical demonstrations, and exhibits of "discoveries in Natural Philosophy." Some of the exhibits and apparatus described are George Strattion's warm air apparatus for heating large apartments (the air was warmed by hot water pipes), models of steam boats, steam guns, steel engraving, and a powerful steel magnet which "fully demonstrates Mr. Faraday's great discovery of the identity of the electric and magnetic fluids."[18]

The Royal Victoria Gallery was quite similar to the Adelaide Gallery in aim and aspect. It opened in June 1840 in the Exchange Dining Room and its ante-room. The collection included technical models and apparatus, antiquities, curiosities, and works of art. Lectures and demonstrations were scheduled, *conversazione* were planned, and a scientific library was in the process of being formed. The gallery was described by early visitors as "elegant, commodious and well-lighted." Especially popular were dial weighing machines, the electrical and philosophical apparatus, William Read's mathematical instruments, fossils from the excavations of the Manchester and Leeds Railway, the model of the "electromagnetic telegraph of Wheatstone and Cooke," the electromagnets, Sharp, Roberts and Company's ball and socket valve, and cast-iron surface plates deposited by Whitworth and Company.[19]

The superintendent of the gallery, the man responsible for the direction of the institution and the presentation of courses of lectures and demonstrations, was Mr. William Sturgeon, a major addition to science in Manchester. Sturgeon came to the city already famous as inventor (of a soft-iron electromagnet, a commutator, and an electromagnetic engine) and lecturer at the East India Company's military academy at Addiscombe. Sturgeon was born in 1783 and raised under conditions of great hardship in Westmoreland. After a difficult period as an apprentice shoemaker, he joined at age 19 the Royal Artillery. He remained in the army as gunner and private soldier until 1820. During his army period, Sturgeon worked assiduously at self-improvement, learning mathematics, ancient and modern languages, and natural philosophy. He left the army at age 37 to take up the trade of bootmaking in a shop at Woolwich, but used his meagre earnings to continue his scientific studies and especially to construct scientific instruments, an interest which brought him into contact with similar practitioners in London and aroused the interest of eminent scientific men such as chemist James Marsh, O. G. Gregory, Samuel Christy, and Peter Barlow.[20] Owing to their influence he was appointed lecturer on experimental philosophy to the East India Company's academy at Addiscombe and lectured at the Adelaide Gallery as well. It was at the Adelaide Gallery that Sturgeon presided over the first meetings of the London Electrical Society and edited their journal, the *Annals of Electricity*. In 1840 he was invited to become superintendent of the new Royal Victoria Gallery, a position which tempted him all the more since the Electrical Society was ending its existence. He took up his duties in March 1840 and, now 57 years of age, began lecturing on electricity, magnetism, optics, and mechanical philosophy.

Sturgeon found many sympathetic colleagues in Manchester, eager to welcome this devotee of electrical science to their circles. One young man from Manchester, James P. Joule, had already made six contributions in the form of letters investigating Sturgeon's electromagnetic motor to Sturgeon's *Annals of Electricity*. By 1841 Sturgeon had arranged for Joule to lecture at the Royal Victoria Gallery; this lecture in February of that year marked Joule's first public scientific appearance and the beginning of his remarkable scientific career. Joule, along with John Davies, at once formed a social as well as an intellectual link with Sturgeon.[21] The little circle expanded as the decade wore on; special friends besides Joule and

Davies included John Just, John Leigh, a young medical man employed as a chemist by the Manchester Gas Works, and Edward Binney, an attorney who already figured largely in the Manchester Geological Society and was eventually to become an important officer and president of the Literary and Philosophical Society.

In a way the coming of Sturgeon added an element of cohesiveness to what had hitherto been the occasional discussions of a small group of friends and acquaintances. Joule had been the chemistry student of Davies, but the disciple (at a distance) of Sturgeon and an avid follower of the *Annals of Electricity.* Binney's interests had been in geology and natural history, as had John Just's. Leigh, a close personal friend of Binney, provided the link, being concerned both with chemistry and natural history. When Sturgeon arrived, he seemed to be another, younger Dalton. Here was a man, self-trained but at the frontiers: a successful applied science inventor; the editor of a reputable journal, and the superintendent of a new and potentially exciting local institution. Hopes were high in 1841 and 1842. Despite the continuing economic depression, the progress of science and, through it, the future of industry must have seemed bright indeed. Joule, in an extended *eloge,* described Sturgeon as "animated . . . his conversation pleasing and instructive. . . . In friendship warm and steady." Joule further described his politics as "conservative" which would have suited Birley and the directors of the Royal Victoria Gallery. He will always, Joule wrote, "be remembered as a distinguished cultivator of natural knowledge."[22]

That his natural philosophy always had the element of the archaic about it would not, in all probability, have disturbed his younger friends and colleagues. As is usual among the self-taught devotees, their views reflected their slow initiation into their subjects. Sturgeon wrote, for example, in an unpublished lecture which reflects his style and his natural philosophy:

The electrical fluid is so universally diffused throughout every part of nature's productions, that every particle of created matter, both animate and inanimate, which has hitherto been contemplated by the philosopher, is full of this surprisingly animated elemental fire. . . . Indeed so universally does the electrical fluid appear to be employed in most, if not all, the grand processes of nature that there is not, perhaps, a plant that grows nor a limb that moves but is in some measure influenced by its powers. Nay, it is, perhaps, this astonishing, this most gigantic of physical agents, which is employed by the GREAT CREATOR to spin the earth and planets on their axes, to sweep them through the heavens in their regular periods of revolutions, and to keep in uniform motion all those massy orbs of matter which compose the countless systems of the universe.[23]

The vigor of the Royal Victoria Gallery and Sturgeon's energies did not wane; however, local support, as so often was the case, turned out to be ephemeral. The Royal Victoria Gallery ended its existence in 1842. Some of its equipment was donated to, some purchased by, the Royal Manchester Institution, whose vice-president for 1842 was H. H. Birley, the chairman of the gallery.[24] Joule maintained, some years later, somewhat bitterly, that "the indifference to pursuits of an elevated character which too frequently marks wealthy trading communities destroyed this, as it has many other useful institutions."[25] Sturgeon found himself once again out in the cold, having to scratch for a living. In January 1843 Sturgeon announced the formation of an entirely new institute: the Manchester Institute of Natural and Experimental Science. Under its auspices, Sturgeon announced new lecture courses and *conversazione.* Yearly subscriptions were the same as at the Royal Victoria Gallery: one guinea for individuals and two for families.[26] The new Manchester Institute failed almost as it began; Sturgeon had to supplement his income from it by lecturing occasionally at the Mechanics' Institution or at the Royal Manchester Institution. Ultimately he became an itinerant lecturer, going from village to village in the Manchester area, drawing his demonstration equipment with him in a cart. In 1844 he was elected, rather belatedly it seems, to membership in the Literary and Philosophical Society. By the late 1840s Sturgeon was destitute. Owing to the strenuous efforts of his friend Binney, and others, the sum of £200 was awarded him in 1847 and again in 1849. Sturgeon received an annual government pension of £50, a sum which was continued to his widow after his death in December 1850.[27] In his eulogy of Sturgeon, commenting upon Sturgeon's straits, Joule took the opportunity of entering a plea for adequate recognition of the scientist's contribution. "The sum which would be necessary to succour the needy man of science, and so to enable him to continue his researches, would appear trifling indeed if regard were made to the important objects to be realized. But he appeals not to his country as a pauper. He asks it to discharge the debt it owes for labours which have contributed to the common weal, a debt which cannot honourably be left unpaid."[28]

The Changing of the Guard: The Evolution of the Literary and Philosophical Society

If the death of Sturgeon went virtually unnoticed in Manchester, the passing of John Dalton, as could be expected, did not. Lyon Playfair fairly commented that Manchester bestowed upon Dalton, in death, the honors due a king.[29] Elizabeth Patterson, in her recent biography,[30] is certainly correct in her judgment that civic pride and civic ostentation took precedence over any real sense of loss. Dalton's death was a "state" occasion; his body lay on display in the town hall, in a large room overhung with black drapery. The windows were covered. Dalton rested in a plain but remarkably beautiful Spanish mahogany casket. The scene was

lit by large gas-fired candelabra. Forty thousand people filed by. The
funeral, arranged by a committee of civic dignitaries, was equally impressive. The coffin was taken from the town hall to Ardwick Cemetery in a
cortege almost a mile long. Manchester had seen nothing like this simple
Quaker burial.

Science, in Manchester, had lost its reigning dignitary; the city itself
had lost the symbol of its scientific-cultural pretensions. W. C. Williamson, then a young naturalist engaged by the Natural History Society (see
chapter 1), wrote that "we apparently entered a scientific interregnum."[31]
The meetings of the Literary and Philosophical Society had already, in the
declining years of Dalton's presidency, become more social than scientific.
The most important officers of the society (the vice-presidents) after
Dalton were Peter Clare, Dalton's friend and alter-ego since his first stroke
in 1837; Dr. Edward Holme, the physician, naturalist and old friend of
Dalton; John Moore, another intimate of Dalton; and J. A. Ransome,
Dalton's physician. Clare and Moore were well-liked but not noted for
their scientific prowess. Clare is described as "genial but fussy," and "not
a profound student nor a frequent contributor."[32] Moore, who took a
major part in the organization of the society, was not an intellectually
strong man.[33] Before his death in 1836, William Henry, Thomas Henry's
son, had been the most prestigious, scientifically, of the old guard; but
business and professional success and social position had eroded even his
connection with the advance of science.[34] The younger, scientifically more
vigorous men who had gathered around Sturgeon (Joule, Binney, Leigh)
in the 1840s had not made for themselves the local or national reputations
to catapult them to institutional leadership. The scientific professionals
(about whom more will be said in the next chapter) likewise were too new
in the town and too tenuous in their local connections to assert themselves.
Interregnum there surely was.

It is no surprise, therefore, that one of Dalton's intimates, almost as a
gesture of respect, would be selected as his successor. In the October
meeting in 1844, Dr. Edward Holme was elected to the presidency; Eaton
Hodgkinson, the engineer, was moved up to the vice-presidency from his
former position of secretary. Holme had been, with Dalton, one of the
oldest members of the organization and had been a vice-president since
1798. Since Dalton's stroke in April 1837, his personal history had been
one of steep decline, mitigated by periods of slight recovery.[35] It is fair to
say, on the basis of several independent accounts, that the last seven years
of Dalton's presidency were far from vigorous. The 1842 edition of the
Memoirs (the first since 1831) contained few papers read since 1839 and
was not a volume to distinguish the society before the visiting members of
the British Association nor to excite the younger members of the Manchester scientific community. Holme's presidency did little to alter the situ-

ation; however, the only volume of the *Memoirs* to appear before Holme's death demonstrated that natural forces were making major changes. The names which dominate volume seven of the *Memoirs* (1846) were Joule (who was elected secretary to the society in 1846), Hodgkinson, J. F. Bateman, the engineer, and Sturgeon.

In a way, it was Holme's death in 1847 even more than Dalton's that unveiled these changes. Holme was the last of the eighteenth-century presidents, the last of the leaders of the Manchester Literary and Philosophical Society who grew to scientific maturity and institutional leadership in the eighteenth century. The accession of Eaton Hodgkinson to the presidency in 1848 was a major break in the traditions of the society.

A cursory examination of successive editions of Slater's *Manchester, Salford and Suburban Directory* gives some hints of the changes occurring in the city and environs and in the society itself. The 1832 edition of Slater's *Directory* lists the names of thirty-three "manufacturing chymists"; about 90 percent of these manufacturers belonged to the Literary and Philosophical Society. The 1852 edition, however, lists fifty-four manufacturing chemists, of whom only eight belonged to the society. On the other hand, the 1852 edition lists the presence of thirty-four civil engineers in the area (none was listed for 1832, although there were nine "engineers and machinists"); about twenty percent of these civil engineers were members of the Lit. & Phil. The 1852 *Directory* quotes Knight's *The Land We Live In* to the effect that "science is better understood at Manchester than at Liverpool. It is certain that the cotton spinners and the calico printers have constant demands made upon their ingenuity—the one to develope mechanical applications and the other both chemical and mechanical applications of the principles furnished by science."[36] In the two decades between 1832 and 1852, then, the engineering establishments in the Manchester area had grown in both economic and social importance. Men such as Fairbairn and Hodgkinson, recognized leaders of the profession in both practical and theoretical fields, were accordingly granted greater recognition in the city itself and in the more important scientific organizations. Eaton Hodgkinson's career is a case in point.

Hodgkinson was born in 1789, the son of a Cheshire farmer who died when the boy was only six years old. His education was left in the hands of a clergyman uncle who had the boy educated in the classics to prepare him for a clerical career. Eaton showed little aptitude for languages and was subsequently sent to a school in Northwich where he learned mathematics. In 1811 his mother, now impoverished, moved the family to Salford and began there a pawnbroker's business. Young Eaton, now twenty-two years old, aided his mother in the shop and eventually took control of the business. His spare moments were entirely devoted, however, to the study of mathematics and science. He read, it is reported, the works of Thomas

Simpson, William Emerson, and Dealtry on mathematics and mechanical science.[37] His interests in science brought him to the attention of the leaders of the Manchester scientific community; in fact, he became a pupil of Dalton, with whom he read mathematics.[38] Other scientific intimates included Dr. Holme, Henry, Ewart, and, later, Fairbairn.

Although as a pawnbroker Hodgkinson was neither a professional man nor, like Fairbairn, a businessman-entrepreneur in a science-related industry, his devotion to scientific matters won him the respect and admiration of the old network which comprised the scientific community. He became a member of the Literary and Philosophical Society in 1826, having read a paper "On the Transverse Strain and Strength of Materials," a paper which brought him national recognition, in March 1822.[39] He was elected to the office of librarian of the society in 1836 and succeeded to the vice-presidency in place of Holme on Dalton's death. He subsequently published six more papers on engineering mechanics in the *Memoirs* of the society until 1844, after which he no longer contributed to the *Memoirs*. His work on strength of materials brought him to the attention of Fairbairn, and the facilities of the firm of Fairbairn and Lillie, a rising engineering company, were placed at his disposal for experiments as early as 1828.[40]

Hodgkinson's national reputation continued to rise, spurred on by his efforts at the British Association and by his election to the Royal Society in 1841 and the award of the Royal Medal for 1841 for his paper on the strength of pillars of cast iron. He discovered that cast iron resists compression six times more than it resists extension; consequently he was able to recommend an optimum cross-sectional area. In 1847 Hodgkinson was appointed Professor of the Mechanical Principles of Engineering at University College, London, and continued to lecture there until 1853. Apparently his appointment at University College did not require Hodgkinson to assume permanent residence in London. He was able to accept the presidency of the Literary and Philosophical Society in 1848.

The character of the society made an abrupt change on Hodgkinson's accession to the presidency. This change was, perhaps, not so much the result of Hodgkinson's personal efforts, but rather the result of the growth in confidence of the younger men, now that many of the old guard had passed from the scene. The battle was joined immediately, and over a trivial issue. After the death of Dr. Holme, several of the members wished to have his paper, *The History of Sculpture,* printed in the next volume of the *Memoirs.* A motion to publish was tendered by Dr. William Fleming and Dr. Charles Clay. Strangely, this seemingly innocuous motion, in homage to a deceased president and long-time officer and member, aroused considerable passion. The issue behind the debate was, of course, no less than the future character of the society. The battle was joined between the devotees, on the one hand, who wished to turn what had become more and more a social club into an active and critical scientific society, and the dilettante remnant. Edward Binney and James Joule, two of the more able of the younger men around Sturgeon, led the insurgent forces. Binney's motion before the council of the society that the paper be read first (a delaying tactic) passed five votes to two. After the paper had been read, Binney and Joule moved that printing be postponed; on this issue they were defeated eight votes to two. Though they lost the battle, the young men won the war. Holme's paper was, with a single exception, the last nonscientific paper read before the society for very many years.[41]

From the beginning of the nineteenth century, the ratio of scientific to nonscientific papers was roughly two to one. In 1848 the ratio dramatically altered: 95 percent of the papers in the *Memoirs* (Holme's posthumous paper being the exception) can be regarded as scientific-technical. Even though it put the society's finances under some strain, the *Memoirs* now began to appear more regularly and with a higher scientific standard of quality than had hitherto been the case. Under Hodgkinson, new names began to appear as officers: Fairbairn, who became a vice-president in 1849; Binney, who joined Joule as secretary when Hodgkinson assumed the presidency; and R. A. Smith, Ph.D., Liebig's student, as member of the Council.

William Fairbairn, the new vice-president, was Hodgkinson's friend, sponsor, and friendly rival. Hodgkinson wrote him in 1840 that "we have both run for some years an interesting race for reputation in practical science, mutually indebted to each other."[42] Fairbairn, a Scot, was born near Kelso in 1789, to a "humble but respectable" portioner and the daughter of a tradesman.[43] At age fifteen he was bound apprentice to Mr. John Robinson, millwright of Percy Main, Northumberland, a colliery engineer. Fairbairn at once began an arduous course of self-education which included a great deal of mathematics and mechanics.

In 1811, after seven years of apprenticeship, he left Northumberland and took up work as a millwright in Newcastle, in London, and in Dublin. At age twenty-four he came to Manchester in the employ of Mr. Adam Parkinson. Of these early days Fairbairn wrote: "I had some leisure for study and made some progress in the first three books of Euclid."[44] In 1817 Fairbairn entered a partnership with a former shopmate, James Lillie. The infant engineering firm began with small orders for machines from fellow Scotsmen like the Murrays; before long it had effected the major changes in J. Kennedy's cotton mill which initially won Fairbairn his reputation. Hitherto all cotton mills had been driven slowly (40 rpm) with large square shafts and heavy wooden drums. Fairbairn suggested using lighter shafts and smaller drums at triple the velocity, thus enabling his engineers to simplify the machinery and make it more efficient. Fairbairn credits this innovation with the prosperity which his company experienced as the leading millwright of the district.[45]

By 1830, then, Fairbairn had won himself a local reputation and, before long, according to Ure, an international one, selling millwork not only in Britain but on the continent as well.[46] The firm had expanded tremendously, indicating a balance of £40,000 and employing over 300 workers. About this time, the firm diversified into such fields as iron shipbuilding; this diversification was apparently the cause of the dissolution of the partnership, and after 1832 Fairbairn carried on from success to success on his own.[47] The Fairbairn success story is one often told and bears no repeating here.

From the very beginning, however, Fairbairn appears to have been driven by far more than the desire for commercial success. "I confess," he wrote about 1840, "that nature had endowed me with a strong desire to distinguish myself as a man of science."[48] In his twin role as man of business and man of science, Fairbairn moved to ease the union of theory and practice. As we have already noted, he supported Hodgkinson's work as early as the late 1820s; he later collaborated with him in scientific work on the strength of materials (see chapter 1). Fairbairn took special interest in the creation of new institutions in Manchester to further both science and science-based industry. In 1824 Fairbairn — along with other engineers and with mill-owners such as Greg, Kennedy, and Murray — took a lead in the establishment of the Mechanics' Institution for the diffusion of useful knowledge and the encouragement of the educable working class. Doubtless Fairbairn saw his own career as evidence of the economic value of encouraging artisan inventiveness; doubtless also the early support of the Mechanics' Institution by the cotton mill owners derived in large measure from their impressive experiences with Fairbairn's ingenuity.

The promotion of scientific industry and the desire for scientific eminence were behind the convening of a scientific circle at his home at Medlock Bank during the early 1830s. In this group was Eaton Hodgkinson; Bennett Woodcroft, the scientific advisor of the Commissioner of Patents; John Elliot, a millwright; and James Nasmyth, the engineer and inventor. "The evenings," Fairbairn reported, "were most agreeably spent — chiefly in philosophical and scientific discussions."[49] Woodcroft and Nasmyth later joined the Lit. & Phil. in the late 1830s and early 1840s. A project which animated the group was the establishment of a quarterly journal to be called *The Workshop,* intended for the working classes. It was to have included articles on industrial and mechanical arts, mathematics, biographies, and so forth, all of which was intended to "interest, stimulate and improve the class." It was further intended to give accounts of practical improvements and to promote moral improvement by including "such matters as would raise the character of the workman, correct dissipated habits, encourage economy, ensure self-respect and render his domestic hearth attractive, instructive and happy."[50] By this time, however, Fairbairn had two establishments, in London and in Manchester, and was employing over 2,000 workers. The others, too, were busy men, and the project withered away.

But Fairbairn never neglected the organizational and institutional needs, as he saw them, of the Manchester scientific and industrial communities. We have already seen in chapter 1 that Fairbairn became a member of the Geological Society and a founder of the Royal Victoria Gallery. In 1848 he assumed the vice-presidency of the Literary and Philosophical Society, and eventually served as its president from 1855 to 1860. In 1861 he served as president of the British Association. The engineering chair at the Owens College owed its existence in part to the efforts of Fairbairn (see chapter 5).

Even after this successful effort, Fairbairn, advanced in years, continued to labor along the same lines. He was the first president and founding member of the Manchester Scientific and Mechanical Society. In his opening address to the society in December 1871 Fairbairn said: "I must endeavour to hold you together, to encourage research and the pursuit of those objects which tend to progress and enlargement of our ideas on mechanical science." With regard to Britain's relative industrial decline since 1851, he continued: "[I] did not think it could be doubted that from the want of sound and first class education amongst the better class of mechanics and artisans we were getting behind. . . . [If] a better system of education was established so that practical mathematics and a knowledge of chemistry were taught there would be raised for the public service a much superior class of men than we possessed at the present time."[51] In 1873, still president of the society, Fairbairn, in an address read by Mr. J. C. Edwards, urged the society to greater efforts along these lines.

I am anxious to impress upon the society the necessity for exertion in every scientific pursuit if we are to maintain our position and cope with the natives of other countries who have equal opportunities and are better educated than ourselves. This is actually the case in France, Sweden and Germany. And in the United States we have to contend with an intelligent and very powerful rival in both scientific and industrial arts. . . . The age of the rule of thumb [is] at an end. . . . The members of this Society will have numerous opportunities of realising these statements if they can spare time and will have the good sense to attend the evening lectures of Owens College under the direction of Professor Roscoe.[52]

In almost fifty years of public effort in the arena of scientific organization, Fairbairn's goals retain a remarkable degree of consistency; he aimed

at nothing less than the promotion of science and industry through education and uplift of the productive classes. However, his public utterances
epitomize the evolution in the conception of science that marked advanced
segments of the Manchester community. In his efforts to organize a Mechanics' Institution in 1824 Fairbairn was concerned primarily with the
diffusion of useful knowledge and in particular with the inculcation of the
elements of science in the advanced workers. By the early 1830s, his abortive project to publish *The Workshop* was intended to supplement the
work of the Mechanics' Institution. By the time of the founding of the
Royal Victoria Gallery however, Fairbairn had placed the emphasis upon
stimulating invention rather than merely inculcating elements; in 1873 the
aims included not only elementary scientific education but the encouragement of research as well. A great deal had happened between 1840 and
1870 to justify this change of emphasis: science-based industry was rapidly
becoming a reality, and the competition of Germany and the United States
increasingly important (see chapter 5).

Fairbairn's rise in eminence nationally as well as locally paralleled a
related phenomenon: engineers began increasingly to figure largely in the
Literary and Philosophical Society. One important participant was John
Frederic Bateman, Fairbairn's protégé and son-in-law. Bateman was born
in 1810 and, after an apprenticeship to a surveying and mining engineer,
struck out on his own as an engineer in Manchester in 1833. He was fortunate enough to catch the notice of Fairbairn, who asked the young man to
join him in a project to improve the water power of the River Bann in
Ireland; the construction of the project was left entirely to Bateman, and
his reputation as an hydraulic engineer was secured. In 1844 he was asked
to investigate the Manchester water supply and in 1846 recommended a
modest plan to bring water from the eastern hills about twelve miles away,
which was rejected as too bold. Soon, however, Bateman was recalled by
the Corporation of Manchester, and this time he recommended a grand
scheme consisting of a series of artificial lakes to be constructed in the
valley of Longdendale, viaducts and advanced hydraulic contrivances
(e.g., one such contrivance was designed to separate clear water from
peaty water). The work was described in a book by Bateman, *The History
and Description of the Manchester Waterworks*.[53] Subsequently, Bateman,
now a recognized authority, was asked to construct Glasgow's supply. In
the Literary and Philosophical Society, his interest lay mainly in meteorology; he became a fellow of the Royal Society in 1860. Like Fairbairn,
Bateman took a vigorous interest in the institutional strength of the scientific community. Bateman was a leading member of the Geological Society, became a member of the Lit. & Phil. in 1840, and was active at the
1842 meeting of the British Association. Bateman was one of those young
men who, like his father-in-law and like Hodgkinson, were convinced of

the importance of the link between science and industry, and, more importantly, of the necessity and value of being up-to-date on the frontier.

James Joule as Devotee

A group more appropriately designated as devotees grew up, as we have seen, around Sturgeon, Joule, and Edward Binney. This circle of friends and scientific colleagues became of importance not only to the Manchester scientific scene, of which they were ultimately leaders, but, in the person of James Prescott Joule, to all of science.

Joule, of course, is the best-known and virtually the only member of the group whose scientific reputation outlasted his life. As a major contributor to the "discovery" of the law of conservation of energy, through his determination of the mechanical equivalent of heat, Joule like Dalton has won for himself a place alongside the immortals in the history of science. It is not our aim here to investigate or to recount the development of Joule's scientific thought; since our goal is to assess the part that the devotees played in the evolution of the scientific community of Manchester, we will restrict our inquiries to Joule's institutional role. By the time of the 1842 British Association meeting, Joule was already acknowledged as an important younger figure in the scientific community.

Joule was born on Christmas Eve 1818 in Salford, the second of five children. His father was a well-to-do brewer, the owner of the Salford Brewery. None of the Joules appears to have enjoyed robust health. James suffered a spinal deformity and in general was a sickly child who spent a great deal of time reading. From age eleven, Joule studied with a series of private tutors and formed an enduring relationship with one, Frederick Tappenden. About 1834 Joule, along with his older brother, began his education with John Dalton. It was James' father who seemed keen for the boys to study with the famous chemist; the physics and chemistry of liquids and gases certainly appeared relevant for budding brewers. They began their studies with elementary mathematics and were just beginning to examine the rudiments of chemistry when Dalton had the first of his strokes in 1837, and their education with him terminated. Of Dalton, Joule wrote in an autobiographical note: "Dalton possessed a rare power of engaging the affection of his pupils for scientific truth; and it was from his instruction that I first formed a desire to increase my knowledge by original researches."[54]

Joule was consumed with research fire. His father gave him a room in the house on Broom Hill and the young man "filled it with voltaic batteries and other apparatus."[55] By the beginning of 1838 he built an electromagnetic engine in an attempt to improve the efficiency of electric motors; he made his debut as a researcher with a series of letters to Sturgeon's *Annals of Electricity* in 1838. In 1839 he began a course of chemistry with John Davies of the Mechanics' Institution and Pine Street Medical School. Joule studied with him privately and also attended his lectures for medical students.[56] When Sturgeon came to Manchester, he found in Joule a young friend and admirer; the nucleus of what may be called the "Sturgeon Circle" began to grow at Broom Hill. Sturgeon invited Joule to appear at the Royal Victoria Gallery in 1840. Joule spoke "to a large audience" on the length of a magnetized iron bar; Joule found that the communication of magnetism to an iron bar increased its length.[57]

Towards the end of the year Joule presented an abstract of a paper before the Royal Society of London entitled "On the Production of Heat by Voltaic Electricity." The Royal Society refused to publish the entire paper, which eventually appeared in altered form in the *Philosophical Magazine* for 1841.[58] In this paper Joule showed that the "calorific effect" produced by the passage of current in a wire is proportional to the square of the magnitude of the current multiplied by the resistance of the wire, "whatever may be the length, thickness, shape or kind of metal."[59] Towards the end of the lengthier published version,[60] Joule examined a theory of combustion which he attributes to Berzelius: "Berzelius thinks that the light and heat produced by combustion are occasioned by the discharge of electricity between the combustible and the oxygen with which it is in the act of combination; and I am of opinion that the heat arising from this, and some other chemical processes is the consequence of resistance to electrical conduction."[61] In short, Joule believed that the heat of combustion is the result of the passage of electricity through some resistance. The analogy between electrical heating and chemical combustion began Joule upon a series of electrolytic experiments that were ultimately to lead directly to his determination of the mechanical equivalent of heat.[62] The paper was refused by the Royal Society; of this rejection Joule later remarked: "I was not surprised. I could imagine those gentlemen in London sitting around a table and saying to each other 'what good could come out of a town where they dine in the middle of the day.' "[63]

At the beginning of 1842 Joule was elected a member of the Manchester Literary and Philosophical Society, having already read a paper before the group during the previous session. Growing in confidence and sophistication, he placed further results before the British Association in June at the Manchester meeting; in his paper "On the Electric Origin of the Heat of Combustion,"[64] he reiterated his view, bolstered by corrected experimental findings that the heat of combustion is an electrical phenomenon, specifically one of resistance to electrical conduction. The question remained: Is resistance to conduction the sole cause of heating effects?

Joule's mode of operation, his scientific style, clearly emerges from these early investigations. Without question, practical, engineering considerations dominate the early papers contributed to the *Annals of Electricity.* Slowly, however, Joule's interests turned from the purely engineering aspects of the electric motor to the physico-chemical character of the forces involved, all the while searching for and weighing the quantitative relationships among them. Rosenfeld aptly describes Joule's work as representative of "the book-keeping mentality of Joule's environment."[65]

After 1842 Joule's scientific career shifted from a relatively private one, centering around his own laboratory and his circle of friends at the Royal Victoria Gallery, to a more public one. Upon election to the Lit. & Phil., he became an active and energetic member and was elected secretary of the organization in 1846. His circle had widened considerably: he had become professionally close not only with his personal friends Binney, Sturgeon, Leigh, and Davies but now also with the Rev. Dr. Scoresby of Bradford, who gave a paper on magnetism at the British Association meeting of 1842 and with whom Joule subsequently joined forces for a paper on that subject. Scoresby seems to have been one of the few outside Manchester whom Joule impressed at the Manchester meeting. Scoresby invited Joule to Bradford for joint work, but parish business pressed, and Joule did almost all the experimental work himself.[66] Another colleague with whom the young Joule was very anxious to form a close working relationship was the young Liebig Ph.D., Lyon Playfair. Playfair was a center of attraction at the British Association meeting, presenting Liebig's results, delivering several papers of his own, and participating actively in the discussions.

The annual British Association meetings were critical for the scientific health and success of the Manchester devotees. These encounters gave the rising young experts a chance to break out of the parochial mold of scientific institutions of the city and to confer, at the frontier, with the British and continental giants of scientific endeavor. The British Association recharged their energies and enabled them to return to Manchester with new spirit for research. It gave them a chance for national reputation; as devotees they saw their scientific horizons spreading beyond the city.

During 1843, while still working in the laboratory in his father's house at Broom Hill, Joule worked steadily on the problem of finding the mechanical equivalent of heat. In a paper read before the Manchester Lit. & Phil. in January 1843, Joule examined heating during the process of electrolysis. He knew from his previous work on electric motors and on the effect of heating, begun in 1840, that the mechanical and heating effects of electricity are proportional to the chemical action of the battery and thus are mutually proportional as well. In the paper Joule supposed the conversion of mechanical energy into heat: "The magnetic electrical machine enables us to convert mechanical power into heat by means of the electric currents which are induced by it. And I have little doubt that by interposing an electromagnetic engine in the circuit of a battery, a diminution of the heat evolved per equivalent of chemical change would be the consequence and this in proportion to the mechanical power obtained." In a note added in February 1843, he records that "I am preparing for experiments to test the accuracy of this proposition."[67]

In August 1843 Joule accompanied Eaton Hodgkinson to the British Association meeting at Cork, where Joule delivered a paper "On the Calorific Effects of Magneto-Electricity and the Mechanical Value of Heat," which supplied the experiments to test the proposition put forth in January. The published version of the paper concludes that heat is generated and not merely transferred from a source in the electromagnetic machine and, more specifically, that a constant ratio exists: "The quantity of heat capable of increasing the temperature of a pound of water by one degree of Fahrenheit's scale is equal to, and may be converted into, a mechanical force capable of raising 838 lb. to the perpendicular height of one foot." A postscript dated August 1843 adds that "I shall lose no time in repeating and extending these experiments being satisfied that the grand agents of nature, are, by the Creator's fiat, *indestructible;* and that wherever mechanical force is expended, an exact equivalent of heat is *always* obtained."[68]

The Cork paper and its published version went generally unnoticed, though it was the first determination of the mechanical equivalent. "With the exception of some eminent men, among whom [were] Dr. Apjohn . . ., the Earl of Rosse, Mr. Eaton Hodgkinson and others, the subject did not excite much general attention."[69] Meanwhile, Joule's father moved from Broom Hill to Whalley Range in the autumn of 1843 and built for the young scientist a greatly expanded laboratory. He began in the interim a collaboration with Lyon Playfair on atomic volumes and specific gravity.[70] The association appears to have been one of collaboration between equal partners, with Joule taking perhaps the more vigorous role. He was not above offering a bit of fatherly advice now and then: "Don't be afraid [he wrote in April 1845] of being called rash, it is a natural thing for successful young theorists to be called so. And remember what Sedgwick said. 'The way to make no errors is to write no papers.' "[71]

The period 1843–47 marked the evolution of the paddle wheel experiments determining the mechanical equivalent of heat, for which he became so renowned. By the time of the 1845 British Association meeting at Cambridge, Joule was sufficiently well known to be elected secretary of the chemical section. During his paper delivered to the chemical section, he exhibited an apparatus consisting of a paddle wheel operated by weights on pulleys. The work expended revolving the paddle wheels produced a rise in temperature of the water, duly measured.[72]

For such delicate temperature measurements as were required, Joule contracted for improved thermometers invented by J. B. Dancer of the firm of Abraham and Dancer. Dancer was a colleague of Joule at the Lit. & Phil.; the latter sent copies of his instrument to Playfair in London and to Thomas Graham.

The year 1847 was a crucial one for Joule. His experiments on the mechanical equivalent were for the most part complete; he had reached in his own philosophical development a point at which the broader implications of his work were becoming clear. On the 28th of April, 1847, Joule gave his famous lecture at the St. Ann's Church Reading Room, "On Matter, Living Force, and Heat." It was in this talk that Joule expressed most clearly and cogently his views on the conservation of what would later be called "energy." The lecture was published in its entirety in the *Manchester Courier* and lay unappreciated until William Thomson's rediscovery of it in 1884.[73]

It was in 1847, too, that Joule's ambitions had for once been kindled beyond his independent, devotee position. Since the age of fifteen in 1834 Joule had worked at the family brewery. From nine o'clock in the morning until six in the evening Joule spent the choicest parts of the day in commerce; scientific investigation remained his passion, but he was able to devote only his spare time before breakfast and after business hours to it.[74] Lyon Playfair suggested, however, that he apply for an academic position at St. Andrew's in Scotland. Playfair reports that Joule was rejected owing to his slight deformity which was, apparently, an objection in the eyes of an elector.[75] A letter from Joule to Playfair in May 1847 seems to indicate that Joule claimed that it was actually he who declined a proffered chair:

I begin to see some fruits from my writing for testimonials *for the chair I have declined.* I am not sorry I wrote for them because it gives me an insight of the opinions of divers people. Graham gave me a very good one. Faraday refused on the alleged ground that one Johnson cheated him some years ago! I received a letter from Becquerel this morning containing a first rate "certificate" which is very gratifying to me. . . . I remain however decidedly of the opinion . . . that a *mathematician* ought to hold that chair. And I shall have enough to do I suspect this half-year in furnishing a house and marrying a wife without bothering my head with infinitesimals.[76]

Academic position did not again tempt Joule. Roscoe's attempts to enlist Joule for the faculty of Owens College repeatedly failed.[77]

At the British Association meeting that year at Oxford Joule made still another attempt to win the attention of the larger scientific community with a paper entitled "On the Mechanical Equivalent of Heat as Determined by the Heat Evolved by the Agitation of Liquids." In it Joule describes the agitation of brass paddle wheels in water and in sperm oil in his attempt to redetermine the mechanical equivalent of heat. The reception was generally cool, as it had been at Cork and at Cambridge. Joule described it, writing in 1885, as follows:

> When I brought it forward again at the meeting in 1847 the chairman suggested that, as the business of the section pressed, I should not read my paper but confine myself to a short verbal description of my experiments. This I endeavoured to do, and a discussion not being invited the communication would have passed without comment if a young man had not risen in the section, and by his intelligent observations created a lively interest in a new theory. The young man was William Thomson [later Lord Kelvin] who had two years previously passed the University of Cambridge with the highest honour, and is now probably the foremost scientific authority of the age.[78]

Thomson, writing in 1882, recalled the incident somewhat differently in the following fashion:

> I made Joule's acquaintance at the Oxford meeting, and it quickly ripened into a life-long friendship. I heard his paper read in the section, and felt strongly impelled at first to rise and say that it must be wrong because the true mechanical value of heat given . . . must, for small differences of temperature, be proportional to the square of its quantity. I knew from Carnot that this *must* be true. . . . But as I listened on I saw that . . . Joule certainly had a great truth and a great discovery and a most important measurement to bring forward. So instead of rising with my objection to the meeting I waited till it was over, and said my say to Joule himself, at the end of the meeting. This made my first introduction to him. After that I had a long talk over the whole matter at one of the conversaziones of the Association, and we became fast friends from thenceforward.[79]

Thomson and Joule collaborated on researches beginning in 1857, culminating in a series of papers, including a number on the thermal effects of fluids in motion, delineating the Joule-Thomson effect.

In 1854, Joule's father sold the Salford Brewery, and James retired from business (at age thirty-six) to devote his entire time to scientific researches. Nationally and internationally his reputation, aided by Thomson's advocacy, was on the rise year by year. The first published notice of Joule's work appeared in a footnote to a paper read to the Cambridge Philosophical Society in June 1848: "In the present state of science no operation is known by which heat can be absorbed, without either elevating the temperature of the matter or becoming latent and producing some alteration in the physical condition of the body into which it is absorbed; and the conversion of heat (or *caloric*) into mechanical effect is probably impossible* certainly undiscovered."[80] The note to the asterisk qualifies the impossibility thus: "This opinion seems to be nearly universally held by those who have written on the subject. A contrary opinion, however, has been advocated by Mr. Joule of Manchester; some very remarkable discoveries which he has made with reference to the generation of heat by the friction of fluids in motion, and some known experiments with magneto-electric machines, seeming to indicate an actual conversion of mechanical effect into caloric."

After this first citation of Joule's work, in the 1850s he began to become well known and well regarded, though not universally. Thomson reports that "I remember distinctly at the Royal Society, I think it was either [Thomas] Graham or Miller saying simply he did not believe Joule, because he had nothing but hundredths of a degree to prove his case by."[81] Still, honors came his way: in 1850 he was elected a fellow of the Royal Society; he received honorary doctorates from Trinity College, Dublin (1857), and Oxford (1860), and was made a member of the council of the Royal Society in 1857.

His reputation in the Manchester scientific community rose commensurately. Almost all his efforts locally went, after the demise of the Victoria Gallery, into the Literary and Philosophical Society. In 1844 he succeeded Eaton Hodgkinson as librarian of the society, and in 1846 he became, along with John Davies, his former chemistry teacher and friend, secretary of the organization. Within a few years of his election, Joule was received into the Council of the society, and his voice was heard in its ruling circles until late in life, when because of failing health he took only an intermittently active role. Joule was elected to the presidency of the society four times before 1880, alternating with his good friends Binney and Edward Schunck, Ph.D., in that post.[82]

In 1872 Joule was elected president of the British Association for its forthcoming meeting at Bradford in 1873; unfortunately, failing health prevented Joule from assuming the position and presiding at the meeting. Among his extant manuscripts, however, is a draft of an address presumably prepared for the British Association meeting on the aims of the scientific life, and its absolute importance. This published address is of particular interest here because it renders insight into the public face which Joule wished to present to his colleagues and to the public—the "vocabulary of motives" for the scientific endeavor[83] which Joule saw fit to justify the scientific life.

> The great object which natural science has in view [Joule wrote] is to elevate man in the scale of intellectual creatures by the exercise of the highest faculties of his nature in developing the wonders of the glorious creation. The second and subsidiary object is to promote the well being and comfort of mankind to increase his luxuries. These objects are closely allied and should not be separated. The benefit to be attained is for the entire man, for his soul, his mind, his body. The importance of this object is measured by the importance of that part of human nature which is beneficially affected. The first object is therefore at least as much more important than the second as the intellect is more noble than the body. . . . And

yet it is evident that an acquaintance with nature's law means no less than an acquaintance with the mind of God therein expressed. This acquaintance brings us nearer to him if such a term may be used in respect of the Infinite.[84]

The religious motive for the practice of science is one which Joule stressed throughout his life. From his early papers and addresses, including the famous talk at St. Ann's Church in 1847, it is clear that Joule expresses his affinity for science in the vocabulary of the devout Christian. It may be, as Rosenfeld has maintained, that one ought not to seek the essential component of his scientific attitude in attitudes and utterances such as these; they are, according to Rosenfeld, "conditioned by uncritical acceptance of social conventions."[85] It is significant, however, to weigh Joule's perception and acceptance of these conventions — Joule is after all no hypocrite — and to assay his interpretation of them. In the quotation above, Joule sees science not merely as the *contemplation* of God's works, but rather as the process by which man elevates himself by (in Joule's words) *developing* the wonders of the glorious creation. Science as *praxis* and not mere contemplation enables him to move smoothly on to the secondary object, the easing of man's material condition: "When the cornerstone of the earth was laid the morning stars sang together and all the sons of God shouted for joy. And although it is not given to mortals to know or to enjoy as do these exalted beings, yet it is our high privilege to learn by patient search and so acquire a body of truth perpetually increasing in quantity and in kind and bringing us nearer and nearer to the fountain of all light."

The utilitarian character of natural science needed less elaboration:

The second and subsidiary object of natural science is to promote the comfort and well-being of man and to increase his luxuries. It is needless . . . to . . . enumerate the thousands of ways in which science ministers to this object, for everywhere may we trace its beneficial influences. In the rudest state of society the bare necessities of life, food, raiment and dwelling are obtained, but whatever comfort or beauty exists it is in application of science. It assists and supplements our natural powers by calling in the forces of nature. Heat, magnetism, electricity, chemistry, geology, optics, mechanics enable man to centre in himself and in a far higher degree the naturally superior advantage possessed by a variety of creatures such as strength, swiftness, extended vision. Natural history in all its branches gives him better food, supplies him with medicine, shields him from disease and so tends materially to lengthen the span of his life.

Unfortunately, according to Joule, science could be misused as well as properly applied:

It is deeply to be regretted that another and more unworthy object has been [instituted] and has periodically and alarmingly increased in prominence. This is the improvement of the art of war and of implements for mutual destruction. I know there are those who think that these improvements will tend to put an end to war by making it more destructive. I cannot think that such an opinion is based on common sense. I believe war will not only be more destructive but be carried on with greater ferocity. Individual campaigns will doubtless be short as well as decisive but this will necessarily cause that rapid rise and fall of states and unsettling of boundaries and constitutions which must eventually deteriorate civilization itself and render peace impossible. And thus by applying itself to an improper object science may eventually fall by its own hand.

It is likely that in this alarmingly prophetic statement Joule was referring to the recent Franco-Prussian war and to the social, territorial, and constitutional upheavals which followed it. He concluded: "In reference to this subject we must also deplore [the] prostitution of science for the aggrandisement of individuals and nations, the result being that the weaker is destroyed and the stronger race is established on its ruins."

Since the scientific endeavor is not without its values, it is clear that Joule may accordingly demand much of the scientific practitioner. If science is to be pursued for the love of God's handiwork and for the relief of man's estate, the scientist must be continually aware of the good, the true, and the beautiful.

One of the chief characteristics of the spirit in which science should be pursued is a love of the wisdom which it unfolds, a love of truth for its own sake independently of a regard to the advantages of whatever kind are expected to be derived from it. The very name *philosophy* shows that this feeling has always been considered the leading one. The pleasure experienced in contemplating the beautiful, the harmonious, the beneficent and the great naturally produces love and this naturally leads us to seek a more intimate acquaintance with these beloved objects.

Secondly, the scientist requires an abiding curiosity, to sustain him through the arduous task before him: "Then there should be certain inquisitiveness of mind to know that which is unknown, a principle which is one of the most important that belongs to our nature and is in fact the principal cause which reconciles us to life which would be miserable indeed if it presented merely the recurrence of the same objects and gave us no hope of varied and fresh enjoyments."

Thirdly, the scientist must be humble before nature. The overbearing or arrogant man is unsuitable for scientific work, for these character defects would of necessity interfere with his scientific judgment:

A third great requirement of him who pursues natural science is humility. Other acquirements may be compatible with pride of heart, to some it may be an essential qualification. A certain [position] may be attained by pride, nay its [votaries] may be the most successful in obtaining wealth and honourable distinction. But science will have none such. The state of mind of the proud man is wholly inimical to success in the pursuit of truth and invariably tends to the dark paths of error. In studying the works of nature a man should commence with the sentiment of his ignorance, that he knows nothing and has everything to learn, he ought to have modesty in writing his opinions and always be ready to modify or retract them.

All this is odious to a proud man. He approaches the subject armed with his pre-conceived and worthless hypothesis, he proceeds to the profane attempt to twist and distort the laws of nature to adapt them to his hypothesis, he ends by accumulating and sometimes unfortunately he publishes a mass of pretended facts and fallacious theories. Thus a humble and teachable spirit are [*sic*] necessary to everyone who wishes to pursue truth successfully. Other characteristics are diligence, effort, patience, yet these [are those] which are requisite for success in any walk of life and no less in that which we are now considering. The latter receive a healthy stimulus from a well regulated love of approbation and the high hope of success.

Anyone acquainted with Joule's personal history must conclude that the above quotation contains considerable autobiography. The career of the young man Joule is marked, both in public appearance and private reality, by those very characteristics of the scientist which Joule is claiming as necessary: humility, diligence, effort, and patience were the watchwords of his scientific life. Joule was himself an essentially conservative man politically, religiously, socially, and temperamentally. Perhaps reinforcement came — if reinforcement was indeed necessary — at Dalton's funeral, when the paradigm Manchester scientist was almost deified for these very virtues. The last sentence quoted above is particularly interesting. It reveals the hallmarks of the unseen and unspoken social regulation of the devotee segment of the scientific community of Victorian Manchester: "A well regulated love of approbation and the high hope of success" provided the cement which strengthened the system of values which held this group together.

It is clear that Joule was strongly committed to the "values" of scientific endeavor which we have come to associate with pure research. He was stimulated and excited by discoveries, and in particular by the process of discovery itself. He was intensely concerned with filling out our picture of the detailed workings of the natural world, and saw his mission to aid in such a grand project. He was committed to the search for what he termed "truth" and to the avoidance of the "dark paths of error." This was his major concern. His daily bread may have derived from his father's brewery, and he may have, as he reported, spent every day there from nine to six until it was sold. But his commercial life was not his real life; his calling was in science, not in beer. As the second son in a family with a successful but not particularly prestigious business, Joule sought his role elsewhere.

Sociologists of science tell us that possession of these values of scientific endeavor is not enough for community coherence in contemporary professional scientific life.[86] Supplementing the socialization process (which for professionals is mainly the result of the graduate training) is some system of internal social control. W. Hagstrom has analyzed competition for recognition among scientists as a form of social control operating among professionals. While the devotees of Victorian Manchester were not yet a professional community, the need for scientific recognition and the desire for success as defined within the limits of the larger scientific community provided the motive power behind the group, impelling them deeper and deeper into the scientific enterprise.

Recall Joule's own career. By his own testimony it was Dalton who instilled within him the desire to, in his own words, "increase my knowledge by original researches."[87] But the desire to increase one's own knowledge does not necessarily invite one into the scientific community, much less stimulate a young nineteen-year-old boy to contribute the results of the first trials to a scientific journal as did Joule. It was rather his "well regulated love of approbation and the high hope of success" that pushed him on. It was a desire for recognition that led him to lecture before the Royal Victoria Gallery in 1840, to speak before the British Association in 1842 and regularly thereafter. It was not merely a love of truth which caused him to be anxious about the publication of his address at the St. Ann's Church.[88] Among his first collaborative researches was that with Dr. Lyon Playfair, the young Liebig Ph.D., who had already become a celebrity at the 1842 British Association meeting. The Joule-Playfair collaboration was a marriage of convenience as well as of division of labor. For Joule, Playfair's cooperation assured a measure of respectability; after all, Playfair was an internationally recognized celebrity, and joint publication would give entrance into circles which included Liebig, Buckland, and Graham. For Playfair, the team effort meant incorporation of Joule's ideas and experimental expertise under the team name.

Very early on in his career, Joule sought recognition beyond the parochial confines of the city. The Royal Society's refusal to publish "On the Production of Heat of Voltaic Electricity" *in toto* in the *Proceedings* set him back, but it did not crush his confidence. He reached a national audience with an altered version of the paper in the *Philosophical Magazine* shortly thereafter. His failure to secure an academic post at St. Andrews (or, if his letter to Playfair is to be believed, his rejection of the proffered chair) caused him to seek institutional recognition in the local community, using it as a safe refuge while expanding his broader reputation. In this context, one may understand his repeated appearances before the British Association, seeking their elusive support year after year; his affinity with William Thomson, who "discovered" him; and finally his collaboration with Thomson in which Joule assumed a subsidiary role, one which placed him almost in the position of experimental assistant.

The national and international honors came of course: the honorary doctorates, the memberships in the scientific societies, the publications in the most prestigious journals. Joule's sense of vocation was amply justified by the recognition of his scientific colleagues and by society at large. Joule was able, late in life, to display his humility. Dining with Lord Rayleigh at the home of Arthur Schuster in 1885 he remarked: "I believe I have done a few little things but nothing to make a fuss about."[89] Joule had labored his entire scientific life spanning almost fifty years to be able to make that statement.

Edward Binney: Career Profile

Osborne Reynolds, Joule's biographer, describes his reaction to the death of Edward Binney in the following fashion: "The death of his oldest friend and colleague in the [Manchester Literary and Philosophical] Society Mr. E. W. Binney, which occurred in December, 1881, was a great shock to him; and the Society was never quite the same afterwards."[90] Binney and Joule were different sorts of men; yet they shared a commitment to science that cemented their friendship. If Joule sought recognition in the larger scientific arena, Binney's strength lay in his vigorous efforts on behalf of the local scientific community and especially of the Lit. & Phil. of which, Joule wrote, he was "a leading spirit,"[91] and indeed, the official *éloge* of Binney's life in the *Proceedings* of the society reports that "for many years little was done without his will."[92]

Binney was the seventh son of a well-to-do corn merchant, born on the family estate at Morton in Nottinghamshire in December 1812. His education was respectably suitable for a middle-class boy, passing through small schools before attending the grammar school at Gainsborough, which at that time was kept by Dr. James Cox, a former master at Winchester.[93] Even as a young man, his interests turned to natural studies. At first he was anxious to study chemistry; family pressures turned him to the study of law. However, natural history was his forte. Always observant during his numerous country walks, Binney was an avid collector of specimens. At school, his son reports, Binney read everything he could find on mineralogy, and here he began seriously to collect geological specimens.[94]

In 1828 Binney was articled to Mr. W. B. Thomas, a Chesterfield solicitor. Chesterfield was then a pleasant country town in the Derbyshire hills, in the midst of the coal fields. Binney's tastes for natural history and for geology were gratified, while his professional training commenced. In his spare time he inspected the countryside, becoming interested in the coal pits, conversing with mining agents, collecting fossil shells and plants. He supplemented his observations with reading. Prominent among the authorities consulted was Bakewell's *Geology*.

When his apprenticeship was completed, Binney was off to London to attempt to forge a career in law in the metropolis. 1835 he was admitted to the bar as a solicitor. London would seem to offer less to the amateur geologist than Derbyshire, but an eager young man finds ways. Binney spent a great deal of time at the British Museum consulting popular and technical works and found site visits to the cuttings of railway tunnels an excellent way to gain insight into the geology of the city.

It was Manchester, however, which he selected as the town in which he would establish himself in his profession. His brother Mordecai had business and social connections there, and the town did, after all, possess a reputation even then as a place where much was happening. Binney took lodgings in Cooper Street — across, as it happened, from the Mechanics' Institution. Since his career was off to a slow start, Binney had much leisure and put it to good use. He became a member of the Institution, attending lectures and using to its fullest the more than adequate library there.[95] One of his first friends was Mr. John Leigh, at the time a physician's clerk at the Royal Infirmary. Leigh, too, was a bright, eager young professional man who shared Binney's abiding interest in natural history and geology and who, as a medical man, appears to have been well trained in chemistry as well.

Binney and Leigh became fast friends. Indeed, Binney often accompanied Leigh on his rounds, waiting outside while Leigh saw his patients. On one of these occasions, Binney inspected a nearby excavation. Cutting the red marls thrown up by the culvert, Binney found casts of shells. He joined with Leigh in collecting further specimens, and the two young men presented a joint paper on the shells to the Literary and Philosophical Society late in 1835. This was Binney's debut in the Manchester scientific community. His career was slowly launched however. In the very next year, he and Leigh wrote a paper on fossils they had found at Newton, which was read before the London Geological Society, of which Binney later became a fellow.

Binney was one of the prime movers behind the formation of the Manchester Geological Society, the background to which has been discussed in chapter 1. Elected secretary in the society's very first year, he ultimately served as president of that group. There may be a good deal of truth in the claim made by R. Angus Smith, who knew him well, that Binney's interest in geology was reinforced by his "belief that by studying questions relating to coal he might largely be employed by coal owners."[96] Certainly his efforts in geology centered upon coal, an interest bolstered by his conviction that "true geological investigation, like charity, should begin at home."[97]

That such eager young professionals as Leigh and Binney would turn to geology is not surprising. Spurred on by the general appreciation of utilitarian social philosophy and by the effects of evangelicalism in religion, such a promising, progressive, and useful science as geology could scarcely fail to appeal. Putting aside for the moment the strong links between geological science and the mining industry, it is clear that of all the sciences geology was the one which most closely reached the position of a "popular" science. As with other natural-historical disciplines, access to the materials of the discipline were readily available to the enthusiast, but academic training was not requisite for serious pursuit of the subject. Moreover, prosecution of geology as a past-time afforded the amateur a welcome opportunity of quitting the smoky, noisy city for quieter, pleasanter surroundings.[98]

At Oxford and Cambridge, Buckland and Sedgwick were establishing national reputations. Both were men of impressive and strong personality and were much honored. In 1830 the first volume of Charles Lyell's *Principles of Geology* appeared; Lyell's works appearing throughout the decade, giving evidence of Lyell's expository gifts, made the principles of the subject accessible to a wide middle class audience. During the 1830s, the Geological Society of London assembled a stellar array of practitioners, including Lyell, Greenough, de la Beche, Sedgwick, Murchison, and Charles Darwin, as well as members recruited from the ranks of the great landowners, peers, bankers, and members of parliament.[99] The discipline was given important popular exposure with the founding of the Geological Survey in 1835 under Henry de la Beche. De la Beche's work with the survey produced his famous *Report on the Geology of Cornwall, Devon and Somerset* of 1839 which excited particular interest owing to its full description of the geology of mineral lodes and of the methods of utilizing them.[100]

It was in fact "economic geology" stimulated by the work of de la Beche and others which provided the motive force behind the founding of the Manchester Geological Society and which drew practical men like Binney into the field. The "cut and try" method of seeking out profitable seams of coal, for example, had led to tremendous waste of money and resources. One wealthy but uninstructed man, disregarding the admonitions of the geologists, had drilled for coal near known iron deposits on his faith "that where God has sent iron-ore he also sent coal to smelt it."[101] Ramsay reported that "throughout the length and breadth of the land, down to this very day, . . .equally fruitless and still more absurd undertakings are being constantly entered upon."[102] In his inaugural address before the Royal School of Mines in 1851, A. C. Ramsay recounted "horror" stories of mining proprietors foolishly digging where no coal could possibly be found and continuing their folly with the excuse that "had [they] gone a little deeper" success would have been theirs.[103]

Ramsay was of the opinion that "even a slender amount of science infused into the general education of the country would strongly tend to prevent the unceasing recurrence of such ruinous absurdities."

The truly practical man [he continued] . . . reasons and advises on very different principles. He is conversant with geological maps and sections; his experienced eye distinguishes the geological relations of the deep and wide-spreading strata of which a country is composed, and as a rule, he knows the utmost limits of the ground where it is safe to adventure; and further if he add to this a general knowledge of the organic forms that characterize the formations, a glance will tell him, however black the shale, . . . that rocks containing graptolites, trilobites, lingulae and pentameri were formed untold ages before the commencement of our carboniferous epoch.[104]

Binney was to become such a "truly practical man."

Binney was therefore already well known in the scientific community when he was nominated for membership in the Literary and Philosophical Society in 1842; he was elected a member along with James Joule and Dr. Edward Schunck in January of that year.[105]

The 1840s were years of success for Binney both in business and in the scientific community. He gave a description of the Lancashire coal-field before the British Association in 1842 and served as secretary of the geology section of the organization. By 1848 Binney was elected an officer of the Literary and Philosophical Society, serving as co-secretary with his friend Joule. Binney was also prominent in these years as a supporter of Sturgeon's Royal Victoria Gallery and, as that organization faltered, Sturgeon's brave attempt at a successor, the Manchester Institute of Natural and Experimental Science. It was in this period also that Binney's studies of the geology of the Manchester district were published widely: in 1843 and 1844, for example, Binney's works appeared in the *Philosophical Magazine,* the *Bibliothèque Universelle, L'Institut,* the *Geological Society Journal,* with translations in German periodicals as well.[106] It was toward the end of this decade and the beginning of the next that Binney entered partnership with James Young for the production of paraffin. The business enterprise made both Young and Binney wealthy men. Young, while still a chemist for Tennant and Company, had consulted Binney for a successful legal defense of Young's stannate of soda patent against a Mr. Mercer. Young often thereafter consulted with Binney on legal and scientific matters and, according to the accounts of Binney himself, Joule, and James Binney, Young was aided materially by Binney's knowledge of coal.[107]

About this time also Binney became associated with the group of young men, devoted to science, who were collecting around the venerable William Sturgeon. Except perhaps for Leigh, Joule became Binney's closest colleague and friend in the community. It was Binney, as was recounted earlier in this chapter, who took it upon himself to rescue Sturgeon from impoverishment, and Binney found himself also the protector of other scientific and literary men in straitened circumstances, such as Butterworth, the geometrician, Bamford, the poet, and Buxton, the botanist. Perhaps Binney's association with "mechanics" at the Mechanics Institution and a very real sympathy for the deserving unfortunate were evidenced in other ways behind what Joule termed "a paternal solicitude."[108]

Despite the press of business and of his legal career, in the 1850s Binney began researches on fossil plants that led to the publication of several papers including a joint paper with J. D. Hooker, the famous botanist and later president of the Royal Society.[109] Shortly after the appearance of this paper, Hooker nominated Binney for fellowship in the Royal Society, to which he was elected in 1856. Binney retained a strong interest in fossil flora. After Sir William Logan, it was recognized that coal seams rest on a rock bed of old soils consisting entirely of vegetable matter. Binney investigated the stigmaria or rootlike fibers in this underclay. He proved that the stigmaria supported a luxuriant vegetation — the sigillaria — which in moist atmosphere at high temperatures formed, through decomposition, fossil fuels.

Partly through force of personality and partly through devotion to science and its institutions, Binney rose to positions of prominence in the scientific community, and particularly in the Geological Society and the Literary and Philosophical Society. From the list of his publications, it is clear that rather than seeking primarily an international reputation, Binney preferred to strengthen his role in Manchester. Joule reports that one of his aims was to maintain the Lit. & Phil. as a publishing institution, and one may surmise that, in the 1860s and 1870s, that became increasingly difficult to do, as many of its more prolific members were by then academic scientists somewhat wary of publishing important papers in a local journal.[110]

Binney became vice-president under Fairbairn's presidency in 1856 and again in 1860 and 1861. Beginning in 1862 he regularly served two-year presidential terms, sharing the office with his good friend Joule and occasionally with R. Angus Smith and Edward Schunck, until his death in 1881. In these years, with Joule, he epitomized the character of the Literary and Philosophical Society. And yet these were difficult years for the devotees. After the founding of the Owens College in 1851 and the arrival among the Manchester scientific community of trained, academic professionals, those who cultivated science — no matter how devotedly — on a part time basis, those without academic connection began to assume a somewhat defensive posture. Binney, his international reputation not nearly so secure as that of Joule, provides perhaps a better case study of the process.

Angus Smith, in his *Centenary of Science in Manchester,* noted that "other men more experienced in vegetable morphology have taken up [Binney's work] and to them, notably to Prof. W. C. Williamson, F.R.S. the task of continuing the observations of Mr. Binney is left, and in continuing we of course expect that improvement is part of the labour. Prof. Williamson has on several occasions objected to the views of Mr. Binney, who certainly did not pretend to a great knowledge of botany and vegetable physiology. This was not pleasant to Mr. Binney but progress must be made."[111] Progress must be made. Although Binney was a liberal politically and admired progress, he was resistant to change in the scientific community and, like Joule, jealous of changes in the form and function of the Lit. & Phil. He aimed, however, at strengthening the society's independent role in the community; he feared that it, like the collections of the Natural History Society, might be swallowed up by the Owens College. He urged the society to retain its rooms, though much enlarged, and offered his assistance in raising the necessary funds for renovation and for the hiring of a librarian and an editor.[112]

Smith was right in suggesting that Binney took umbrage at the efforts of the academics to transcend his work.[113] Matters came to boiling point at the annual meeting of the Manchester Geological Society on 27 October 1868. The subject under discussion was the proposal, violently objected to by Binney, to transfer the Museum of the Society to the Owens College. The candid outburst by Binney reveals his views:

I hold that Geology is to be advanced as a science, not merely by young boys in classes at college, but by practical amateur cultivators. Manchester has always stood pre-eminent for working scientific men. What have we to hope for from Owens College with regard to our Society? Look back to see what its Professors have done for us. When we are in the city of Dalton, the Henrys, Hodgkinson, Roberts and a host of men who have raised it to its present position, why should we look to universities or colleges? What help has been received from them in building up this great hive of industry? It has nearly all been done by amateurs and practical men.

Manchester runs risks, he claimed, in threatening the independence of vital segments of the scientific community. He continued:

I should like to ask the gentlemen connected with the Coal trade of Lancashire what assistance they have had from Professors of science, either at Owens, or any other college? or what they expect for the advancement of the sciences of Geology and Mining from colleges, or young boys brought up at colleges? A college certainly can educate a man, no one doubts that; and a good man when well educated will be better than one without education; but still, it is impossible to have men made to order by colleges and other similar institutions. I consider that a city and a district like ours should have a Geological society with a museum independent of a college. The latter may give a taste for Geology, but for the advancement of the science a society is needed.[114]

The words are harsh and imply a philistinism which was not actually there. Binney realized the value of academic training in geology and mining; he advocated, however, a symbiotic relationship between theory and practice, between classroom and field operation, which did not then exist. Just two years before, in a presidential address before the same group he advocated the establishment of a school of mines in Manchester, similar to the successful one in Jermyn Street, London: "Is it not possible to have in Manchester an institution connected with our Society for the teaching of geology, mining, metallurgy, engineering, and chemistry, where our officers may have the advantages of a scientific training, in addition to their sound practical knowledge in a mine?" Indeed, he continued, if the construction of a wholly new institution along these lines is not feasible, perhaps the existing structure at the Owens College could be utilized:

If this is not practicable at present, and I must confess that I do not like to multiply societies and institutions in Manchester, cannot we establish in connection with Owen's College Professorships of practical mining and engineering and geology? These would be steps in the right direction, and the creation of other chairs might follow. What is much wanted is, that our scientific men should be more practical, and our practical men more scientific. We need not fear over-educating our mining officers.[115]

Was Binney in fact promoting for himself the hope of an academic position? In the absence of any positive evidence to indicate that he was interested in such a prospect, we may dismiss it as a motive. A more likely source for his advocacy is Binney's conception of the role of scientific education, which he held to be vital not only for the young men of the district, but for the working men as well.

We should do all we can, as a Society, not merely to encourage and support the education of the rising generation, which can only be done effectually by a broad and liberal system of national education, but by frequently lecturing to under-lookers, firemen and colliers, on the nature of the gases found in mines, their origin, and likely places to be met with, under what circumstances they become dangerous, the different lamps now in use, and how and when they become perilous. Nothing appears to me more alarming than to place a scientific instrument like a safety lamp into a poor ignorant miner's hands, without giving him some knowledge of the nature of the gases found in coal mines, the principles of the construction of the lamp, and the circumstances under which the lamp becomes a dangerous lamp, and not a safety-lamp.[116]

Binney's view—that no academic organization alone could suffice for the great hive of industry that was Manchester—exacerbated his fears for the independence of the organizations for which he had worked so long and so hard. But he was, ultimately, fighting a losing battle.

John Leigh (1812 or 1816-1888): The Surgeon as Devotee

One of Binney's closest associates and oldest Manchester friends was Mr. John Leigh, surgeon and, after 1868, Manchester's medical officer of health. By the time Binney met and shared quarters with Leigh, the latter had already established himself as a knowledgeable and able scientific man. In the early 1830s, before he had reached the age of twenty-one, and before he himself had obtained a qualification in surgery, Leigh held the chairs of chemistry and forensic medicine at the Pine Street and Marsden Street Medical Schools.[117]

Born at Middleton of an old and highly respectable Cheshire family, Leigh was educated at the Moravian School near Dukinfield and afterwards attended the Pine Street School, served as clerk at the Royal Infirmary, and studied at Guy's Hospital, London, where he received a diploma. Before he received his accreditation, he was already training students; when he himself appeared before Sir Astley Cooper and the accrediting board, Cooper asked the young man about his namesake, John Leigh in Manchester, from whom the Board of Examiners had been receiving certificates. "I am that lecturer," said Leigh. "You have been receiving my certificates and I am proud to say you have never yet rejected one of my pupils." Further questions were dispensed with.

Besides chemistry, Leigh shared a number of interests with Binney, especially geology and botany. His debut as a scientific researcher occurred with the paper he co-authored with Binney, "Observations on a Patch of Red and Variegated Marls, containing Fossil Shells, at Collyhurst, near Manchester," which was read before the Geological Society and which appeared in the *Philosophical Magazine* for 1836.[118] By the beginning of the 1840s, Leigh joined with Binney in the group gathering around Sturgeon at the Royal Victoria Gallery; Leigh was much taken with Sturgeon and, like Joule, wrote a eulogy of him for the *Manchester Examiner and Times,* 14 December 1850, in which he too lamented the sorry state in which England leaves its discoverers.[119]

By this time Leigh had been hired by the Manchester Gas Directors as a chemical analyst. The date of his connection with the company is as yet unknown. The Directors advertised for an "operative chemist" towards the beginning of 1838, and Leigh may have filled this post.[120] Manchester was one of the first towns to make extensive use of coal gas. Phillips and Lee's spinning mill in Chapel Street, Salford, was the first firm to use it in 1805. Gas works were built in Gas Street, off Water Street, in 1817. George William Wood, Thomas Potter, and Thomas Fleming were instrumental, as police commissioners, in bringing the city into the gas business, and a municipal enterprise it remained.[121]

The production of coal gas had, however, several problems which only the science of chemistry would be able to solve. Quality control was very important, for some of the impurities involved in the production of illuminating gas from coal are noxious as well as unpleasant. Coal varies in constituents, and some varieties are more suitable for the production of illuminating gas than others. The chemist, therefore, is useful in testing samples of coal for their suitability and in analyzing samples of the gas produced from them. Furthermore, all coals utilized in the production of gas produce certain impurities, most notable among them being hydrogen sulphide gas, cyanogen, and carbonic acid. Hydrogen sulphide is malodorous, toxic, and when burnt yields corrosive oxides of sulphur. Oxygen and ammonia also can prove troublesome, the former forming water with hydrogen and carbon dioxide with carbon, and the latter causing injury to the distributing apparatus. The gas chemist would therefore be engaged in devising or employing suitable purification measures, that is, employing chemical means of separation using substances which would react with, say, hydrogen sulphide gas and not with the illuminating gas itself. The most common of these, first suggested by Dr. Henry of Manchester, utilized dry lime. Later, Richard Laming devised an "oxide method" using a hydrated ferric oxide to oxidize hydrogen sulphide to sulphur.[122] The gas analyst might also be concerned with gas testing—that is, valuing gas by comparing its light with that yielded by a standard.

Much of Leigh's subsequent published work related to his expertise in gas analysis and to the experience gained in his connection with his service for the Manchester Gas Directors. At the British Association meeting in Manchester in June 1842 Leigh exhibited considerable quantities of benzene, nitrobenzene, and dinitrobenzene, all coal tar derivatives, to the chemical section.[123] The description of his work in the *Report* for that year did not permit identification of the substances. Later, in 1863, Leigh patented his process of extracting benzol by passing coal gas through nitric acid or a mixture of nitric and sulfuric acid.[124] Leigh has some claim, therefore, to the discovery of benzene in the destructive distillation of coal.

In 1851 Leigh presented to the Manchester Literary and Philosophical Society (of which he had become a member in 1849) three papers relating to his work as gas analyst. The first, a lengthy account of the findings of his determinations of the constituents of coal gas entitled "On the Analysis of Gaseous Mixtures," was read on 7 January. In it, Leigh recounted the usual methods which gas makers have employed in determining the quality of coal gas. One of the most common was simply the determination of specific gravity. "The heavier the gas, . . . it is said, the better is its quality." Leigh showed that a gas containing a great deal of carbon dioxide and deficient in light hydrocarbons, and therefore deficient in illuminating power, might weigh heavier and seem better than a gas of higher quality. A second means often used was optical comparison. "After a few trials," Leigh concluded, "the eye fails to appreciate any but large differences." Thirdly, some estimate the illuminating power by measuring the amount of a gas's constituents condensible by chlorine, a method first suggested by Henry and recently recommended by Dr. Fyfe. Leigh rejected this method also, because the olefinic gas is not the only kind which will react with the chlorine, and therefore at best affords only a crude approximation. The method which was most highly recommended by Leigh was likewise suggested by Henry. In this process, the gases are fired by electricity in graduated tubes, and the oxygen consumed in producing carbonic acid is calculated.[125]

Leigh's experience as a gas chemist was to stand him in good stead professionally in the decades to come. Spurred by Edwin Chadwick's *Report . . . on an Inquiry into the Sanitary Condition of the Labouring Population of Great Britain* (1842), Leigh began in the mid-1840s to concern himself seriously with the public health problems of the city (see chapter 4). In October 1844 Leigh spoke before the Manchester Royal Institution on what the *Guardian* described as a subject of the "deepest importance to the whole community." His paper "On Some Circumstances Affecting the Sanitary Condition of the Town of Manchester" was well attended and evoked some lively discussion. Leigh described how he had first come to the subject in his years at the Royal Infirmary. Leigh was convinced the condition of the air, especially as Liebig found the presence of ammonia and ammonium salts and of considerable numbers particles suspended in it, was in large measure responsible for the unhealthiness of the town. The smoke nuisance, he claimed, was destructive of "vegetation, goods, furniture, and health." In postmortems which he had performed as surgeon at the Royal Infirmary he often noted the presence of "black carboniferous material in the lungs" of the deceased. It is no accident that Manchester suffers inordinately from respiratory disease, which he found to be much commoner in Lancashire and Cheshire than in the rest of England. Heavy fuel consumption in these areas was the primary cause of the polluted and unhealthy state of the atmosphere of the town.[126] Of course, proper ventilation of homes was obligatory and "health required a due supply of nitrogenized food—fresh meat and bread," as Liebig had shown; but above all the city must rid itself of smoke and the heavy concentration of particles suspended in the air.[127]

Leigh's views were opposed at once; self-interest is usually narrowly conceived. Alderman Hopkins rose to object that the heavy use of coal, on which the economy of the region largely depended, actually *purified* the air. By locally heating the air, the large number of fires causes the heated air to expand and ascend to higher regions, the purer "country" air flowing in to take its place. In the Jews' quarter in Rome—the lowest, dirtiest, most crowded quarter—"malaria was scarcely known" while in other parts of the city, the disease was a serious problem. The salubrity of the Jews' quarter was obviously, according to Hopkins, the result of the greater number of home fires and the resultant local heating of the air. Hopkins' remarks were greeted with animated applause.[128] Leigh made little reference in his reply to Hopkins' theory. In 1845 and 1846 he made reports to the town council on the condition of the air and the possible unhealthy character of its polluted, smoky state.

Leigh also served on the committee appointed by the Literary and Philosophical Society to inquire into the nature and extent of the potato disease and to find a means of preventing or containing it.[129] By 1849 he was considered by the town fathers one of the "chemical experts" on civic matters to be consulted as needed on matters of importance. In May of that year, for instance, Leigh was called upon to confirm J. F. Bateman's recommendation of Angus Smith's method of preserving water pipes by coating them with a mixture of pitch and oil, which Leigh confirmed would prevent oxidation at only a moderate cost per ton.[130]

In 1850 Leigh teamed with N. Gardiner, superintendent registrar of Manchester, to produce a study of the cholera epidemic in the city: *A History of the Cholera in Manchester in 1849.* The slim volume is of special interest; it presents a vivid picture of the visitation of cholera in an industrial city in the late 1840s and of Leigh's intelligent and sober attempts to analyze the causes and inhibit the spread of the dread disease. It marks an early attempt to use the vital statistics to map the progress of cholera and to deduce its causes. Leigh's aims involved nothing less than the scientific assessment of disease and the public health measures taken against it:

By abstraction or accumulation are general laws attained and the iterations of a single phenomenon may become the general expression of one of nature's constant operations. If, instead of mere guesses concerning what Cholera may possibly be, or may possibly be dependent on, actual investigations were made by men of science concerning what it is, or is dependent on, or even if positive determinations were obtained as to what it is not there would be a better hope of our arriving at something satisfactory in its history, for every negative fact, by limiting the sphere of inquiry, conducts us nearer to the true path. To discover what a thing is not, is next in value to the discovery of what it is. But unahppily the Baconian system so fertile of discovery in physics, has been all but discarded in physic, and every wandering of the imagination that the non-worker has put forth has been hailed and applauded as the true arcanum of thought.[131]

It is as one of the "men of science" he respected that Leigh sought to ensure his reputation, using his "Baconian system" to ferret out true causes and right methods. He began his Baconian review with an account of the physical character of the city: its geology, climate, and drainage. Only then did he proceed to the elaboration of the spread of cholera throughout several parts of the city in 1849.

The first recorded death was in the Market Street district, that of a young prize fighter recently returned from Liverpool "from the haunts where Cholera had prevailed, and which place he had visited for purposes of dissipation" (p. 10). After two weeks numerous cases began to appear — in the St. George's district, in the Ancoats district, and more in the Market

Street area: "After the occurrence of a few cases, the disease seemed to radiate in all directions, and to increase with frightful rapidity" (p. 14).

Leigh and Gardiner found that the disease, in each district, was generally confined to those areas in which the worst sanitary conditions prevailed:

In Manchester, at all events, whatever may have been observed in other parts of England, it is proveable then, that Cholera has almost entirely confined its ravages to those localities or particular streets in which noxious exhalations proceeding from putrifying or otherwise decomposing matters in rivers, canals, stagnant water, etc. prevail; in which the inhabitants breathe a polluted air, arising from bad drainage, overcrowding, and bad ventilation, whose employment is precarious and not very remunerative, and whose means, therefore, are insufficient to procure a necessary amount of suitable and nutritious food. (p. 15)

Still, these circumstances existed before the onslaught of cholera and exist still in the fetid districts afflicted in 1849. "It cannot therefore be assumed that any one of them, or all of them associated, have been sufficiently potential [*sic*] for the production of the disease, but that this disease has arisen from something superadded to all these" (p. 15).

It is true that the aforementioned conditions tend to enervate, weaken, and impair the body and deplete its resistance, yet some outside agency must bear responsibility for the epidemic. Certain diseases, like small pox and scarlet fever, Leigh knew to be "propagated by specific poisons of an organic nature . . . capable when received into a living organism of producing specific trains of symptoms . . . and that these particles of organic matter are capable of being conveyed by the atmosphere as a vehicle from one individual to another; . . . and that thus diseases existing in one individual may be produced in another" (p. 18). The spread of these toxic "organic particles" is facilitated by climatic conditions. As a gas chemist, Leigh was particularly aware that

the quantity of any gas that water is capable of retaining in solution varies with the pressure to which the water and gas are subjected. If the water of sewers, rivers, canals, etc. then be saturated with a number of gaseous matters when the barometer is high, on a decline in the column a portion of these would escape into the atmosphere rendering it offensive. It is partly on this account . . . that before rain, sewers and the vicinities of stagnant waters everywhere become exceedingly offensive. (p. 19)

However, comparing the meteorological tables of 1848 and 1849, nothing unusual is noted; "no relation, so far as we are able to judge or can legitimately deduce, [exists] between any ascertained condition of the atmosphere and the advent or continuance of Cholera" (p. 19).

Several theories, however, have been proposed respecting the general cause of the disease. One group, Leigh reports, believe the cause to be a minute fungus infesting the air and water. "This theory, having received its quietus in the Microscopical Society need not occupy us," Leigh acidly remarked. Another group suggested that the disease might originate in a deficiency of ozone. "At present," Leigh wrote, "we know nothing of the general existence of ozone in the atmosphere at all, and nothing of its functions in the system of nature, nor do we know that it has existed in less than usual quantities in the past year" (pp. 19-20). Thirdly, M. Audrand of Paris and his medical followers have suggested that the cholera outbreak may have its origin in a deficiency of atmospheric electricity. During the prevalence of cholera in Manchester, Leigh enlisted William Sturgeon, his friend and well-known electrician, to make kite experiments to test the theory. Sturgeon reported that nothing unusual about the atmospheric electricity occurred during the epidemic (pp. 20-21).

Putting out of view, then these three latter, as untenable hypotheses, we have seen that in Manchester, at least, the Cholera has almost exclusively affected the inhabitants of ill-drained, ill-ventilated districts, within or adjacent to which are bodies of excessively foul half-stagnant water. . . . Yet it is equally well proved that these alone are incapable of actually producing the disease, and that the latter is the result of some special cause superadded to all these. (pp. 21-22)

We know further that the progress of cholera has been westward, from Asia to Europe, the study went on to state. Had the poison been subject to the laws of gaseous diffusion it would have been extended in the atmosphere equally in all directions. The effects of the disease would be radial and all the inhabitants of each country affected would have been subject to the disease. But this has clearly not been the case. If the disease were brought by the atmosphere alone, it would have reached us first in a dilute and then in an ever more concentrated state. But this is contrary to the actual case. Rather than a general atmospheric phenomenon, Leigh concludes that cholera, apart from its origins, is communicated from one individual to another, "not by mere contact only, but by infecting the atmosphere or clothing, or any medium or media adjacent to, in contact with or extending to a certain distance from the affected subject. . . . The first cases of Cholera which appeared in Manchester . . . were all ascertained to have been imported cases, one from Glasgow, the others mostly from Liverpool. . . . The earliest cases in England occurred in sea port towns" (p. 25). Leigh then provided maps to show the progress of the disease through the city, demonstrating that a family in a particularly hard hit district left the neighborhood and brought "the seeds of the disease" with them to their new surroundings. Regarding treatment, Leigh suggested active attempts to arrest the vomiting and to check the diarrhea, to "restore the secretions" (i.e., to maintain fluid levels as well as possible), and to maintain surface circulation "by the outward application of stimulants." Leigh recommended large doses of opium, in some cases hydrocyanic acid, followed by the administration of powerful astringents such as the nitrate of alumina, along with calomel. "The treatment, whatever it be, must be active; for there is little time for the operation of comparatively inert medicines" (p. 29).

Leigh's synthesis of the medical and scientific man, of the clinical experience with experimental knowledge of chemistry, fashioned his approach to public health problems. His reputation as scientific investigator and analyst, and his undisputed interest in and qualifications for sanitation and public health made him an obvious choice for Manchester's first medical officer of health. On the fourth of September 1867, the General Purposes Committee laid before the city council a recommendation to appoint an officer of health for Manchester, to investigate the rate of mortality, to inquire into the condition of the various districts of the city, to examine the waterways for pollution, and to examine the condition of the sewers. Alderman Vertegans moved acceptance of the report, which extracted heavily from the writings of R. Angus Smith and John Leigh. Vertegans also suggested that the council accept the recommendation of the committee and appoint Mr. Leigh as officer. Mr. Ashmore rose to object to the appointment of an officer who, he assumed, would be powerless to act in any important case, and to object to Leigh.[132] Consideration of the report was delayed until January; at that meeting, concerned councilmen insisted that "Manchester held the unenviable position of being the most unhealthy town in the kingdom," a disgrace it ought to be anxious to remove.[133] By March the position was approved and numerous applications had been received from all over Great Britain, including some from respected medical and military men. The Building and Sanitary Committee recommended four for interview: Dr. C. W. Philpott of London; Dr. E. J. Syson of Wath-on-Dearne; Dr. Reed of the Manchester Royal Infirmary; and Mr. Leigh. Interviews were held and the committee recommended Dr. Reed, a thirty-five year old physician. *The Free Lance* reported that many members of the council wished to choose a candidate[134] whom they could keep under their thumb. When the matter came before the whole council, however, Leigh was championed by James Heywood, his colleague in the Literary and Philosophical Society. Leigh's age, Heywood held, was being held against him and ought not to be. At age fifty-six Leigh was still vigorous, and his experience could not be denied. In 1859 he had been appointed by the Ardwick Nuisance Committee to investigate air pollution among manufacturers, and a great improvement had been noticed. The nomination of Leigh was seconded by Alderman Robert Rumney, a manufacturing chemist, who rose to stress Leigh's ability as a chemist. Reed, Rumney reported, had never published an article in a single scientific or medical journal, whereas Leigh continued a laboratory practice as an analyst, had lectured on organic chemistry, and was an authority on the chemistry of gases. Leigh's scientific qualifications carried the day, and the council overrode the interviewing committee's recommendations, voting thirty-three to twenty-six for Leigh.[135]

Leigh began his term vigorously, to say the least. He soon prepared several reports to lay before the council on the subjects of air pollution and sewage treatment.[136] In May of his first year of office, Leigh declared that respiratory diseases occurred with far greater frequency in Manchester than in other towns and that the fault lay in "impure air," about which something must be done.[137] Alderman Pochin conceded that while he hoped manufacturers would conduct their business "so as to cause the least possible annoyance," he feared lest "anything be done by the Health Committee to cause the removal of important manufactures from the borough." Mr. Barningham, a rail manufacturer, for example, was head of a concern which emitted considerable black smoke into the atmosphere. He employed, however, two thousand workers and paid five thousand pounds a week in wages and indirectly supported even more in the community. Still, the Council voted seventy-six to fourteen for an amendment which would permit the Health Committee to deal with Barningham as it saw fit.[138]

It is clear that it was Leigh's scientific reputation, even more than his medical one, which was responsible for his appointment as medical officer, and that Leigh conceived it his duty to bring to bear upon public health problems all the scientific expertise which was in his power to muster.

Joseph Baxendell, F.R.S. (1815–1887)

In his work on cholera Leigh disdained the view, which continued to be a popular one, that the amount of ozone in the air was intimately connected with the health of the town. In 1881, Joseph Baxendell, another devotee, laid before the Literary and Philosophical Society a paper in which he examined the vital statistics of Southport, a town outside Manchester, over the previous decade and concluded "variations in the actual amount of free ozone exercise a very sensible influence upon the state of the public health."[139] Baxendell's expertise was not, however, in the area of sanitation or public health; he was, for the most part, well known as an astronomer and meteorologist.

Baxendell was the son of an intelligent Manchester tradesman who had raised himself from "humble life" into the "respectable" ranks by his own exertions. Joseph received some tutelage from Mr. Thomas Whalley of Cheetham Hill, but on the whole—in science—Baxendell was self-taught. His interests lay in the observational sciences, and he exhibited a mathematical bent which was to serve him well in the subjects of his choice. At age fourteen, he became a seaman, returning to Manchester in the mid-1830s to aid his father in his business. About this time he befriended Robert Worthington, a scientific amateur who had built himself an observatory at Crumpsall Old Hall near Cheetham Hill. Since Worthington was partially sightless, he was pleased to share with Baxendell the use of the thirteen-inch reflecting telescope, the speculum of which was cast, ground, and polished by the young devotee himself.[140] When the observatory was moved from Crumpsall Old Hall to Altrincham, Murray Gladstone permitted Baxendell to use his own private facilities.[141]

Baxendell is described by his associates as an amiable and simple character, with a retiring disposition not much given to thrusting himself forward in scientific debate nor concerned with personal advancement. His involvement with the Manchester scientific community was therefore slow but, once commenced, steady and fruitful. His earliest observations at Worthington's observatory concerned variable stars, and his first publication appeared in the *Astronomical Society Monthly Notices* for 1848–49.[142] But it was not until 1858, at age forty-three, that Baxendell made his first communication to the Manchester Literary and Philosophical Society. It was probably Worthington who introduced him to the society, and to membership in 1858, but it was Baxendell's devotion to his subject and acknowledged expertise which permitted his election to the council of the society in the following year. By 1861 he was elected secretary, a post in which he served for twenty-four years, and was responsible for the publication of the *Memoirs* and the *Proceedings.* The year of his election to the Lit. & Phil. appears to have been a watershed: in the previous year he had been elected to the Royal Astronomical Society, again probably under Worthington's sponsorship, and he succeeded the Rev. H. H. Jones as Manchester's municipal astronomer in 1859. In 1884 he was elected to fellowship in the Royal Society. By this time he had published over seventy papers and a well-known catalogue of variable stars.

Besides Worthington, the members of the Literary and Philosophical Society most closely associated with Baxendell were his fellow members of the physical and mathematical section, including Edward Binney and J. B. Dancer. Baxendell served as president of the section ten times and was without question one of its moving spirits. James Bottomley has recorded a vivid description of these meetings:

Most of Mr. Baxendell's papers were in the first instance communicated to this section; the quiet social character of the meetings seemed quite in harmony with his retiring disposition; first there was the pleasant half-hour spent in pleasant conversation over tea, and when the cloth was drawn, the half-dozen members or thereabout, who ordinarily composed a meeting would draw up to the green baize table with the President of the Section at one end, a cheerful fire would give a pleasant glow to the old council room, and four wax candles, if they could not compete with modern methods of illumination, were not without a certain charm, reminding us that we belonged to an old society, and recalling the days of Main-waring, Massey, Percival, Henry, Dalton and other old worthies of the Society. To such an audience Mr. Baxendell would read his papers in a low-toned voice, but marked with earnestness.[143]

In a way, Baxendell typifies the devotee-scientist who dominated the society in the period after Dalton and before the professionals assumed leadership. Baxendell's papers, his obituary in the *Monthly Notices of the Royal Astronomical Society* reports, "evince his great mental activity and the earnestness with which he pushed his investigations"; he had "a mind which never seemed to regard anything as not worth knowing."[144] Self-educated, devoted to his subject, yet wide-ranging in his interests, Baxendell's contributions clearly reflect his approach to his science.

Besides his work on variable stars and his catalog of them, he is best remembered for what Balfour Stewart in his eulogy considered his pioneering contributions in meteorology; Baxendell's paper "On Solar Radiation" concluded that the maxima and minima of the heat intensity given off by the sun correspond with sunspot frequency.[145] He discovered the fact that the faculae accompanying the sunspots are thrown behind them with respect to the rotation of the sun.

Baxendell's value to the community, and his effective role in it, is better evinced by his other activities. He often spoke before the society on meteorology and its effect on public health; he took an active interest in storm warnings, and, according to Balfour Stewart, he was the first to propose the use of a universally adopted meteorological system; he warned the community of the forthcoming summer drought of 1868 and enabled the municipal water works to prepare for it; he was correct, it seems, in his warning the town of Southport against a coming small pox epidemic. In all these matters, irrespective of the bases for his conclusions, his reputation was greatly enhanced. Beyond these substantive matters, his assiduous devotion to official duty in the Lit. & Phil. made him a very valuable officer of the group, and it was largely owing to Baxendell that the subjects of meteorology and astronomy received the weight in the society that they did in the 1860s and 1870s. Upon his death, an editorial in the *Guardian* summed up municipal feeling quite neatly: "Baxendell was one of the old school of scientific investigators; it is impossible to imagine him as ever having been tempted by a feverish desire for either fame or wealth. His love of knowledge for its own sake was essentially the delight and solace of his life."[146]

His work, however, did not go without criticism. Like Binney and most of the other devotees except Joule, his career overlapped the entrance into the scientific community of Manchester of the academic scientists; and as with most of the others, his wide range of interests, the "amateur" cast to his writing, and the occasional lapse subjected him to severe questioning. But Balfour Stewart perhaps put the subject in its proper perspective:

We have heard it objected that Baxendell generalized from a comparatively small number of observations; but in a question like this such a procedure is essential to the pioneer. His task is to deduce, with a mixture of boldness and prudence, something of human interest out of the observations already accumulated and thus to stimulate [others] not only to go on with their labour, but to cover more ground in the future than they have covered in the past.[147]

Devotee Science: A Social Analysis

In a paper presented before the Royal Manchester Institution in 1845, "On the Study of Physical Philosophy," John Leigh created a minor scandal by saying straight out what a good many in his audience liked to think privately. After platitudinizing on the "elevating" character of the study of the laws of nature and the beneficial character of natural philosophy in developing the powers of observation, Leigh went on to remark that "nearly all great discoveries had been made and great projects conceived by men sprung from the middle and lower classes of society."[148] He pointed specifically to Peel, Arkwright, Davy, Faraday, Sturgeon, and Mary Somerville. He was only reflecting what a goodly number of commentators all over Britain had been saying for some time. Baden Powell put it equally strongly, though in a different vein: "Scientific knowledge is rapidly spreading *among all classes* except the higher, and the consequence must be, that that class *will not long remain* the higher."[149]

The devotee scientists of Manchester reflect the social changes occurring in the ranks of scientists since the eighteenth century: the slow upward diffusion from the lower segments of society into positions of prominence and importance. Everett Mendelsohn has pointed to the effects of this process on British science;[150] we may in some measure corroborate it for the Manchester community. The devotees were largely recruited from middle and lower segments of the *bourgeoisie:* Sturgeon was a shoemaker and private soldier; Hodgkinson and Joule entered the family businesses of pawnbroking and brewing and Fairbairn and Binney were successful, self-made men of business; Leigh, Bateman, and to a lesser extent Sturgeon capitalized on their expertise to forge careers for themselves. All were self-educated in science and reached a position on the scientific frontier only by dint of arduous labor in the moments spared them by their careers. For these men, of whom none was naturally of the city's cultural, social, or political elite, science was a means of obtaining the *recognition* and the acceptance of gentlemen of accomplishment that others sought in business, politics, or polite literature.[151] All, in a sense, were self-made men, benefiting from Manchester's justified fame as a congenial city for careers open to talent.

The conception of science which these men shared was similar to those held by many of their group. They shared the utilitarians' view of science as vital to the progress of Britain and the health of its economy; they saw their work as contributing, even if in a small way, to that noble end. They also saw science as worthy in its own right, the proper study by man of God's handiwork. As we have seen earlier in this chapter, Joule's genuinely religious conviction justified the study of science in lofty terms; even Edward Binney, much more the materialist, would declare before the Manchester Geological Society:

As students of geology, we hold, in common with all who are working in other sections of natural science that we ought to know as much about the universe we are in as our body, mind and soul can possibly bring within our ken. We hold it as a right and duty that we may search as close, as deep, and as far after knowledge as ever our intellects will carry us, and that there is nothing either in mind or matter that we may not legitimately try to fathom and understand.[152]

The devotee's position as serious but still part-time and independent investigator of nature permitted this dual conception of science to be elaborated in actual practice. Although seriously devoted to work in a single specialty, their interests were still relatively wide-ranging. Theory and application provided no real dichotomy. Their roles as scientists and citizens could be, and often were, united by their application of their science to problems of immediate concern to their townsmen. The independence provided by their "outside" incomes permitted them this. Unlike the academics of the decades to come, there was no push to produce books or papers in a regular fashion. They could and, as in the case of Joule, did concern themselves with the largest issues, the "big questions" that professionals like to avoid. They were able, given the circumstances, to work at their own pace and present their work in a fashion which their peers would appreciate. The literary flourishes which mark the works of Sturgeon, Binney, Leigh, and even Joule were soon to disappear from the scene.

The taint of amateurism remained with them. Binney and Baxendell were often criticized by the academics in the Literary and Philosophical Society. Binney's view that the advancement of science had best be left to societies and not colleges[153] is a reflection of these confrontations. Even Joule, with an international reputation, was not exempt from the occasional patronizing attitude. Joule wrote Henry Roscoe, professor of chemistry at Owens College, to the effect that it would be desirable to appoint a nobleman president of the British Association for 1873. "It seems to me" Joule wrote, "a most desirable thing to get at the sympathy and influence of the upper ten." Roscoe prints the letter in his autobiography, because "it is interesting as showing that he was not at all averse to the appointment of eminent noblemen who are not especially scientific as presidents, an opinion which is not now generally held, it being felt that the first requisite for the President is a distinguished scientific position."[154]

In the case of the engineers, Fairbairn, Hodgkinson, and Bateman, their business was related to their scientific work and offered them new perspectives on it; the problems addressed by them grew out of engineering situations. But all the devotees valued research. All were "frontier-minded" and regarded being up-to-date as a necessity. Hagstrom indicates that transcendent value placed upon research by contemporary scientists, their undoubting view that research is "natural" originates in the socialization process of their professional training.[155] The devotees of Victorian Manchester were self-trained, their valuation of research was more dearly won.

It is no injustice to view devotee science, at least in part, as a search for recognition within the scientific community and for status within the larger society. For most of the devotees, the role of savant drew them out of their "tradesman" status and elevated them into a world in which they could hope to consort with university men, professional men, prime ministers and peers. Doubtless the "inspiration" so necessary for the successful man of science was present; their professed love of science for its own sake was no hypocrisy. As the *Guardian*'s obituary stated for Binney: "It is not, however, as the successful man of business that he will be remembered, but as a genuine lover of science for its own sake."[156] Yet, in social terms, strong reinforcement was present, too. None began as a member of the city elite, either social or political or cultural. All would find their way.[157]

Above all, what linked the devotees was their common consciousness of their "calling," their sense of mutual concern for the advancement of science which went far beyond the desire for cultural elevation and for cosmopolitanism of the founders of the Literary and Philosophical Society towards the end of the eighteenth century. To those who did not share their inspiration, this common bond often translated itself as elitism, as an exclusiveness based upon expertise which violated the amateur traditions of the community. The increasing friction between the dilettante and the devotee, the interested amateur and the scientist, translated itself into sharp differences, not only in the discussions on the floor of the Lit. & Phil., but also in the makeup of the *Memoirs* and the composition of the membership. After the presidency of Eaton Hodgkinson, the name Literary and Philosophical became a misnomer. Henceforth, virtually all the papers published were scientific papers; the *literati* were excluded in all but name. It was on the recommendation of Joule in 1851 that the society voted to publish the *Memoirs* annually, and in the same year the rules were altered to note the date of submission of papers so that the "scientific world will be assured of the precise date of each paper and that proper credit will be given to the Author in case of disputed priority of discovery."[158] Surely a new age had dawned.

The open conflict between the devotees and the dilettantes had been broached as early as 1848 in the dispute over the posthumous publication of Dr. Holme's paper on sculpture; as we have seen earlier in this chapter, the devotee party was led in this instance by Joule and Binney. A more serious confrontation, the Grindon Affair, surfaced in the 1860s and concerned a membership issue. It threatened to cause a serious and perhaps permanent split in the society.

On 7 October 1862 Mr. Leo Grindon, noted amateur botanist and literateur, was proposed for membership in the Literary and Philosophical Society. He had expressed a desire to join the microscopical section; prior membership in the society was required. In part owing to the opposition of the society president, Edward Binney, Grindon's candidacy could not muster the required three-fourths of the membership, and he failed to gain admission. Recently the rules on membership had been altered. Previously a vote of two-thirds the membership was sufficient for the election of new members; the new rules required three-fourths. The new rules helped the cause of the devotees and professionals of the society in making it more selective and converting it increasingly into a society of scientists. Opponents of the rule change charged with much justice that the new rules transferred responsibility for membership to the council of the society, usually dominated by the "scientifics"—that is, the devotees and academics. The council, when it desired, could easily muster over a quarter of the membership to oppose any candidates it deemed unfit.

Grindon was a well-known amateur botanist in the city; the local newspapers remarked that he had done more for the popularization of the subject in Manchester than any other man. As president, secretary, and founder of the Manchester Field-Naturalists' Society, he was the "life and soul" of a popular scientific group. Born in Bristol on 28 March 1818, the son of a solicitor, Grindon was educated at Bristol College, and while still at school, as a teenager, he founded the Bristol Philobotanical Society. As a young man of twenty, he came to Manchester as an apprentice in a warehouse and in the following year became cashier in a cotton firm, a position which he held for over twenty-five years. In Manchester he continued his botanical interests at the Mechanics' Institution and ultimately was appointed lecturer on botany at the Royal School of Medicine (Pine Street) and taught the subject privately. In the spring of 1860 he founded with his friend Joseph Sidebotham the Manchester Field-Naturalists' Society, which had over five hundred friends and patrons.[159]

The *Proceedings* of the Field-Naturalists' Society describe the founding in the following fashion:

The Manchester Field-Naturalists' Society owes its existence to a few gentlemen of the town and neighbourhood who, impressed with the belief that an association for the outdoor study of Natural History would be highly useful and agreeable, formed themselves into a temporary Committee . . . in order to give practical shape to the idea. . . . The Right Hon. the Earl of Ellesmere was invited to accept the office of President. His lordship kindly acceded to the request and Thomas William Tatton Esq. . . . and James Aspinall Turner, Esq. M.P. were then invited to become Vice-Presidents.[160]

The constitution of the group declares that the society is "composed of ladies and gentlemen who are specially interested in Natural History. . . . It is open also to those who, without paying minute scientific attention to the objects of nature, delight to ramble in the country, and find pleasure in the contemplation of its loveliness."[161] The inaugural address was given by Thomas Turner who acknowledged the efforts of Mr. Grindon and who reiterated, "I dare say there are many present who do not profess to know much of any of the sciences we cultivate, and these are the very people we want for members."[162]

These were certainly *not* the kind desired by the council of the Literary and Philosophical Society, and Grindon all too unhappily appeared to appeal to this group. His *Life: its Nature, Varieties and Phenomena* (1856) received "some very antagonistic reviews,"[163] but this did not discourage him from publishing *Manchester Flora* in 1859. In his biography of his friend Sidebotham Grindon provides a glimpse of his approach to his studies: "Can I ever forget the summer evenings when we were accustomed to stroll, almost hand in hand in the lovely Reddish valley . . . getting moon wort and curious sedges, and the great white cardamine, and many another charming plant, helping one another, challenging one another to new discovery, sharing everything."[164]

On 18 November 1862, Grindon was proposed a second time and for a second time rejected for membership. His friends, however, refused to see the defeat as final. David Morris, Sidebotham, and Thomas Nevill, all associates in the Field-Naturalists' Society attempted to push through an alteration of the rules, to reduce the required number of affirmative votes to two-thirds the membership. Nevill admitted that their motive was to insure Grindon's admission. The attempt succeeded, and a major split in the society was precipitated. On 8 April 1863, Joseph Baxendell and Prof. Henry Roscoe, secretaries of the society, circulated a letter unanimously approved by the council censuring the pro-Grindon group for their attempt at "coercion":

[We] believe that a continuance in such a course will materially interfere with the usefulness of the Society and lead to a lowering of the high position which it at present occupies. . . . In conclusion, we express our opinion that the members must henceforth choose whether the Society is to be conducted for the advancement of Literature and Science, or is to be an arena for the tedious discussion of party disputes and personal matters.[165]

Privately, during the month of April the acrimony reached a new peak. The pro-Grindon group, led by Sidebotham, Nevill, and Morris, determined to oust Binney, on whose head they heaped much of the blame for Grindon's rejection. The alliance between devotees and most of the professionals threatened reprisals. There were statements, more than mere hints, that the leading scientists would resign. The reply of Sidebotham, Nevill, and Morris, dated 18 April, deplored that "the Society is openly threatened with a withdrawal of the services of some of its eminent scientific officers in consequence of this independent action of the majority of the members."[166] They asserted that the more relaxed admission rules had served the Society well in the past; it had "worked well under the presidency of such truly eminent and honoured men as Dalton, Fairbairn and others." Privately they accused the devotees and professionals of snobbery and elitism.

The Sidebotham-Nevill-Morris letter evoked still another response signed by sixteen members, led by most of the Owens College faculty in the society, James Joule, James Baxendell, and two of the society's Ph.D.'s, Edward Schunck and R. A. Smith. This letter, undated, strongly supported the position of Binney, extolled his "high standing in the scientific world," the "services rendered by him to the Society," and his "many valuable scientific communications" and deplored the "active canvass" and "regular meetings" of the pro-Grindon group. Then followed a significant statement which clearly delineates the underlying issues:

With regard to general interests, we hold, that this Society which is instituted for the *advancement* of Literature and Science, and not for the *diffusion* of knowledge already acquired, should be conducted with the view of preserving the character thus given to it by its founders; and that a shortsighted desire of popularising the Society would be detrimental to the true interests of the Society, and ultimately lead to its utter ruin.[167]

The devotees and professionals were in no mood to risk the gains which they believed they had made in the preceding decade: "We are convinced that members cannot have failed to observe within the last few years a very decided progress, and a marked improvement in the position of the Society. This position we contend must be maintained and the administration of the Society's affairs entrusted to Council able and determined to carry out the system which has been attended by so much success." (Ibid.) The issue could not have been more clearly stated. The scientific men of the council were determined to retain control of the society's affairs and maintain a "professional" character for it.

The Grindon group responded to what they pointedly referred to as a "Circular without date . . . signed by James P. Joule, R. Angus Smith, E. Schunck and thirteen others" in a letter dated 18 April 1863 in which they provided a complete slate of nominations for the offices and for the council of the Society, headed by Fairbairn.[168] We do not know whether Fairbairn was consulted about the nomination.

On 20 April 1863, the day before the election, the council members supplied their own slate headed by Binney, the present president, accompanied by a letter from Joule, Smith, and Roscoe reaffirming their goals. They asserted that the society, like all other scientific bodies, must be and had always been, "of a limited and exclusive nature." Furthermore, from the openly expressed statements of the opposition, they concluded that it was the latter's aim to divert "the Society from its proper object — the *advancement* of Science" and feared lest "the administration of the affairs of the Society may eventually pass into the hands of those who, however sincere their professions, may not be able to maintain the scientific status of the Society."[169]

It was more than concern for the scientific status of the society which prompted some of the devotee-professional group to publish in the form of a circular an acerbic parody of the pro-Grindon faction, obviously identified as dilettante-amateurs. The parody, dated 20 April 1863,[170] is unsigned but a postscript lists by name "the sixteen snobs" who signed the letter defending Binney; these self-designated "snobs" included Joule, Schunck, Angus Smith, Baxendell, Roscoe, and a number of others. Because the circular so pointedly evokes the devotee-professionals' contemptuous view of the dilettantes, it will be pertinent here to quote from it *in extenso*:

If we were asked to name the distinguishing feature of the age, we should unhesitatingly say PROGRESS. The triumphs of the Railway, the Telegraph, and the Photograph, are illustrations of the general spirit which now pervades mankind. But this feeling is not universal; there are individuals whose only aim appears to be obstructiveness; alone, these persons would be powerless for evil, but when they combine, as at our Universities and some of our so-called learned Societies, they cease to be harmless. (p. 1)

If we use the parody to reconstruct the private charges of the Grindon faction, we easily discern their resentment against the serious scientific practitioners both at the Owens College and within the society itself which, the circular continues, has adopted "bigoted conservative principles, [and] become absolutely effete."

We long observed with disgust the haughty attitude adopted by this Society towards our less antique institutions; and having managed to obtain admission within its exclusive precincts, we found ourselves in a stronghold of intolerance and pride. . . . The Council have evinced a bias for abstruse and therefore useless papers on Chemistry, Geology, Natural Philosophy and Mathematics to the exclusion of more interesting subjects. (p. 1)

The language reflects quite clearly that employed by the diminishing and defensive dilettante group: "haughty," "exclusive," "intolerance," "pride," "abstruse." We have in this document a fascinating illustration of the tensions which marked the transition periods in professionalizing scientific societies.

And to show the want of common fairness, it is a fact they have uniformly refused to print the most valuable papers by the Patroness elect ["Josephine Chesapeake Dyer," a crude parody on Joseph C. Dyer, a defender of the caloric theory and critic of Joule and Rankine] who has shown *without a single experiment* that the recent theories advocated by some of the members of the Council are quite untenable. (p. 1)

Dyer, an American-born inventor and entrepreneur, was a very active civic leader. Involved in the founding of the *Guardian* and in Anti-Corn Law agitation, Dyer served as chairman of the Reform League. He was elected to the Literary and Philosophical Society in 1818 and had been involved in the founding of the Royal Manchester Institution and the Mechanics' Institution in the 1820s. By the 1860s, Dyer was an elderly but still alert man whom the progress of science had largely passed by. His papers against the mechanical theory of heat and Joule's views were always respectful and not polemic,[171] but the council and the secretaries (Baxendell and Roscoe) did not see fit to publish them in their entirety in the *Memoirs,* although accounts of them are printed in the *Proceedings.*[172] Dyer had been a vice-president of the society, mainly by dint of seniority, since 1851, and the devotees were not prepared to add injury to insult. He appears, therefore, for re-election on their slate of candidates under Binney.

The parody continues, addressing itself to the Grindon affair:

We submitted to many slights from the men in office, and might have been still in patience possessing our souls, until the blackballing of the venerable founder of the Field and Flirtation Society [Grindon and the Field-Naturalists] a man who as Professor Williamson justly observed, had done more for science than any man that was ever connected with the Society, aroused us to action. . . . We succeeded in increasing the number of blackballs necessary to exclude a candidate, and we shall not desist from our endeavours until all restrictions as to admission are abrogated, and the Society shall be free to all. Among other improvements which Mr. Grindon intends to propose . . . is the admission of Ladies, who will grace the Coffee Tables by their presidency and enliven the hitherto dull proceedings by occasional playing and singing in addition to which their presence will afford an opportunity of enjoying the festive dance. (p. 1)

Williamson, a candidate for the vice-presidency on the pro-Grindon slate, was professor of botany and zoology at the Owens College. He was no friend of Mr. Binney and probably welcomed the opportunity to embarrass him further. The slate "nominated" by the signatories of the satirical circular, "Joseph Sidebottom, T. H. Navel and David Morish," included David Morris as president, "Sidebottom," Nevill, and Williamson as vice-presidents, with "Lion Hyena Grindon," and some of his associates in the Field-Naturalists' Society as members of the council.

The appointment of President [the circular continues] has occasioned much discussion, but it was felt that the claims of the gentleman named could not be passed over, as he is the Chairman of the Reform Association, and he has pledged himself to permit only such papers to be read as he can himself understand. A guarantee is thus given as to the character of the communications. (p. 2)

The list of papers promised to the society is a ripe spoof of the interests and qualifications of the dilettantes and their allies:

1. On the Union of Science with Utility and Economy, exemplified in a plan for the construction of a building which shall serve as an Observatory in Winter and a Smoke Room in Summer. The Astronomical Instruments not to cost more than £5, by [Morris]
2. An Account of an Expedition of the Field and Flirtation Society . . . wherein it is shewn how he drew upon his wits in naming the plants brought to him, and how he assured the country people that his followers were not foreigners, but simple naturals, who would do them no harm if not touched, by [John Lee]
3. On the Effect of Wind upon Crinolines, by Alderman Polly Hopkins [Thomas Hopkins]
4. Curious Comparative Statistics of Cats and Old Maids by [David Chadwick]
5. On the Sort of flattery commonly known as "Soft Sawder" by T. Turner
6. On Extreme Unction, by the Rev. Professor Williamson
7. Also a conjoint paper on the History of the Three Tailors of Tooley Street by [Sidebotham, Nevill and Morris]

The letter was "signed" by Sidebottom, Navel, and Morish, and to it was appended a postscript: "Observe the date. The circular signed by the sixteen snobs [Joule, etc.] is dateless" (p. 3).

The "snobs"—the devotees, the professionals, and their allies—carried the day. On 21 April 1863, the entire Binney slate was elected. Every officer and member of the council designated in the letter of 20 April signed by Christie, Clifton, and Heelis was elected by the society at large. Mr. Grindon never became a member of the Literary and Philosophical Society. The Manchester scientific community had passed a landmark; the progress of winnowing out the dilettantes was prepared to enter its final stages.

3

The Coming of the Civic Scientists

The New Professionals

The 1840s marked the infusion into Manchester of new scientific blood. These immigrants are recognizably a new scientific type which can with considerable justice be labeled "professional" by virtue of their advanced training, their special outlook, and their determination to pursue a career in science. Lured to Manchester by its reputation as a "boom town" and by its renown as Britain's city of opportunity, they came seeking to establish the utility and relevance of science—and thereby themselves—in a way which had few precedents. It was an exciting time, and they were, for the existing scientific community, exciting people.

They came into a city which had grown helter-skelter in a remarkable fashion, which was scarred by amazing squalor and unrestrained profit, and whose atmosphere and rivers had within living memory become open sewers. Yet despite periodic economic depression, it was a city of unmatched industrial vigor, with a reputation for progressive thinking and for receptivity to new and fresh views.

Of the small band of professionals who came to Manchester in the 1840s, by far the most important was Lyon Playfair (1818-1898), the scion of a distinguished Scots family, born in India and educated in Scotland and Germany. His father, George Playfair, a medical man, was inspector-general of hospitals in Bengal. At age fourteen Lyon entered St. Andrews University, but within a few years was sent to Glasgow as a clerk for his uncle James, a merchant. Chemistry had interested Lyon at St. Andrews (or so his autobiographical sketch relates), and as part of his training for a medical career he determined in 1835 to pursue the subject further. He chose to study not at the Glasgow University, where the celebrated Thomas Thomson, then well past his prime, was the professor of chemistry, but rather at Anderson's University where a much younger and more vigorous researcher, Thomas Graham, directed a laboratory and lectured on chemistry. The "Andersonian" was more technically oriented and aimed its curriculum at the lower-middle and working classes, offering part-time and evening courses of study. Playfair probably studied part-time while continuing as a clerk for his uncle.

Graham had excited a great deal of interest and enthusiasm upon his coming to Glasgow. He lectured first at the Mechanics' Institution; later, upon Dr. Andrew Ure's leaving, Graham succeeded him at the Andersonian. His popular lectures were well attended and without exaggeration can be said to have renewed interest in a subject with a long and distinguished history in Scotland. A future Manchester chemist, R. Angus Smith, has provided a description of Graham in this period: "He did not cause enthusiasm by brilliancy of address, but a certain reserve and a certain feeling of power, as well as of ambition . . . so acted on his demeanour, that students became attracted, and were ready to work beside him and devote themselves to his service."[1] Graham established at the Andersonian, in a modest way, a practical teaching laboratory after the fashion, but quite independent, of Liebig's at Giessen. In it Graham began the training of a number of young men who later were to distinguish themselves as industrial and academic chemists, including Young, Playfair, Gilbert, Stenhouse, Crum, Harvey, Blyth, and Thom. "Graham began his [laboratory] very naturally," Smith later wrote. "His pupils wanted work, and he gave them his to do."[2] Playfair was, he himself reports, one of Graham's three favored pupils; the others were James Young, a carpenter first employed to repair equipment but later an eager assistant and student, and David Livingstone, later to win fame as explorer in Africa.[3]

In 1837, when Graham accepted a professorship at University College in London, Playfair resumed his medical studies at Edinburgh—studies which were cut short, according to his own account, by his allergy to "the atmosphere both of the dissecting rooms and the hospital." After a brief attempt at a mercantile career in India, he rejoined Graham in London in 1838. Young was also there at the time as an assistant to Graham, and Playfair was offered the post of "private assistant in his researches."[4] After a year or so of this close relationship, Graham suggested that Playfair go to the University of Giessen to study under Liebig, whom Graham greatly admired and with whom Graham had been forming close ties.[5] According to Playfair, almost all the students at the Giessen laboratory were engaged in extending Chevreul's work on fats. Playfair himself was to find a new fatty acid in the butter of nutmeg—"myristic acid"—and a new crystalline substance in cloves which he termed caryophylline.[6]

During Playfair's stay at Giessen, Liebig was completing his treatise on agricultural chemistry. Eager for a British audience, Leibig asked the young man to translate it into English and to report for him on another subject at the British Association meeting in Glasgow in 1840. These tasks Playfair agreed to do and returned to Britain for the British Association meeting in June 1840.

The Glasgow meeting was an important event in Playfair's professional life. As an advanced student with some original research to his credit, Playfair already enjoyed some status. He was selected to serve as secretary of the chemical section, presented Liebig's paper on poisons for him, and reported on his own work on fatty acids.[7] At the meeting he formed professional friendships which were to stand him in good stead in the near future. Among them, Playfair singled out two: Henry de la Beche, of the Geological Survey, and, according to his memoirs, Dr. Buckland, the famous geologist, as well.[8] After the meeting he returned to Giessen to receive his doctorate.

In the spring of the following year Playfair received an offer of a position as manager of the dyeing works of the calico printers James Thomson and Brothers, of Primrose, Clitheroe, in Lancashire. Thomson had consulted both Graham and Liebig; both had come up with Playfair's name.[9] Thomson was reputed to be a man of wide culture, both literary and scientific, and the possessor of a fine library and a well-furnished personal laboratory. Playfair would later refer to Thomson as a man of "singular talent and cultivation."[10] His works manufactured expensive prints for the well-to-do.

Thomson was himself a man of some scientific training and accomplishment. He was born at Blackburn in February 1779 of a family that boasted close family connections with the Peels. In 1794 young Thomson left Blackburn to pursue his studies in Glasgow, where he met and befriended Gregory Watt (son of James Watt), with whom he remained close after Gregory's death. About 1795 he began working for his cousins' firm, Joseph Peel and Company, in London. Here he remained for six years, during which time he numbered among his intimate friends Humphry Davy and W. H. Wollaston. It was, in all likelihood, through Gregory Watt that Thomson had first come into contact with Davy; Watt had provided the link between Davy and Dr. Beddoes at the Pneumatic Institute at Bristol, and Thomson was to mediate between Davy and the managers at the Royal Institution of London, Davy's major base.[11] Thomson, Watt, Underwood, and the poets Coleridge and Southey were among the friends whom Davy had visited on his first visit to London in December 1799; Davy had had a scientific correspondence with Thomson in the previous months. In his "Experiments and Observations . . . on Nitric Acid,"[12] Davy credits "my ingenious friend, Mr. James Thomson" with the view "countenanced by all the facts we are in possession of" that nitrous acid ought not to be considered "as a distinct and less oxygenated state of acid, but simply as nitric or pale acid . . . loosely combined with nitrous gas." Thomson was one of the first to inhale nitrous oxide gas, the account of which is preserved in Davy's *Works.*[13]

Thomson himself published two scientific papers: "On the Analysis of Sulphate of Barytes," which appeared in *Nicholson's Journal* for 1809, and "On the Mummy Cloth of Egypt," which appeared in the *Philosophical Magazine* for 1834, as well as a pamphlet *Notes on the Present State of Calico Printing in Belgium.*[14] Thomson was elected a fellow of the Royal Society in 1821 and was an original member of the Chemical Society upon its founding in 1841. Edward Baines, in his *History of the Cotton Manufacture of Great Britain* (1835), wrote that Thomson combined "in an eminent degree scientific with practical knowledge" — an accolade he well deserved.[15] Thomson and other printers such as John Mercer served as corroboration for the widely held notion that the Manchester vicinity owed a great deal to, and uniformly respected, chemical and mechanical knowledge.

Accordingly, given Thomson's scientific accomplishments, his interest in the application of chemistry to dyeing, and his local reputation, it is not altogether surprising that he chose to hire a Liebig Ph.D. to aid him in his dye works, despite the fact that the job he was offering the young man was not, after all, a research position.[16] Perhaps Thomson hoped his chemists (he hired Mr. George Steiner as well as Playfair) would develop patents, as he himself had done. In 1813 and 1816 Thomson had patented processes for discharging Turkey-red by printing tartaric acid on cloth and passing it through bleaching powder or chloride of lime.[17] The second patent involved a process which combined the acid with a mordant (a metallic oxide) which, after the dyed color was removed, was capable of imparting to it some other color.[18]

Almost at once upon beginning work at Primrose, Playfair began to seek to improve the "chemical life" of the district. He interested a small group of young men, ten at first, in attending regular monthly scientific meetings at his home at Whalley. Most of the young men were employed in the chemical and calico printing and dyeing works in the area. Among those attending, besides Playfair himself, were John Mercer (1791-1866), a calico printer and partner at Fox and Brothers, whose chemical treatment of cotton fiber (mercerizing) made him famous, and, in all probability, James Young, Playfair's old school-mate at the Andersonian and in London under Graham. It was at these meetings that Mercer's theoretical interests flowered, and his interesting work on catalysis, which Playfair later extended, originated there.[19] For men like Mercer the Whalley meetings provided their first and exciting encounter with the wider world of chemical research. As a recent Liebig student and Giessen Ph.D., Playfair brought to them the exhilaration of what they took to be research at the chemical frontier; Mercer spoke approvingly of "men *trained* at our Whalley meetings."[20]

The meetings were held monthly; each of the members of the group was expected to make a presentation in rotation, one per meeting, with an alternate kept in reserve in case of the non-appearance of the designated contributor. The majority of papers were chemical, although not exclusively so. The group later grew to about thirty and formed the nucleus of the first chemical section of the Lit. & Phil.[21]

Playfair was not satisfied, however, to confine his efforts to this small and relatively informal group. He began, almost immediately upon his arrival in Lancashire, to turn his attention toward that magnet for ambitions, Manchester. In 1841 he had already published a paper in Sturgeon's *Annals of Electricity,* a journal to which he continued to contribute.[22] In March 1842, under Sturgeon's auspices at the Royal Victoria Gallery, he made his debut before Manchester audiences, on the subject of "The Chemical Relations Now Subsisting Between Plants and Animals."[23] Furthermore, to Playfair's good fortune, the British Association chose to meet in Manchester that year. As at the Glasgow meeting (1840), Playfair was selected to serve as secretary of the chemical section.

At the Association meeting, Playfair was able to assume the role of intermediary between the "locals," of whom he was beginning to know an increasingly larger number, and the major figures of British and continental science who gathered in the city in June 1842. In his role as Liebig's representative, Playfair attracted the attention of the scientific community (see chapter 1); in Manchester, as at Glasgow, it was he who served as Liebig's surrogate and who presented an abstract of Liebig's report to the British Association on "Organic Chemistry applied to Physiology and Pathology."[24] Moreover, Playfair also gave two papers of his own and took a leading part in the discussions. Thereafter, he became somewhat of a scientific celebrity. The local secretary for the British Association was James Heywood, M.P., fellow of the Royal Society, founding member of the Geological Society, and moving force in the Royal Manchester Institution.[25]

It was, perhaps, due to the efforts of Heywood that Playfair received the position of honorary professor of chemistry at the Royal Manchester Institution early in 1843. The Royal Manchester Institution had at its inception provided in its aims for the prosecution of lectures in science and literature before its proprietors. Indeed each year, such lectures were given by distinguished gentlemen contracted for the purpose. At the beginning of 1843, however, the council of the Royal Manchester Institution appointed two honorary professors: Playfair in chemistry and Thomas Turner, a local surgeon, in physiology. At the general meeting in March 1843 the secretary of the council reported: "The council feel persuaded from the very high character and great scientific acquirements of [Playfair] that [his] appointment will give universal satisfaction to all connected with the institution and will be appreciated by the scientific world as an honourable testimony of respect . . . [to one who has] done so much for the promotion of science." By the time of the meeting, Playfair had already completed one course of lectures on chemistry for the Institution, for which he also was commended by the secretary: "the clear and lucid manner in which the subject was treated gave instruction and delight to the large and attentive audiences which attended them."[26]

The Royal Institution professorship was itself unpaid. Playfair received, however, rooms which he fitted up as a laboratory and compensation for courses of lectures delivered throughout the year. The prestige of being the only "professor" of chemistry in the city was a distinct advantage in drawing private pupils to his laboratory. Playfair recalls, in his memoirs, that "pupils came to me in numbers greater than could be accommodated, and I had to secure the services of assistants, one of whom was Dr. Angus Smith, so well known afterwards for his researches on air and disinfectants. . . . My lectures at the Royal Institution were certainly popular and attracted crowded audiences. . . . Dalton was a frequent visitor to my lectures."[27]

Playfair had already been tempted by an offer of the professorship of chemistry at the University of Toronto; Faraday, on the suggestion of Graham, had written to Playfair about it in October 1842. The story is well known. Buckland and de la Beche interceded with the prime minister, Mr. Peel; Playfair was invited to Tamworth and decided to seek a career in Great Britain.[28] De la Beche suggested first that some post in agricultural chemistry be invented for him and, failing that, that Mr. Richard Philips be moved from his post at the Museum of Economic Geology to some other and Playfair be induced to accept it. Peel accepted the idea and broached it with Playfair at their meeting. However, Philips refused to accede to the request, and the Agricultural Society was unwilling to enter into complex negotiations about a new post for Philips. Everything fell through, although Playfair cleverly saw that the makings of a British career in chemistry were spread around him in bits and pieces and that with a smattering of intelligence and confidence he would be able to fashion them to his advantage. He remained in Great Britain and accepted the honorary professorship at the Royal Manchester Institution. There he remained until 1845, when de la Beche arranged a post for him at the Geological Museum and Playfair removed himself from the Manchester community to settle in London. But 1843 and 1844 were active years for him. He lectured at the Royal Manchester Institution and served also on the Health of Towns commission, reporting on the large towns of Lancashire (see chapter 4).

In March 1844 Playfair was seeking testimonials to support his application for a chair at Edinburgh; Dalton wrote on his behalf.[29] The post went, however, to another Liebig student, William Gregory, whom Playfair eventually succeeded upon his premature death in 1858. The year 1844 also saw his famous collaboration with Professor Bunsen of Heidelberg on the gases evolved from iron furnaces. Playfair and Bunsen analyzed the gases at various points in furnaces utilizing cleverly contrived collection apparatus. They concluded, rather startlingly, that almost 82 percent of the fuel escaped as gases from the mouth of the furnace and thus was wasted. They showed as well that an iron furnace may be "dissected" and its operations anatomized as if in a laboratory.[30]

In general, however, Playfair was beginning to regret — at least a little — the declined Canadian professorship. His career was not advancing with the rapidity he had hoped for and which he felt he deserved. Declaiming at the Royal Manchester Institution on the recent progress of chemistry in July 1844, Playfair lashed out against the "class of man who cannot see why years should be spent investigating laws which may after all lead to no practical end. Such are the men who can only see beauty in the infinity of divine wisdom when it shows how to cheapen the yard of calico by a diminution of labour."[31]

The death of Dalton in 1844 only served to enlarge Playfair's status as professor of chemistry in the city but yielded no broadening of his prospects. He had been elected, along with James Joule and Edward Schunck, a member of the Literary and Philosophical Society in January 1842, and at the first succeeding meeting in April 1843 he was elected to the governing council of the society. Without question, the translator and favored student of the renowned Liebig was a rising and acknowledged star in the Manchester community by the time of the passing of Dalton from the scene.

Within a week of Dalton's demise, letters began to appear urging a suitable memorial for the great man. Writing to the *Guardian,* "D" proposed that "every man, woman and child have the opportunity of contributing something however small no matter if it be only a penny."[32] In October, a meeting convened to discuss a suitable monument heard encomia to Dalton by John Davies and John Moore. James Heywood urged the establishment of a Dalton professorship:

While it would be of direct benefit in improving our knowledge of chemistry in connection with manufacture, in which respect we were inferior to the French manufacturers, it would also be of great value and assistance to the students in the medical school and to students in chemistry. Dr. Hope's lectures in Edinburgh were attended by a class of 600 or 700 young men; he [Heywood] was sure that in a pupil of Liebig now resident amongst us, we had a gentleman most competent to fill the chair of chemistry.[33]

Heywood urged not only the establishment of a Dalton chair of chemistry but also the filling of it by Dr. Lyon Playfair, a young man whose services the council of the Royal Manchester Institution (of which Heywood was chairman) had secured the year before. Heywood's suggestion was not without opposition. John Davies, for example, preferred the erection of a monument; he had little desire to elevate Playfair's position in the community at the expense of his own. Playfair was, after all, his competitor in chemical tuition.

A letter soon appeared, signed only by "One of 'Young Manchester'," which openly attacked both the proposed professorship and Playfair himself. The writer pointed out that Dalton and Henry had both flourished in the city without any endowment. Already, he continued, there is a professor of chemistry at the Royal Manchester Institution who "though he has not yet distinguished himself as a chemist is known to have ample opportunities of gaining a knowledge of chemistry. There are besides several other teachers of chemistry who have spent considerable sums of money on chemical apparatus and who have settled here as teachers, viz. Dr. Smith (Liebig's pupil), Dr. Davies (Dalton's successor in the chair of chemistry at the school of medicine), Mr. Bowman, Mr. Day and Mr. Stone. If any one of these be selected for the endowed professorship it will be an injury to the rest."[34] Furthermore, if an outsider be named to the post, the locals will be unable to compete with him for private pupils. "The advantages derived . . . from free and fair competition would be lost. . . . It would hardly be consistent for a community asking for free trade and prohibition of monopolies to create an endowed professorship when there is already plenty of talent in the town. . . . Chemistry like commerce should be unfettered."[35] "Lover of chemistry" replied to "Young Manchester" that a professorship "raises and benefits all who are connected with it."[36]

But nothing, at least for a while, came of the Dalton professorship. Playfair left the city early in 1845 for London and by the time a *Guardian* editorial supported the professorship notion, much of the steam had gone out of the movement, which was to be revived a few years later by John Owens' bequest.[37]

James Young (1811–83) was another Scot and former student of Graham who may have attended the Whalley meetings. The son of a Glasgow cabinetmaker, he began attending classes at the Andersonian while still apprenticed to his father. Like Playfair and his fellow student David Livingstone, Young was attracted by Graham's earnest and dedicated approach to chemistry. Graham found work for him repairing equipment and eventually hired him as his assistant. When Graham was offered the post at University College London in 1837, Young followed him there and remained with him until about 1842 when, on Graham's recommendation,

Young was hired as a chemist by Messrs. Muspratt of St. Helens. About 1843 he moved to Manchester and served as a chemist for the firm of Charles Tennant and Company. It was as a chemist for Tennant that Young renewed his acquaintance with Playfair and he may have attended the circle of chemical workers that Playfair had formed at Whalley.

It was Playfair, on most accounts, who called Young's attention to the Derbyshire oil pit from which Young began his career manufacturing illuminating and lubricating oils. When the spring began to fail, Young turned to the distillation of cannel coal to produce paraffin, and it was the production of paraffin from "Boghead coal" for which Edward Binney, Edward Meldrum, and Young formed their company and made their collective fortunes.

It was only after the removal of Playfair to London in 1845 that Young began to assume a more active role in the Manchester scientific community. By the end of 1845 we find that Young, along with Thomas Graham's brother John and John Thom of Chorley (another Andersonian student of Graham) was serving on a Literary and Philosophical Society committee to investigate potato blight, although Young was not apparently elected to the parent body. After Playfair left, it appears to have been Young who assumed leadership of the chemical group which afterwards assimilated into the Literary and Philosophical Society as a section.[38]

Playfair was responsible for the introduction of yet another Scotsman into the Manchester scientific community, Dr. Robert Angus Smith. For Smith, as for Playfair and Young, it was Thomas Graham who initiated an interest in chemical science. Smith attended his popular lectures, and he and his friends were enthralled by the possibilities opened to them by the earnest young lecturer.

Smith was born near Glasgow in February 1817, the son of a manufacturer.[39] His older brothers were already devoted to scholarship, and one, John, had developed an interest in science which he shared with Angus.[40] Angus, however, showed some talent for classics, and his mother hoped for a career in the church for him. After preparation at the Glasgow High School and a short time at the university, he spent several years as tutor to several families, eventually traveling to Germany with the Rev. H. E. Bridgeman. During his German travels, Smith's interests in chemistry were reawakened, and he began working in the laboratory. His fellow students were Playfair and Schunck, soon to be colleagues in Manchester. Before leaving in 1841, he took the degree of Ph.D. Soon after his return to Great Britain, Smith published a translation of a paper by Liebig under the title "On the Azotised Nutritive Principle of Plants,"[41] but his chemical prospects were dim. He does not appear to have been one of Liebig's most luminous pupils. Never strong as a chemical theoretician, Smith stressed the useful and practical side of Liebig's teaching. T. E. Thorpe's *éloge* perhaps put it best:

His chief point of contact with Liebig lay in his recognition of the utilitarian side of his science; for upwards of forty years he laboured unceasingly to show how chemistry might minister to the material comfort and physical well-being of men — not in the manufacture of new compounds useful in the arts, or in the establishment of new industries — but in raising the general standard of health of communities by checking or counteracting the evils which have followed in the train of that enormous development of the manufacturing arts which is the boast of this country.[42]

What more appropriate city to enter upon the study of public health chemistry than Manchester? After leaving Giessen, without prospects, Smith for a short while toyed with the idea once again of entering a career in the church. But in 1843 Lyon Playfair, whom he had known at Giessen, was in need of an assistant, and Smith accepted the post, occasionally lecturing before the Royal Manchester Institution and aiding Playfair with his pupils. When Playfair accepted Sir John Graham's offer of a position to examine the health of towns in Lancashire, Smith joined him as his assistant in this project as well, and thus began his long and distinguished career in sanitary chemistry.[43] Smith was introduced into the Manchester Literary and Philosophical Society in 1845 and by the end of the 1840s was established as a leading member of the scientific community. In 1852 he was elected secretary to the society and eventually succeeded to the vice-presidency and presidency.

The density of Liebig students in Manchester was extraordinary. A third Liebig student, also trained at Giessen in the period 1839–41 was Henry Edward Schunck (1820–1903). Schunck was born in Manchester in August 1820, the son of a wealthy merchant who had emigrated from Germany to Manchester in 1808. Although the family fortune was made in the foreign export trade, Edward's father had also acquired a calico-printing and dyeing works near Rochdale. After preliminary schooling in Manchester, Edward was sent to Germany to learn chemistry as part of his preparation for managing the dye works. Working for a short while with Rose and Magnus at Berlin, Schunck found his way to Giessen, where he began serious efforts towards a doctorate in chemistry under Liebig. By 1841, under Liebig's direction, he published his first paper *"Ueber einige durch die Einwirkung der Salpetersäure auf Aloe entstehende Producte"* in Liebig's own journal *Annalen der Chemie und Pharmacie*.[44] The paper included his account of the discovery of a new nitro acid, chrysammic acid. By the action of reducing agents on this acid hydrochrysammide is obtained, a substance resembling indigo blue.

Schunck returned to England during the course of 1841 and continued his work on coloring matter. In January 1842, along with Playfair, Binney, and Joule, Schunck was elected to the Manchester Literary and Philosophical Society, certainly the most potent single "class" inducted into the society in its history. On January 4, 1842, he read a second paper before the Chemical Society in London entitled "On Some of the Substances Contained in the Lichens Employed for the preparations of Archil and Cudbear," which was published in the *Memoirs* of the society for 1842 and in the *Philosophical Magazine.*[45] Acting upon the suggestion of Liebig, Schunck had reinvestigated the work of Heeren, Robiquet, Kane, and others on coloring matter dervied from these colorless plants of the Lecanora and Variolaria families. "Our knowledge," Schunck wrote in the 1842 paper, "concerning that department of organic chemistry which embraces the colouring matters . . . is of the most imperfect kind. Though many other branches of organic chemistry have been so thoroughly and accurately investigated that little or nothing remains to be known concerning them, this may be called an unexplored field."[46] The budding dye-chemist succeeded in isolating a crystalline surface which he named *lecanorin,* and, using the combustion technique of organic analysis made famous by his mentor, he fixed the formula as $C_{18}H_8O_8$ for what he termed "one atom of lecanorin" (i.e., for the lecanorin molecule).[47] During the course of the 1840s Schunck continued his work on coloring matters, investigating next the coloring principles of the madder root (rubia tinctorum); Colin and Robiquet were the first to isolate alizarine from it in 1826, and in the period 1846-48 Schunck produced a series of papers in which he demonstrated that the active coloring principle in madder, alizarine, was not present as such in it, but was present as rubian which when decomposed yields alizarine and sugar (i.e., that alizarine is formed only after the death of the plant).

Schunck at once formed close relationships with Lyon Playfair and Angus Smith, whom he knew at Giessen. Playfair arranged for him to lecture before the Royal Manchester Institution in June 1844, and within a few years Schunck was making the rounds of the lecture circuit of the city, speaking for example at the Mechanics' Institution on the chemistry of plants.[48]

By the time Schunck published his first paper in the *Memoirs* of the Manchester Literary and Philosophical Society in 1855 he had already published fourteen papers elsewhere. His reputation established (he was elected a fellow of the Royal Society in 1850) and an officer of the society (secretary with Smith) starting in 1856, Schunck began more and more to turn to the affairs of the society and to publish in its journals. After 1862 he was elected president or vice-president of the society for the next twenty years; as his obituary in the *Proceedings* stated, "thenceforward [he] was seldom out of harness."[49] Within a few years of his return to Manchester he "retired" from his industrial career; his independent means permitted him to devote full time to his researches. It is likely that he found himself overtrained for the post of manager of the Belfield dyeworks; it is also likely that his father, who had an interest in science, concurred.[50]

Before his retirement, however, perhaps in an effort to relieve the tedium of the color-shop of the printworks, Schunck formed a chemistry class, composed of the young men working there. About fourteen men gathered for reading and study under Schunck's direction. His plan called for each to have a copy of some text; after a paragraph or two was read from the text, Schunck introducd an illustrative experiment, and so it went. "I look back," Schunck wrote many years later, "with pleasure to the many hours spent with these honest fellows — they were so genial and unceremonious."[51]

Other Liebig students, associates of Playfair at Giessen, came to Manchester after 1841. James Allan (1825-1866), a Scot, studied with Liebig and the Berlin physicists. Allan was an analytical chemist in Manchester during the 1840s and taught at the Pine Street School of Medicine until 1854. J. H. Gilbert (1817-1901), later associated with Lawes at Rothamsted, knew Playfair at Giessen in 1840 and came to the town as a calico-print chemist in 1841-42.[52]

Playfair and Graham were responsible for yet another trained chemist's entry upon the Manchester scene. In 1846, on the suggestion of both Playfair and Graham, whom he presumably consulted in London, Dr. Frederick Crace-Calvert moved to Manchester where, he believed, a career might be forged in consulting and industrial chemistry. Crace-Calvert was born near London in November 1819 but in 1835 left England for France where he began the study of chemistry, first in a private laboratory in Paris and with Gerardin at Rouen and later under Chevreul and Dumas. After two years in Paris, Crace-Calvert accepted the position of manager of the chemical works of Robiquet, Pelletier and Company, and shortly thereafter became assistant to the lecturer on chemistry at Gobelins.[53] In 1841 he assumed the post of preparateur of chemical lectures applied to the arts and manufactures under M. Michel-Eugène Chevreul, the celebrated chemist. After a number of years with Chevreul, he studied for a short while with Dumas in 1846 and then returned to England.

After settling in Manchester in 1846, he began eagerly to set about making his way in the scientific community. He had already published a number of papers, the first, "Sur l'extraction de quinine et cinchonine," appearing in 1842 in the *Journal de Pharmacie*[54] and others in Liebig's *Annalen* and the *Comptes rendus*. He was an extremely prolific researcher; in the period 1842-63 he produced no less than fifty-seven papers, ten of them jointly, in the major journals of Europe. A dark, animated man of middle height, Crace-Calvert kept a trace of the French accent acquired in over ten years' residence in France. This was no handicap in Manchester; shortly after his arrival he succeeded to the honorary professorship of chemistry at the Royal Institution vacated by Playfair and to the chair at the Pine Street Medical School, once held by Dalton.

Manchester was the likely place of residence for him. Trained in a variety of chemical specialties, Crace-Calvert retained a major interest in the chemistry of color, following in the footsteps of Chevreul, whom he always referred to as "my old master." In addition to his private practice as consulting chemist and his public professional roles, Crace-Calvert ultimately became a successful manufacturing chemist, building a small scale operation producing purified carbolic acid in 1859 to a much larger scale firm producing the greatest volume of it after 1865.[55]

Crace-Calvert was elected to the Literary and Philosophical Society early in 1847 not long after his arrival. He did not, however, rise as the other trained professionals into the inner leadership circles. He never became close with Smith and Schunck nor with the professors at the Owens College in the 1850s and the 1860s. He sought instead to create his own constituency; the shortlived Manchester Chemical Society was virtually his creature. He lectured before it nearly forty times in 1848 and 1849 on such subjects as "Animal Black and Coals," and "On Sulfuric Acid," "On Chlorine," and "On Metals," lectures to which a shilling admission was charged.[56] The Royal Manchester Institution became virtually his preserve — at least for chemistry. Smith and Schunck turned for a platform to the Mechanics' Institution, the program of which was thereby revitalized.

An early lecture by Crace-Calvert at the Royal Institution on the subject "What is Chemistry?" is particularly revealing of the subtle changes in attitudes that the 1840s had wrought in the scientific community of Manchester. He began with what by then must have almost platitudinous: the proposition that science is the base of "all real and substantial progress." "The scientific man," he continued, "did not merely theorize and discuss the abstract laws of nature but looked with strict attention to the application of chemistry to the arts; and also it was essential that practical men should have a sound knowledge of science which was the fulcrum of true and productive practice."[57] He then proceeded to recite the litany of the major benefits of applied science, paying special attention to those effected by chemistry, electricity, heat, light, hydraulics, and so forth. Replying implicitly to the argument that much technological progress had been brought about by practical and not scientific men, Crace-Calvert insisted that the question was not what had been done, but what will need to be done in the future. After 1815 Britain stood virtually alone; now other nations are attaining the mechanical skill which was the root of British success. "Practical men [are] useful but we must give them sufficient knowledge to conduct their works with discernment and to render them capable of appreciating with correctness the discoveries made by their scientific contemporaries." As we have already seen was the case with Fairbairn, Crace-Calvert no longer stressed the education or the laborer and the artisan in the elements of science but now concerned himself with the more difficult problem of educating the managers, the owners, and the investors. It was in this sense, he continued, that we must recognize that if we are to maintain Britain's economic position and success "we must generalise science." The discussion which followed Crace-Calvert's address is likewise revealing. Thomas Turner insisted that "an institution for the teaching of chemistry as applied to the useful arts is a great desideratum in a city like Manchester." J. A. Ransome concurred: "We should appeal to the utilitarian spirit of the time; chemistry was actually the creator of wealth."[58]

There were, however, increasing opportunities for young men to learn practical chemistry in Manchester by the late 1840s. With the coming of the civic scientists to the city, the small number of local science teachers (men such as Davies and Sturgeon) were greatly augmented by trained and ambitious young chemists who were determined to make their living not merely as hired hands in local chemical or calico-printing works, but in a manner befitting their training. Mr. Henry Day, Jr., as early as 1845 advertised private courses in elementary, practical and analytical chemistry for "medical students, engineers, bleachers, dyers, miners, agriculturists, emigrants and others going abroad."[59] Day was president of the Mechanics' Institution; his brother-in-law, Daniel Stone, Jr., was secretary and managing director. In July 1849 Stone resigned his position at the Manchester Mechanics' Institution and determined to make a go of it as a consulting chemist and teacher. His case will provide some insight into the character of the enterprise circa 1849. In the same year that he was elected a member of the Lit. and Phil. Stone opened his laboratory in Oxford Street. The laboratory was well fitted up. It had a working room with furnaces, sandbath, "descending flues for removing noxious products," warm air chambers, steam apparatus, and blowpipe table. The students were to have separate work places with tests, reagents, and apparatus available for their individual use. The premises had as well a private class room and apparatus and drying rooms. Stone offered private and class tuition, in addition to his consulting practice which concentrated upon the lighting, warming, and ventilation of private and public apartments. He hired David Yates as his laboratory curator and was open for business by September 1, 1849.[60] Stone was the son of a successful local tradesman who was educated at Edinburgh and returned to Manchester to pursue a career in chemistry. In addition to his analytic practice, his heating and ventilation consulting and his lecturing, Stone taught at the Manchester Mechanics' Institution and at the Pine Street Medical School.[61]

Crace-Calvert and Smith followed a similar pattern; Schunck had independent means. Each attempted to form a wide network of institutional connections: Smith lectured at the Manchester New College and at the Mechanics' Institution in addition to his consulting work; Crace-Calvert held the honorary professorship at the Royal Institution, lectured for fees before the Manchester Chemical Society, consulted and later began his own chemical firm.

The question remains, in what fashion do the civic scientists form a group or class? This relatively small group of men represents a relatively high concentration of trained chemists, even for so vigorous and important a city as Manchester. Almost all of them (Playfair, Young, Smith, Gilbert, Allan, Schunck, and Crace-Calvert) received extensive practical and theoretical training with some of the most celebrated chemists of Europe, including Liebig at Giessen, Graham, Chevreul, Gerardin, and Dumas. Each was determined to make a career for himself in chemistry and chose Manchester for it. Manchester was a magnet for the chemically trained; the Lancashire area had for some time drawn young men into the works, and the vital cotton industry was a natural attraction for those like Schunck and Crace-Calvert who were interested in dyestuffs.

It is interesting to note that Playfair, Young, Allan, and Smith were Scotsmen; Crace-Calvert came on the advice of Playfair and Graham; Schunck was a local. There was much precedent for a Scottish immigration. One often notes the remarkable large number of Scottish names in the scientific and technical institutions of the city. In the case of the Playfair group, this northern migration may be of some real significance. The 1830s was a time of disastrous depression for the Scottish cotton industry; many mills were closing down and the economic hardships reached across all the industries dependent upon cotton.[62] At the same time, in Glasgow at the university and at the Andersonian under Graham and in Edinburgh fairly large numbers of young men were attending the chemical courses and were convinced of the value of scientific education, especially in chemistry. These young men, students of Thomas Thomson, Hope, or Graham, saw little chance of carving out a career for themselves in Scotland and began to look southward. Graham's own brother, a calico-printer-chemist, settled himself in the Manchester district. Thomas Graham's students followed. Alfred Binyon reported in 1849 that he noted that the demand in the Manchester area for chemists had been met not by locals but by students of the Andersonian, although as of late "foreigners" were being hired as the Scotsmen stopped coming.[63]

The foreigners were, of course, German, and it is possible that the flow of Scotsmen began to dwindle after Graham's removal to London in 1837. Why Manchester? It was, after all, still the most explosive economic area in the country, despite its own hardships in the 1830s. Moreover, there had been much Scottish migration into the cotton towns of Lancashire since the late eighteenth century. Leading cotton firms, the M'Connels, the Kennedys, the Murrays, were all Scots immigrants. There is evidence that successful Scots employers brought families from Scotland to work for them. One finds anxious references to "the vast numbers of Scotch artisans" suggesting contemporary consciousness of a northern technical invasion. The career of Fairbairn, for example, demonstrates the value of Scottish connections and how far a bright Scots boy could go.

Another striking fact which the profile of the civic scientists accentuates is the imprint of Justus von Liebig (1830–73). Playfair, Schunck, Smith, Gilbert, and James Allan, the five Ph.D. chemists resident in Manchester during the 1840s, all received their training at Giessen under Liebig, and all were stamped in some fashion in the Liebig mold.

It is far beyond the scope of the present inquiry to examine in any reasonable detail the life, work, and achievement of Liebig, one of the nineteenth century's most important and innovative chemists. It will be necessary and useful, however, to sketch in briefly the impact which he had on the British, and in particular upon Manchester's, scientific community.

Liebig and the British Chemists

Liebig was born in Darmstadt, the son of a dealer in drugs and colors. First experimenting upon colors in his father's shop, Liebig eagerly developed his interest through wide reading at the Darmstadt library. Subsequently he as apprenticed to a druggist before entering the universities of Bonn and Erlangen, where he received his doctorate. The chemical frontier was, however, in Paris, where the laboratories of Gay-Lussac and Thenard were considered among the most exciting. It was in Paris that Liebig became the protege of Alexander von Humboldt who, along with Gay-Lussac, was responsible for Liebig's appointment as extraordinary professor of chemistry at Giessen in 1824, and it was at Giessen during the next three decades that Liebig founded his school and established his reputation as one of the leading chemists of Europe.[64]

Liebig's contributions to pure chemistry are both wide-ranging and deep and again it is beyond the scope of this inquiry to examine them. His most impotant area of investigation was organic analysis, which before Liebig's time was made only with great difficulty. Liebig's combustion method made it possible for advanced students to make such analyses with relative ease, and it is no exaggeration to claim that the rapid development of organic chemistry after the 1830s was made possible by Liebig's methods and the apparatus that he developed.

As discoverer of new facts and innovator in theoretical chemistry, Liebig ranked among the leaders of his time. Among the areas of his extensive research were investigations into the character of organic acids (he introduced Graham's ideas of polybasic acids into the organic realm), into aldehydes (which he introduced), chloral, and with Wöhler into the chemistry of benzoic compounds and urea. Liebig's work on the benzoyl and ethyl radicals provided the experimental basis for the radical theory with which his name was closely associated.

It is, however, for the Giessen laboratory and the perfecting of the chemical laboratory as a teaching instrument that Liebig is best re-membered. Liebig virtually created the nineteenth-century chemical profession; products of his laboratory staffed the major research centers in Europe. Liebig was not considered a great lecturer. He lectured only infrequently and even his direct instruction in the laboratory was only desultory. But Liebig exerted upon his students what can only be called a charismatic influence; he set the problem and inspired the young men towards a solution. An anonymous eulogy in the *Proceedings* of the Royal Society by a British student of Liebig provides a vivid description: [65]

What unexampled activity reigned in those Chemical Halls! There met and worked from morning to evening the future professors of chemistry, the future manufac-turers. All the dialects of Germany might be heard there, nay every European language, and in one hall, somewhat predominating, the English language — all mixing and yet in order, for every one felt that he was striving for a noble object; he was serving science; in other words he was a pupil of Liebig.

The sense of community and the vigorous striving towards a luminous goal overcame the welter of the laboratory:

And how inspiriting was this meeting together of earnest youths under such a master! Many a noble and life-long friendship was founded by community of work and aims. And Liebig had mostly a bit of good advice or a happy thought which helped one who had got into either a scientific or experimental difficulty.

What can only be described as Liebig's charisma is verified by one of his most influential pupils, A. W. Hofmann:

We remember his fascinating control over every faculty, every sentiment that we possessed; and we still, in our manhood now, remember how ready we were, as Liebig's young companions in arms, to make any attack at his bidding, and follow wherever he led. We felt then, we feel still, and never while we live shall we forget Liebig's marvellous influence over us. [66]

Liebig's army, Liebig's companions in arms: this was the essence of the *esprit de corps* which marked the body of chemical warriors to issue from the Giessen laboratory. It was not, however, merely Liebig's contribution to pure chemistry that invigorated his students, not merely the sweeping away of old hypotheses (which "fell, like captured forts before him"[67]) that captured the imagination of his young disciples. For Liebig and his stu-dents, chemistry was the key to the new industrial age, a potent weapon in the struggle for a new age. Increasingly, in the late 1830s, Liebig turned away from pure chemistry and towards the applied. From the late 1830s until his death, his efforts were mainly devoted to the application of chemistry to agriculture, physiology, and manufacturing. By 1840 he claimed he was tired of laboratory work and that "only the applications attract me, and these must be the object of the later periods of life."[68] He told Berzelius in the same year that he was sick of theoretical discussions.[69]

Yet precisely at this time the new Giessen facility, opened in 1839, was attracting relatively large numbers of British students. Liebig had traveled to England in 1837 to deliver an address before the British Association, and had met during his trip a large number of British chemists; he reported to Wöhler that "among the older [chemists] Thomson is still the best, among the younger, Graham."[70] After this time, increasing numbers of aspiring young British chemists, especially Scotsmen, found their way to Giessen, including Playfair, Smith, Gilbert, Allan, and Schunck.

Liebig's new turn of interest was apparent as early as his Liverpool address, which marked the first exposure of the well-known continental chemist to a British audience. In it, Liebig invited British participation in advancing the frontiers of organic chemistry, a science which was about to lead to amazing progress in all areas of modern life:

Organic chemistry has made its first step, and already its field has been extended to a surprising degree. We meet every day with new and unexpected discoveries. It is, however, remarkable, that in the country in which I now am, whose hospitality I shall never cease to remember, organic chemistry is only commencing to take root. We live in a time when the slightest exertion leads to valuable results, and, if we consider the immense influence which organic chemistry exercises over medicine, manufactures, and over common life, we must be sensible that there is no problem more important to mankind than the prosecution of the objects which organic chemistry contemplates. I trust that English men of science will participate in the general movement and unite their efforts to those of the chemists of the Continent.[71]

The area which excited the most interest in Britain was Liebig's work in agricultural chemistry. It was Playfair's edition and translation under the title *Organic Chemistry in its Applications to Agriculture and Physiology* published in 1840 which introduced the British public to Liebig's views. The book owed its origin to the British Association, which had requested Liebig to produce a report on the state of organic chemistry; the book grew out of his efforts to fulfill this request. The reception of it was astonishing, and it went quickly through six editions in six years.

The "humus theory" was still, by 1840, generally accepted, although attack had been made on it for a number of years. It was held that the land possessed powers which enabled crops to be produced; these powers were exhausted by the growth of plants and replenished by rest or manuring, just as animal strength is replenished by rest and food. The source of fertility was thought to be *humus*, a dark decay product of organic matter.

In short, organic matter, such as manures, was required to replenish the soil, for crops were nourished by the organic ingredients of the land. Liebig's views startled the agricultural and scientific communities, for, if he were correct, a *science* of agriculture was possible. According to Liebig, the foods of plants are inorganic substances, such as carbonic acid, water, and ammonia, and such complex substances as albumen, starch, and fat; animals live on these complex products and reduce them to simpler ones. Manures do not aid crops by providing organic substances for them simply to assimilate, but rather by providing decomposition products for them. For example, manures provide carbon and nitrogen compounds, which form carbonic acid and ammonia or nitric acid in the soil.

One startling conclusion that Liebig drew was that artificial manures were possible; organic manures may be replaced by fertilizers which provide the inorganic substances into which natural manures decompose.[72] Thereafter, Liebig and others began experimenting with the production of artificial fertilizers according to his views.

Almost instantly Liebig became a celebrity in Britain. The *Chemical News* later reported that "the application of chemistry to agriculture and to many of the wants of daily life received so powerful an impulse from Liebig that the popular mind has taken him for the representative of the science in its application to practical purposes."[73] Part of the explanation for the instant success of Liebig lay in the changing structure of the British economy. By the end of the 1840s, the import to export ratio had risen to eighteen percent, with food imports providing as much as thirty-four percent of the whole. Grain alone had risen to ten percent as compared with four percent in the 1820s.[74] Rising imports, especially of grain, and the artificial supports provided by the Corn Laws were, as is well known, a critical political issue. The Malthusian specter picturing an ever-expanding population pressing relentlessly upon limited means of subsistence was one which was increasingly haunting all segments of British society. Liebig seemed to offer a way out. As Hofmann, Liebig's student, wrote later: "Science, meantime, speaking with Liebig's voice, overruled these gloomy forebodings, and showed the collective organism of mankind to be a self-supporting institution."[75] It is no wonder then that the public as well as the scientific world grasped Liebig's position with enthusiasm. Liebig himself induced the alkali manufacturer James Muspratt of Liverpool to prepare mineral manures for wheat, rye, oats, clover, and potatoes.[76] In 1843, after a tour of Great Britain, he wrote to Robert Peel concerning the absolute necessity for agricultural improvement:

> I have arrived at the conclusion that the most indispensable nourishment taken up from the soil is the phosphate of lime. In the course of a Journey through England last year I have become convinced that all fields of that country are in such a state

[i.e., of soil exhaustion]. . . . There is no Question that in a very short time England could entirely [dispense] with the importation of foreign grains if a rich and cheap source was opened to her for the supply of phosphates. . . . The reason of my writing this letter is to acquaint you with the fact that there exist in England layers of fossil guano in a quantity sufficient to provide her with phosphates for centuries to come. . . . These layers are the copralithes discovered by Dr. Buckland. . . . As this material must acquire great value I was led to consider that I might draw from my discovery the advantage for myself. . . . [But] as a man of science and not of Commerce I have thought it therefore preferable to communicate my discovery to her Majesty's government.[77]

Liebig's claim to be a man of science and not a man of commerce was only partially true; as the years passed he permitted himself to be lured into business. The most important such excursion followed upon his work upon animal chemistry. In 1842, before the Manchester meeting of the British Association, Lyon Playfair delivered an abstract of Liebig's report, "Organic Chemistry applied to Physiology and Pathology," which introduced Liebig's ideas to the British audience, and W. Gregory, another Giessen student, edited and translated Liebig's manuscript under the title *Animal Chemistry* in the same year. Gregory was also responsible for the introduction of Liebig's *Researches on the Chemistry of Food* in 1847. Essentially, Liebig divided foods into the *plastic* and the *respiratory,* the former (nitrogenous in character) serving to produce blood, build organs, and replenish the muscles debilitated by use, the latter (largely fats and carbohydrates) providing body heat. About 1865 "Liebig's Extract of Meat Company" was formed, and Liebig himself announced in 1866 that he had assumed the directorship of the scientific department of the company and would work closely with his former assistants Dr. von Pettenkofer, Dr. Finck, and Mr. Seekamp.[78] The famous *extractum carnis,* or meat extract, was to provide in a convenient form the plastic food required in large amounts by active men and women, using the wasted meat of the cattle of the Rio de la Plata region, which were being killed for their leather.[79] Within a few years the company had become a success, and the British government was supplying its troops with the meat extract as a substitute for fresh meat on their Abyssinian expedition.[80]

Shortly thereafter Liebig introduced methods of supplying "artificial mother's milk" (or baby formula) and alternative means of making an improved bread "to replace the old fashioned fermentation process."[81] In short, Liebig himself carried out the promises made in Liverpool in 1837, and, as Hofmann stated, he merited the classical encomium *"illustrans commoda vitae."*[82] His zeal of extending the empire of organic chemistry over the problems of everyday life extended downward even to the utilization of sewage: he was concerned that the sewage system of cities in England brought about the loss of valuable elements necessary for the fertility of the soil. "Of all the elements of the field," he wrote in the late 1850s, "which in their products in the shape of corn and meat, are carried into the cities and there consumed, nothing . . . returns to the field. It is clear that if these elements were collected without loss, and every year restored to the fields, these would then retain the power to furnish every year to the cities the same quantity of corn and meat; and it is equally clear that if the fields do not receive back these elements, agriculture must gradually cease." Mindful of the difficulties, he then urged the use of the contents of sewers as a source of manure elements. British engineers and scientists working in harness doubtless could manage the task. "Intelligence in union with capital, represents a power in England which has rendered possible and practicable things of much greater apparent difficulty."[83] In Manchester, Joule, among others, took up the Liebig standard on this issue and suggested means of sewage treatment with an eye to recovery of the valuable elements; it was "filthy," Joule claimed, not to remove sewage to the soil.[84]

It is no accident, therefore, that the British students of Liebig and especially those who settled in Manchester in the 1840s devoted their careers to the application of organic chemistry to the relief of man's estate, both in industry and in public health; they were, in a sense, Liebig's chemical missionaries. Most came to him at a critical period of Liebig's own career, at that time at which he himself turned his seemingly unbounded energies toward applied chemistry. The charismatic figure provided an ethos and held out to them the promise of glittering careers as scientists both willing and able to tackle the major problems of British society. This was a time, it ought to be remembered, at which the profession of chemistry virtually did not exist; the wish, in large measure, fathered the deed.

They carried the Liebig message back to an England which was in Liebig's words "nicht das Land der Wissenschaft" but in which existed "nur ein weitgetriebener Dilettantismus."[85] Smith and Playfair, as Gardner and Gregory, began with translations of the master; all the Liebig students and their allies would thereafter work out the ideal. In sum, Liebig provided for them the basis for a view of science's role in society which marked the early stages of the professionalization of science in Manchester. The Manchester civic scientists identified themselves as a group, that is, they thought of themselves as professionals.

Professionalization of science involves something far beyond merely earning one's living as a practising scientist. Professionalization involves the consciousness of one's exclusive status, along with an acceptance of what comes to be recognized as professional values and responsibilities. It is important, therefore, to examine how these "new professionals" looked at the scientific enterprise.

For them, as for Liebig, the search for scientific truth, for knowledge, was paramount. Since, however, all knowledge, all truth, is in a very basic sense practical, it is science which, in the words of Liebig, "gives nourishment and life to industry." Training and continued research in the pure sciences is valuable for the applied; experience in the practical in return nourishes the researcher.

Thus, for them, there is no real distinction between the pure and the applied. Liebig wrote in his *Familiar Letters on Chemistry*: "It may well be said that the material prosperity of empires has increased manifold since the time oxygen became known and the fortune of every individual has been augmented in proportion. Every discovery in chemistry has a tendency to bring forth similar fruits. Every application of its laws is capable of producing advantages to the state in some way or other or promoting its welfare."[86]

Liebig's disciples among our civic scientists would act upon these words. Combining the German notions of *Wissenschaft* and state service with a British utilitarianism and concern for individual effort, our civic scientists evolved a coherent set of views on the relationship between science and society. The basic elements are simple and straightforward:

1. Science is *pro bono publico*. Science can, and must, increase both the wealth and the welfare of the people.
2. Science must have a conscience, for it has a public responsibility
    (a) to solve critical problems facing all citizens;
    (b) to educate the public to its own responsibility;
    (c) to foster institutions for the advance of science and for the fulfillment of those responsibilities.
3. The fruit of the scientists' labor should be presented to the public in the form of lectures, in journals, even in the public press.

To be accurate, most of these values of and about science — the science of civic virtue if you will — had already emerged in other groups responding to the same problems and sharing similar aspirations. This earlier group of men, including James Joule, the physicist, Edward Binney, the geologist, and John Leigh, the chemist-physician, attained a high level of competence in their respective sciences and exhibited the originality befitting their roles as devotees of science. What sets the civic scientists under discussion here apart from the similarly motivated devotees is the intensity with which they followed through and the consistency with which they adhered to their goals. The practitioners of the science of civic virtue were intent primarily in forging alliances with existing power groups and demonstrating to them the necessity and worth of their mission. With the German experience and a sense of destiny, their confidence in their own professional role was reinforced. Placed at the chemical frontier, they realized that *research* in science was the key to progress and, once again through the German example, realized also that the organization of research was both possible and necessary.

Much of the foregoing is expressed eloquently in a rather paternal letter Liebig wrote to Playfair upon the occasion of the latter's accepting a post with James Thomson at Clitheroe. The letter clearly delineates Liebig's sympathetic concern both for the requirements of practice and for *Wissenschaft*:

In the first place my dear friend let me express my sincere delight at the fine and pleasant post which you are taking up. . . . In all that you do, do not forget science and keep fresh and vigorous your taste for mental work, for unless a man is making progress in that which gives nourishment and life to industry he is scarcely in a position to fulfill the demands of his times.[87]

Liebig goes on to advise Playfair about his higher interest and, in doing so, may give us insight into the values of our civic scientists:

After satisfying the claims which Mr. Thomson makes on your attainments for his business, you ought not to trade on your experience and your chemical knowledge, but to give advice and help where they are useful. . . . You will have sufficient time for this, and it will raise you higher in the eyes of sensible men and be more really beneficial to your interests than if you tried to make money. Be true to yourself and true to science — This is all I wish to ask of you.[88]

We have here, in a pithy and human statement, the civic virtue which these scientists brought to Manchester with them: the reconciliation of profit and *Wissenschaft,* the identity of interest among the citizen, the scientist, and the productive classes of society.

Several years later, in his *Familiar Letters on Chemistry,* Liebig prophesied of the fruit of this union:

A more vigorous generation will come forth powerful in understanding, qualified to appreciate and to accomplish all that is truly great, and to bring forth fruits of universal usefulness. Through them the resources, the wealth and the strength of empires will be incalculably increased; and when, by the increase of knowledge, the weight which presses on human existence has been lightened, the difficulties of obtaining subsistence lessened, and man is no longer overwhelmed by the pressure of earthly cares and troubles, then, and not till then, will he be able to devote his intellect, purified and refined to the study of higher objects of investigation, and finally to the highest of all.[89]

This, too, was the credo of Manchester's civic scientists.

Science and Public Health

In the decade just before 1840 Manchester itself had undergone still further significant changes. In 1800 the population of Manchester and its sister city, Salford, was together about 90,000. This population is described as residing in the "midst of smiling meadows and well growing plantations, interspersed with ponds stocked with trout." To be sure, the public and private nuisances afflicting cities were beginning to be felt, but they possessed nothing like the intensity they were to have by mid-century, when the growth of the city had in the words of Sidney and Beatrice Webb "defiled the atmosphere, polluted the streams and destroyed the vegetation."[90] By mid-century the population of Manchester and Salford had grown more than fourfold—to 400,000; the population of England as a whole had merely doubled—from about eleven million to about twenty-two million.

This booming population reflected the changing economic and social position of what was in all but name a city. Yet the place was ruled by the relics of a medieval manorial system—an antique *court leet*. In 1838, under the provisions of the Municipal Reform Act, Manchester became an incorporated borough, though it did not become a city until 1853. Underlying both the tremendous population growth and the push for political restructuring of the town was the industrial might of Manchester and its environs. Cotton and its ancillaries were and remained, of course, the leading industries.

All these aspects of modern industrial life had come upon Manchester quite suddenly. In 1840 it found itself rich, vital, a true "boom town." As Asa Briggs has put it: "All roads led to Manchester in the 1840s. It was the 'shock city' of the age."[91] For Engels "it is the masterpiece of the Industrial Revolution."[92] But it also began to see itself with the eyes of others: as a graceless, dirty, dingy town virtually without cultural amenities, with not even a public park, let alone a symphony or a university. It was replete with smoky factories; its three rivers, the Irk, the Irwell, and the Medlock were by this time stinking, foul open sewers. Its air was a national disgrace. Alexis de Tocqueville, on visiting the city in 1835, reported: "A dense smoke covers the city. The sun appears through it like a disc without rays. . . . It is here that the human spirit becomes perfect and at the same time brutalised, that civilization produces its marvels, and that civilized man returns to the savage."[93] The King of Saxony in 1844 displayed an unfortunately typical reaction. "I could not help being forcibly struck by the peculiar dense atmosphere which hangs over [this town] in which hundreds of chimneys are continually vomiting forth clouds of smoke. . . . [The atmosphere] is not like a mist nor like dust nor like smoke but is a sort of mixture of these three ingredients, condensed moreover by the particular chemical exhalations of such towns."[94]

The damp, smoky, crowded conditions produced, naturally enough, a public health crisis which had been unknown in Great Britain since the seventeenth century. Cholera, for example, struck Britain in 1831–32 and returned in epidemic proportions in 1848-49, 1854, and 1867. These epidemics struck down tens of thousands and dramatically raised local death lists. "It was," in the words of a noted historian of the period, "the clearest warning of the lethal propensities of the swollen towns of the new industrial era."

In sum, the 1840s and the 1850s were for many of the British "boom towns," and for Manchester in particular, a time of environmental crisis. The awareness of the crisis was extreme, exacerbated by the rapidity in which it came about. The malaise felt by most Mancunians extended to all classes of society, in all sectors of life. Manchester men were in fact just beginning to see the dimensions of the crisis in housing, sanitation, food adulteration, fuel supply, paving, sewerage, water supply, open spaces and the public provision of these services. As one might expect, when Edwin Chadwick's famous *Report on the Sanitary Condition of the Labouring Population of Great Britain* was delivered to Parliament in July of 1842, the effect upon leading elements of the large northern cities, and especially Manchester, was tremendous. The rapid growth of population of the town and of Salford (in the single decade 1821-31 the population increased by a staggering 47 percent) was not only highly visible but was causing anxious reappraisals on the part of numerous citizens, many of whom still remembered a more pastoral Lancashire. The number of houses per acre was on a steep rise, owing to the increased practice of building of back-to-back houses. The quality of life as seen by the influential bourgeoisie was rapidly becoming a vital political and social issue. The questions of sewage disposal, water supply, sanitation, housing, ventilation, open spaces were not ones which would just go away. They had to be resolved.[95]

The sense of crisis was heightened by the mortality figures. A report of the Manchester Statistical Society in 1842 (commonly known as the Chadwick Report) showed that the death rate in the inner city of Manchester was 35.2 per 1000, whereas in the suburb of Broughton it was only 15.8 per 1000.[96] Only? The Chadwick report had some unpleasant news even for the upper classes. Chadwick compared the average age of death in the period 1837-40 for Rutlandshire and for Manchester. The figures are the following:[97]

Average Age of Death

	Manchester	Rutlandshire
Professionals and gentry	38	52
Tradesmen and families	20	41
Mechanics and laborers	17	38

J. R. Wood of the Manchester Statistical Society noted that fifty-seven percent of the laboring class die before the age of five, that is, "before they can be engaged in factory labour."[98] The historian may underestimate, only at his peril, the effect of these statistics upon the residents of the city. Indeed, mortality figures were widely discussed in Manchester for the next half century.

Thus when Engels wrote (perhaps a bit dramatically) of the city in the mid-1840s, he was raising issues already before concerned citizens: "At the bottom the Irk flows, or rather, stagnates. It is a narrow, coal black, stinking river full of filth and rubbish which it deposits on the more lowlying right bank. In dry weather, this bank presents the spectacle of a series of the most revolting blackish green puddles of slime from the depths of which bubbles of miasmatic gases constantly rise and create a stench."[99] The miasmas! When Southwood Smith wrote in his *Treatise on Fever* in 1830 that the cause of fever was the "putrefaction of animal and vegetable matter,"[100] he was expressing a widely held view that the health problems of congested areas, with their decaying offal and refuse, owed largely to the "miasmas" given off into the surroundings. Chadwick's famous report and the public health and sanitation movement spurred by it assumed the miasmatic theory. The historian Michael Flinn underlines the importance of it: "The eradication of miasma — not entirely achieved even by the midtwentieth century — was a sound instinct and could do nothing but good."[101]

Not only were the air and water called into question, but even the wholesomeness of the food in the town was under critical examination. Ever since 1820, when Frederick Accum, the chemist, published his *Treatise on Adulterations of Food and Culinary Poisons* ("There is death in the pot") there had been increasing concern and public complaints about millers, brewers, and bakers. The available evidence seems to indicate, however, that adulteration and public agony over it reached its peak in the decade 1840-50.[102] John Mitchell's *Treatise on the Falsifications of Food* (1848) and A. H. Hassall's landmark investigations for the *Lancet* in the 1850s demonstrated the surprising extent of adulterations. Bakers' bread contained large quantities of alum, flour contained chalk as a filler, a common adulterant of beer was iron sulphate which aided in frothing, milk was diluted with water, oatmeal contained barley-meal, even the evening tea was fraudulent.[103] Eventually, public agitation led to the passage of the Adulteration of Food Acts of 1860 and 1872; F. Filby has recounted the importance of these for the rise of the profession of analytic chemistry.[104] In addition, the citizenry had to concern themselves about drug adulteration, which Accum likewise called to their attention.[105]

Even before Chadwick's alarming report reached Parliament, residents of Manchester on their own initiative met together in what was advertised as a town-wide smoke abatement meeting at the Royal Victoria Gallery on May 26, 1842. The meeting was called "for the purpose of directing attention to plans for the prevention of smoke, with the ultimate view of bringing the subject before the British Association," which was to meet in Manchester during June. A committee was appointed to pursue the matter and included, as its scientific contingent, men associated with the old Mechanics' Institution and later with the Royal Victoria Gallery: John Davies, Peter Clare (Dalton's friend), Richard Roberts, the engineer, William Sturgeon, Director of the Victoria Gallery, Alfred Binyon, and that inveterate organizer, William Fairbairn.[106]

The *Guardian* two days later commented favorably upon the townsmen's efforts in a leader entitled "The Smoke Nuisance." Smoke was, it reported, "an opprobrium under which Manchester labours more heavily perhaps than any other town in her majesty's dominions." Not only, the *Guardian* lamented, "is the greater density of smoke in this town palpable to the senses," but even in the few open spaces remaining "the flowering shrubs will not grow at all" and the trees are dying. "Why should the inhabitants of Manchester continue to be annoyed and to have their clothes and their habitations defiled by the dense clouds of smoke with which improvident and careless people will persist in polluting the atmosphere?[107] By the fourth of June, the anti-smoke vigilants had organized themselves into the Manchester Association for the Prevention of Smoke and had planned an exhibit to be shown before the British Association. Another meeting was held on June 8th at the Royal Victoria Gallery, and Fairbairn reported on the progress of the society and insisted there was "no doubt that smoke could be consumed." At the June meetings of the British Association, Fairbairn spoke on the possibility of reducing smoke from steam boilers. C. W. Williams, who had addressed the Manchester association at their June 8 meeting, reported on plans to reduce the smoke nuisance. Within a year, the Manchester Association had petitioned the House of Commons, insisting that "excessive smoke [was] . . . a great evil."[108]

Of course, the smoke menace, like sewage, sanitation, overcrowding, ventilation, and water pollution, was not merely a local Manchester problem. Other cities like Liverpool and Leeds were likewise forming environmental protection organizations. A Leeds lecturer on chemistry, William West, published in 1842 *An Account of the Patent and Other Methods of Preventing or Consuming Smoke* in which he examined statistically the rise in respiratory ailments in heavily industrialized areas, and, like his Manchester colleagues, he was particularly concerned to effect a more complete process of combustion.[109]

In the wake of Chadwick's report and petitions from civic organizations such as the Manchester Association, in 1843 Parliament appointed a select committee to investigate smoke prevention. The Manchester *Guardian* happily reported that the committee "received the evidence of the most eminent men in the science of chemistry, of practical engineers . . . and of leading master manufacturers," and that the committee recommended a smoke prevention bill.[110]

The Select Committee to inquire into the "smoke nuisance" called before it those like C. W. Williams and John Juckes, who held patents on "smoke consuming" devices and scientific consultants like William West, Andrew Ure, W. T. Brande, and the redoubtable Michael Faraday. The emphasis of the testimony and of the inquiry was on searching for the most efficacious means of providing more complete combustion. All the methods discussed were similar; the C. W. Williams device was typical. Dr. Andrew Ure advocated Williams' patent, claiming that in the cases where it had already been applied "there is seldom a particle of smoke to be seen."[111] Very simply, Williams' device was a unit with a whirling device and numerous small holes which would draw air into the combustion engine and mix it with the gases therein. A workman would operate a vent to admit air in quantities sufficient to consume the smoke. Williams based his invention on "all the first chemical authorities," whose principles indicated to him the necessity of a mechanical mixture of the gases with air so "the atoms of the one must come into contact with the atoms of the other." West and Ure endorsed the device, the former considering soot the greatest evil, the other gases emitted during combustion of "extremely little consequence." Ure testified that the Williams patent was already utilized in over twenty great steamers, among them that of Henry Houldsworth of Manchester, and that the eminent William Fairbairn has expressed favorable comment on it. Houldsworth, a Manchester factory owner who had been a student of Ure, maintained that although the device worked, some manufacturers would be tempted to save money, not fuel, by overworking the engine and thereby reducing the effectiveness of smoke prevention.[112]

Michael Faraday, F.R.S., testified that "there can be no difficulty as a natural effect to obtain perfect combustion of smoke," but that the production products required a trade-off. Getting rid of black smoke meant invariably the greater discharge of carbon monoxide and carbon dioxide gases, both of which were injurious to health. But, he claimed, the problem as it existed was not "so much a question of health as of cleanliness and comfort."[113]

In the wake of Chadwick's report, however, the smoke nuisance became inextricably intertwined with the health meance. Even the Tory *Quarterly Review* was alarmed: "When we reflect that the air the labouring classes breathe—the atmosphere which by nuisances they contaminate—is the fluid in which rich and poor are equally immersed—that it is a commonwealth in which all are born, live and die equal—it is undeniable that a sanitary inquiry . . . becomes a subject in which every member of the community is self-interested."[114] The Borough of Manchester was moved to action and on 4 July 1844 passed an ordinance (see below, chapter 4) stating that beginning on the first of January 1845 all smoke from steam engines or furnaces be consumed by itself, under penalty of a forty-shilling fine each week the annoyance continues.[115] Without adequate enforcement provisions, however, the ordinance did little to mitigate the problem.

What was the answer? Where did it lie? The *Quarterly Review* was optimistic. The great evil lay in the "disregard of scientific precautions," for the supply of "air in [nature's] laboratory" is great and "an ounce of science" would go far to ameliorate the situation. The *Review* went further: "The purification of our great cities, and a watchful search throughout the land we live in for every removable cause of disease, are services which science should be proud to perform."[116]

The call for a massive environmental effort on the part of science was echoed by the sanitary reformers and by the scientists themselves. R. D. Grainger, lecturer on physiology at St. Thomas' Hospital, London, and author of *Observations on the Cultivation of the Organic Sciences,* commented that "we are on the eve of a great social amelioration," since "for the first time in our career science is about to be brought to bear on the various questions which exert so direct an influence on every inhabitant."[117] Before a large audience at the Manchester Athenaeum in May 1845 Grainger spoke "On the Causes of Insalubrity of Towns and on the Means of Their Removal," and dramatized anew the supposed effect of the shameful conditions in manufacturing towns. Quoting statistics demonstrating Manchester's shocking disadvantage in mortality rates and age of death, and indicating the worsening situation in the 1840s, Grainger insisted that "of all the causes which are productive of diseases in towns, infinitely the most influential is the vitiated state of the atmosphere," fixing upon smoke, and invisible effluvia from decaying animal and vegetable wastes — the hated miasms — as the worst offenders. The remedy, however, is at hand — and it is *Science.* "There is no great art or business in which a knowledge of science is not necessary. . . . There are scientific questions concerned in all the business of drainage, a knowledge of geology and chemistry is almost indispensable." The great evil is ignorance. "All persons charged with the public health are expected . . . to acquire much scientific knowledge."[118]

Shortly thereafter, Chadwick himself sharply maintained much the same position. In an address before the College of Civil Engineers, Putney, and given ample coverage in the *Manchester Guardian,* Chadwick spoke movingly of "The Want of Science in our Public Works." Although Great Britain was preeminent in practical science, Chadwick maintained, such knowledge was not put to use in the construction of public works. For example, "large masses of capital were expended in works for the distribution of water of towns without such a competent knowledge of hydraulics and other branches of science applicable to the [problem] as a good scientific course of instruction must afford. The task [of improving the health of towns] was only to be achieved when such practical science . . . was duly appreciated and received by the public and properly applied." Lyon Playfair, also at Putney, corroborated Chadwick's views and remarked upon "the deplorable absence of competent scientific knowledge."[119]

Such calls from the leaders of sanitary reform did not go unheeded. Indeed, Manchester was fertile soil; it possessed a willing and able scientific community and, in the students of Liebig and their allies, men already ideologically conditioned and professionally in need of such a mission. As we have already seen, John Leigh was in 1844 lecturing before the Royal Institution on the sanitary condition of Manchester and scientific approaches to its improvement, and he had urged upon his audience, among many other things, the need to be rid of smoke and suspended particles in the air.[120] Edward Binney, in a letter to the *Guardian,* insisted that the population must "be provided with as pure an atmosphere as the circumstances of a manufacturing town will allow."[121] Letters came in to the offices of the *Guardian* on a regular basis lamenting that "still the mischief continues unabated" despite the rapid strides that "chemical science has been making" or ironically commenting upon the contamination perpetrated upon the community by "our *scientific* manufacturers."[122]

The lesson was not lost upon Manchester's small group of professional scientists. Robert Angus Smith was one of the first to throw himself fully into the question, one upon which he would later make his reputation. Writing from the laboratory of the Royal Institution, Smith published two long letters on the air pollution menace which deserve quotation at length both for their content and their impassioned sense of concern.

Coming in from the country last week on a beautiful morning, when the air was unusually clear and fresh, I was surprised to find that Manchester was enjoying the atmosphere of a dark December day. This is not seldom the case. . . . We see that the atmosphere of our streets in winter is frequently of the deepest black.

Does the smoke do harm? Such a question is an insult to physiology; an insult to the tastes of mankind; and we may even say to the great Maker of the atmosphere. We must learn to obey more implicitly the dictates of natural laws without constantly inquiring if nature knows how to act or not. That which she gives us we had better use, until we find out how to act or not. . . .

We know very imperfectly the effect of carbonic acid on the lungs; we cannot well explain why its effects are so very powerful, still less do we know the reason of the deadly effects of a very small percentage of carbonic oxide, that most poisonous product of imperfect combustion. We know that the lungs are injured by the entrance of solid matter; but we do not know . . . how little may gradually act . . . on the general health. . . . The life of man is not so long that the loss of even one year is to be considered as trifling. . . . Those who would defend such evils, who would remain careless as long as any probable cause is un-removed must surely be devoid not only of mercy but of clear perception and of good taste. The gloominess and uncleanness is everywhere around us; the depression of filth on the spirits and on the pockets is continually before our eyes; the destruction of our landscapes and of our town views is undoubted, and can we fail to look upon the cause of this as a small evil?

We must make men feel that, although they employ workmen, and thereby do good, they pollute the air and partly balance good by evil. . . . True it is, that when necessity arrives, we shall manage it better. Is it not a necessity now?. . . . At all events, let an exertion be made lest we be found guilty of the charge that we seek wealth at the expense of happiness.[123]

The almost Biblical quality of the denunciation invoked by Smith—"the gloominess and uncleanness is everywhere around us"—was doubtless heartfelt; Smith had, perhaps, already discovered his mission. He was nonetheless a young chemist attempting to make a career for himself in a city of manufacturers and their dependents. His second letter on the subject, several weeks later, was more muted and perhaps more orthodox in the city of Manchester economics: "If smoke is a nuisance which cannot be cured without loss to the country, then it is a nuisance which must be permitted. . . . But if we prove that manufacturers may flourish at less expense, and the community have the benefit, then one may with a more determined voice—Let coal be perfectly burnt! you shall cease to poison the air. . . . Give men freedom and an inducement to improve, and then men will improve." And if it is said that we need facts, we shall have them: "No class of men are so able to confirm theories by practice . . . as the manufacturers and merchants of Britain."[124]

Whereas Smith, in his first letter, was on the point of drawing a dichotomy between purity and profit, the second letter is more representative of the early work of the civic scientists. An almost pristine liberalism ("Give men freedom and an inducement to improve") is combined with Liebig's scientific zeal—profit reconciled with *Wissenschaft*—to produce the idea of the identity of interest among scientist, citizen, and the productive classes of society.

4

The Science of Civic Virtue: Chemistry, Health, and Industry

The very air seems to have changed in quality, and to tax the powers of Sanitary Reform to the uttermost. In Pickwick *a bad smell was a bad smell; in* Our Mutual Friend *it is a problem.* —Humphry House, in *The Dickens World*

The Environmentalist Attack

All this environmental turmoil and all these calls upon Science to rescue the manufacturing towns from environmental disaster could not have come at a better time for Lyon Playfair. It was only through the efforts of de la Beche, Buckland, and the prime minister, Robert Peel, that Playfair was prevented from considering more seriously a vacant professorship at the University of Toronto. Although "saved" for England, Playfair had little professionally to remain for. He returned to his honorary (and unpaid) professorship at the Manchester Royal Institution late in 1842.

With the appearance of Chadwick's report and the widespread concern in official circles which the report stimulated, however, Playfair's fortunes took a rapid turn for the better. The British Association had voted a small sum for the investigation of the chemical operation of blast furnaces in the manufacture of iron; it was then that Playfair began his famous joint researches into the matter with Professor Bunsen. After Chadwick's report, in 1843, a Royal Commission was convened to inquire into the state of large towns. The president was to be the Duke of Buccleuch, and it was to be manned by such scientists and engineers as Henry de la Beche, Prof. Richard Owen, and George Stephenson. Also selected by Peel to serve as a commissioner was Playfair, who chose for his special area of inquiry the large towns of Lancashire so that he could continue with his chemical work

117

in Manchester.[1] At once he asked R. Angus Smith, his assistant at the Royal Institution, to serve him once again, as his assistant commissioner.

Playfair was an obvious choice. As a student of Liebig he capitalized upon the tremendous impact that Liebig was having, in part through Playfair's own efforts, upon the British medical profession. Chadwick was anxious to contact Liebig, and it was through Playfair that he did so.[2] Playfair was able to report before long that Chadwick's "work was met with much eclat here and will produce most beneficial results."[3] Chadwick subsequently sought advice from many Manchester scientists. Angus Smith, John Leigh, and William Fairbairn were frequent correspondents. With the appearance of Liebig's views on animal chemistry in the versions of Playfair and Gregory of 1842, Liebig began to attract considerable attention. Henry Ancell outlined Liebig's views on health and disease in a series of articles for the *Lancet* in which Ancell exulted in the "new era in medicine," a "truly rational" approach to be ushered in by Liebig's chemical methods.[4] By 1844 the *Lancet* was publishing in extenso Liebig's lectures on organic chemistry at Giessen, reporting upon Liebig's triumphant tour through Britain, and answering such queries as "Can you inform *A Beginner in Chemistry* what is the expense of attending a class with Liebig at Giessen?" Answer: six pounds, ten shillings for the term plus materials and apparatus costs.[5] By 1845 Liebig became an almost mythic figure, and the appearance of articles such as "Anecdotes of Liebig" suitably marked the baron's entrance into the *Lancet's* Valhalla.[6]

It was, in fact, Playfair who had introduced Liebig's ideas on health and disease into Britain. In 1840, at the Glasgow meeting, Playfair reported on Liebig's views "On Poisons, Contagions and Miasms," which discussed the effect of inorganic and organic "poisons" on the human body. The principle was simple: morbid poisons, such as are emitted by decaying animal and vegetable, induce a similar state of decay in the body. "A body in a state of . . . decomposition is capable of inducting or imparting its own state of . . . decomposition to any body with which it may be in contact."[7] These miasms, in short, cause a chemical fermentation in the blood. When Playfair made his debut before the Manchester scientific community in a lecture "On Miasmata" at the Royal Victoria Gallery the very next year, he expanded upon his version of Liebig's notions of contagion. Playfair distinguished between *gaseous contagious matter* which is generated from the blood and is capable of reproducing itself there, and *miasm* which is the product of the decay of animal or vegetable matter and acts upon the blood without being reproduced. Some miasms are purely chemical; two of the more pernicious are hydrogen sulfide gas and carbonic acid. It is noteworthy that Playfair remarked that miasms are produced in abundance even in Great Britain and will continue to be present until "better drainage and cleanliness be afforded to our towns."[8]

Liebig's chemical pathology supported the generally accepted miasmatic notions of the medical profession and reinforced the views of the sanitary reformers, like Chadwick and Grainger, whose zeal for sanitation involved the attempt to eradicate the cause of the miasmas, the filth which polluted the town. It also drew a connecting link between the smoke nuisance proponents and those concerned about the sewage and drainage of towns, for not only was decaying organic matter harmful to health but also inorganic substances such as hydrogen sulfide gas, "carburetted hydrogen," carbonic acid (CO_2), and carbonic oxide (CO) were potentially pernicious as well.[9]

Thus it was with sterling qualifications and a prior interest and experience with miasmas and contagion that Playfair entered upon his new duties as Royal Commissioner in the Health of Towns investigation. Not surprisingly, it was his colleagues in the Literary and Philosophical Society and around the Royal Victoria Gallery to whom he turned for expert advice on the condition of Manchester. Edward Binney, whom Playfair describes in his report as a "well known geologist," testified as to the wretched condition of the Irk, Irwell, and Medlock rivers owing to the sewage problem:

> In all the streams above described an abundance of dead dogs and cats are to be seen in the various states of decomposition. Bubbles of gas, chiefly light carburetted hydrogen, rise up to the surface, and although offensive smells are met with at all times, they are by far the most annoying when the barometer has experienced a sudden depression. . . . Sulphuretted hydrogen is the gas which chiefly causes the odour.[10]

Medical men whom Playfair consulted, such as John Leigh, were "unanimously of the opinion that the emanations from the putrid streams . . . are a cause of disease and mortality." Leigh testified that the "ravages of [cholera] were so great on the banks of these putrid streams that . . . one court (Allen's court) in Long Millgate was entirely pulled down" (p. 348) Thomas Ashton, a manufacturer and member of the Lit. & Phil. testified on the beneficial effects of supplying clear water to his workers' cottages (p. 410), and Playfair's colleague in the Lit. & Phil., John Graham, a partner in the Hoyle and Sons print works, reported on their experiences in filtering water (p. 412). Playfair commented on the smoke nuisance, supporting the notion that "they who occasion the nuisance do so in ignorance of the benefits to be derived to themselves from a consumption of smoke," quoting Leigh to the effect that the pecuniary loss to the city each year is twice the amount gained from its poor-rates (p. 377).

In his conclusion, Playfair argued eloquently for all-out war on the causes of disease: too much effort had been frittered away upon mere amelioration, whereas prevention of disease was not only possible but necessary. The "excessive mortality" which he found in large towns of Lancashire, "is due to adventitious causes, in almost every instance removable" (p. 474). Enforcing legislation is necessary, for although "capitalists can calculate on the intelligence of the middle and higher classes," they "must doubt the progress and conviction and voluntary change amongst the lower classes" (p. 486). Upon the appearance of the report, the *Manchester Guardian* gave it considerable publicity, providing an outline for its readers and endorsing its conclusions.

Playfair immediately proceeded to another parliamentary investigation, this time with Henry de la Beche, on the subject of the "Means of Obviating the Evils Arising from Smoke." Their report generally supported earlier views on the consumption of smoke, although claims for the great economic benefit of smoke reduction were now moderated: "There is frequently [de la Beche and Playfair concluded] a loss of heat in the prevention of smoke, but with careful management an economy of fuel may be produced."[11] By this time, however, Peel had offered Playfair the post of chemist to the Geological Survey, which he accepted with alacrity, and he joined de la Beche in London. Except for his continuing collaboration with Joule, Playfair severed most of his relations with the Manchester scientific community.

The sanitarians, Playfair and Smith, interested others in the Manchester community in application of chemistry to public health. James Young, for example, tied to both Playfair and Smith both by friendship and professional concern, served on the Literary and Philosophical Society's committee along with Smith and Leigh to inquire into the potato blight in 1845; Young suggested a weak solution of sulfuric acid to retard decay. A more important foray into public health chemistry by Young came two years later, in December 1847, when he read a paper before the Literary and Philosophical Society "On the Deodorization of Manures." The subject was of particular importance in two respects, as Liebig had shown in his essays on agricultural and animal chemistry: (1) the preservation of refuse matters for agricultural purposes, and (2) the prevention of the decomposition of such matters in urban areas before their removal can be effected. Young wrote: "Our great sanatory problem, the solution of which has occupied much attention is the prevention of decomposition in organic accumulations in towns."[12] Young at the time was engaged in manufacture of chlorine, a waste product of which was manganese chloride, mixed with iron perchloride. One manufacturer was throwing away thirty-six tons of it per day. Young was convinced that "this solution has, in a high degree, the property of preventing decomposition in organic matter." Young added that "I mentioned the matter to Dr. R. A. Smith and Dr. Lyon Playfair, both of whom fully agreed in the views I had taken on the subject."[13]

Young's paper brought out sharply two important points: (1) the desirability of chemical treatment of wastes to preserve them for agricultural use and (2) the disinfection of such wastes as part of the public health effort in the larger towns. The Towns Improvement Clauses Act of 1847 and the Public Health Act of 1848, by recognizing two methods of disposal, the profitable and unprofitable, implicitly drew the lines for further work in both areas.[14] "Disinfection" in the pre-Pasteurian use of the term meant largely the prevention of decomposition of organic matter; for agricultural purposes this meant the retention of important soil nutrients, and for sanitation it implied either the destruction or the retardation of the production of miasmas.[15]

A colleague of Young in the Manchester Literary and Philosophical Society, Alexander McDougall (or sometimes M'Dougall), was lecturing by 1856 "on the Preservation of Natural Manures" and joined with Angus Smith and Frederick Crace-Calvert in experiments on the preservation and deodorization of sewage from the "foul River Medlock" in Manchester. At first, the team used lime, which, they reported, deprived the water of most of its characteristic foul smell, but the odor soon returned. It was then that they developed and utilized their own disinfecting powder, a patent for which was taken out in January 1854 in the names of Smith and McDougall for "improvements in treating, deodorizing and disinfecting sewage."[16] A second patent was taken out in 1859 by McDougall himself and marketed widely under the name "McDougall's Powder." Smith did not enter into the manufacturing of the powder and drew no profit from it.[17] Before the widespread use of the powder, the lime process was the most prevalent means of disinfecting sewage; McDougall claimed that his process improves the quality of the manure elements by preventing the extrication of ammonia from it, whereas the lime process precipitates phosphates and thereby weakens it.[18]

By 1861 a report before the British Association was able to state that "the manufacture of disinfectants has now become a regular and constant one; and since the inquiries instituted on the subject by [Angus Smith] and Mr. M'Dougall of this city [Manchester], the use made in this district has been enormously increased."[19] McDougall manufactured his disinfectant, which used carbolic and sulphurous acids, near Oldham, outside Manchester, and sold it in powder form for the prevention of decomposition of manures. He also marketed a liquid employing carbolic acid and lime-water for use in sewers. The powder was a mixture of calcium and magnesium sulphites and carbolates.[20] Already by 1861 it was being used in dissecting rooms and in the treatment of sores, dysentery, and foot-rot.[21] P. H. Holland, the medical inspector for the Westminster Burial Board, followed closely the application of McDougall's process for the deodorization of sewage which was adopted by the town of Carlisle; according to

Holland, the process consisted mainly of the addition of crude carbolic acid at a cost of under eleven pence per day, with results regarded as "completely successful."[22] It is noteworthy that Joseph Lister claimed to have taken the hint for his use of carbolic acid for surgical purposes from its use as a disinfectant for Carlisle's sewage.[23]

Crace-Calvert followed his confrères by producing his own patented disinfectant, mainly carbolic acid, and pursuing the subject in a number of publications.[24] Later, Crace-Calvert became a major manufacturer of carbolic acid and disinfectant products; his firm, Crace-Calvert and Company (later Crace-Calvert and Thomson) flourished until his death in 1873.[25]

He had been drawn into sanitary reform not long after his arrival upon the Manchester scene. By the end of 1848 Crace-Calvert was lecturing on the necessity for sanitary reform and on drainage and disinfection.[26] His major contribution in the public health area, and one in which he reached some eminence as an authority, was in the field of food adulteration. The 1840s was a critical decade in this as in other areas, and apparently actual adulteration and public awareness of it began to climb alarmingly during the course of it (see chapter 3). Crace-Calvert chose this issue as suitable for demonstrating the chemist's civic value. In December 1848 he lectured before the Royal Institution "On the Adulteration of Food," in which he stunned his audience by reports of massive adulterations of sugar, tea, coffee, and flour, but ended optimistically, citing the chemical tests available to detect the frauds. The Board of Guardians of the city took action against offenders, using Crace-Calvert as an expert witness.[27] Despite angry letters and denials of the charges, Crace-Calvert continued to perform adulteration analyses both locally and nationally. By 1856 he was a nationally recognized expert and was called to testify before the parliamentary commission on the adulteration of food, along with such other widely known investigators as A. H. Hassall and Alphonse Normandy. Crace-Calvert testified that he had "several times been called upon, as a chemist, to inquire into cases of adulteration taking place among tradesmen in Manchester, and . . . to examine the quality of various products used in large chemical works. Besides that I have had my special attention called to the question of the adulteration of food as supplied in large quantities to public establishments." He explained that he found significant frauds in such household items as flour, oatmeal, butter, soap, sugar, and bread. He had found flour ranging in purity from zero to seventy-four percent adulterated. It is no wonder that the police often consulted him, and poor people who could not bake bread with their alum-adulterated flour sought his help.[28]

Professionals and devotees came together to form the Manchester and Salford Sanitary Association whose directing committee included Stone,

Smith, and Crace-Calvert, and whose vice-presidents included the Heywood brothers, Fairbairn, and Whitworth. A chemical and geological committee was formed which included Alexander McDougall, Crace-Calvert, Edward Binney, Allan, Stone, Smith, and Edward Frankland, professor of chemistry at the new Owens College.[29]

Sanitary Chemistry: Career Profile of Robert Angus Smith

Of the civic scientists, the one most devoted to the subject of sanitary chemistry—a field which he virtually created and in which he became a leading authority—was Robert Angus Smith. Along with McDougall and Crace-Calvert, Smith was centrally concerned with questions of disinfection and sewage treatment; by the time of the appearance of his book *Disinfectants and Disinfection* in 1869, the *Chemical News* was able to report that "by common consent Dr. Angus Smith has become the first authority in Europe on the subject of disinfection."[30] Yet his primary interest was the detection and analysis of impurities in the air. A lover of the countryside, Smith was aesthetically as well as professionally disturbed by the condition of the town air which he encountered as Playfair's assistant in the health of towns inquiry. His first major publication following the heartfelt letters of 1844 to the *Manchester Guardian* on air pollution was a paper delivered before the Chemical Society and later published in the *Philosophical Magazine* for 1847.[31]

Having given considerable attention to the inquiry into the causes affecting the health of towns, I was anxious to find what the real evil in their polluted atmosphere consisted of; the air has been frequently examined, but the differences found do not sufficiently account for the differences perceptible in breathing for the first time in entering a large town from the country, or for the very great difference in the colour and appearance of both when contrasted side by side, by an individual outside a town, having both before his eyes.[32]

Most of the adult residents of the city could remember a time when the air was purer, the town less crowded, and the countryside more accessible. The rapidity of the demographic and industrial change affecting Manchester prodded urban nuisances into an urban crisis. Smith was eager to get at the root, the "real evil" in the "polluted atmosphere." He collected and examined rain water from cisterns and from special porcelain vessels. During a concert at the Mechanics' Institution he collected condensed moisture from a window, and he amassed samples from rivers, churchyards, and cesspools. He found an "increase in organic matter on approaching towns" in the air and water.[33] In 1848 he reported before the British Association on his researches in a paper entitled "On the Air and

Water of Towns." Contrary to Dalton, who insisted that no difference could be determined between town and country air, Smith, in what was for him a remarkably polished exposition, insisted that the subjective view of air and water pollution in the contemporary city ought not be disregarded.

Men accustomed to experimental inquiry are apt to forget the value and force to be attached to those apparently less rigorous observations which the senses are constantly and unconsciously making, and to believe only that which can be demonstrated by the grosser processes of a laboratory. Most men would be satisfied of the impurity of an atmosphere through which a blue sky could never be seen of a blue colour.[34]

His work for the Health of Towns Commission had convinced him that "crowded towns are dangerous places," and that it was imperative that Britain act "in self-defence" to avert disaster. He concluded that, contrary to some opinion, the pollution of air in crowded places derives largely from organic matter, and not merely carbonic acid (carbon dioxide); that this matter has the power of supporting the life of possibly noxious "'animalcules"; and that water purifies by filtration through soil—a process which may be replicated in the laboratory with a sand filter.[35]

The character of Smith's early work on air analysis is best summarized in a report prepared for the Condition of Mines Inquiry of 1864. In his *Report on the Air of Mines*, Smith reviewed the work of Priestley, Gay-Lussac, Regnault, and others who had made determinations of the percentage of oxygen and carbon dioxide in air. Using a nitric acid test and other procedures, Smith found that "variation is really so great in some place that we must admit some powerful local cause. . . . When carbonic acid increases is it wonderful that oxygen should decrease? . . . If emanations arise from foul places, they must occupy space."[36] He compared street air with "midden or closet air" in Manchester and found the oxygen level of the latter considerably lower (20.7 percent compared with street air's 20.943 percent). He found that carbonic acid outside Manchester averaged 3.69 percent whereas in "close places" in the city he found the level to average 16.04 percent, the worst in mills.

The *Report of the Metropolitan Sanitary Commission* of 1848 contained testimony by Smith, as well as by Thomas Graham, Playfair, and Young. Young testified on his manure preserving process, and Playfair and Graham on disinfectants. Smith tied together the work he had been doing on airs and waters with the need for disinfection. "The prevention of infection," Smith testified, "will depend upon the prevention of putrefaction." For Smith, the natural elements provided the most obvious and safest means of disinfection. "Air I consider the great disinfector, partly by its removal of matter and partly by its oxidizing powers."[37] Water and earth, too, can contribute to the purification of the urban environment if preserved and used properly. In short, proper condition of air, water, and drainage in towns are requisite for a healthy urban condition.

By the beginning of the 1850s, Smith was investigating the influence of land filtration, comparing favorably the water obtained from springs with surface and rain and town-well water, the last of which is disturbed by nitrates and chlorides left after the removal of impurities. Smith urged, in a report to the British Association in 1851, consideration of the advantages of a porous soil in removing offensive materials. "As an agent for purifying towns this oxidation of organic matter is one of the most marvellous, we might almost say, and necessary." Moved by the study of the moment, Smith went on to claim that "we may very correctly look on the soil as the greatest agent for purifying and disinfecting." Smith also made extensive inquiries on sulphuric acid in the air and water of towns and presented the results to the British Association in the same year. Liebig had found ammonium carbonate in the air; Smith demonstrated the existence of relatively large quantities of sulphuric acid in the atmosphere of large manufacturing towns.[38]

In 1862 the House of Lords appointed a select committee to inquire into "Injury from Noxious Vapours." The committee took expert testimony from a number of witnesses, including Lyon Playfair, who outlined the vast problems: muriatic acid gas (hydrogen chloride) from soda manufacture; nitrous acid from the manufacture of sulphuric acid; and sulphuretted hydrogen gas (hydrogen sulfide). Playfair was of the opinion that injury from the alkali trade was "entirely preventable" simply and cheaply owing to the solubility of muriatic acid gas in water. John Leigh of Manchester testified that he was retained as borough chemist for the corporation of Manchester and in that capacity had inspected several factories but found no health hazard beyond an occasional nausea induced from malodorous products. Peter Spence, like Leigh and Playfair a member of the Manchester Literary and Philosophical Society, and a manufacturer of ammonia alum, testified regarding his experiences as producer and as defendant in a nuisance suit. Spence admitted that he had been found "guilty of a nuisance" from his Pendleton works but insisted that it was "proved not detrimental to health."[39]

The committee's report marked an important stage in the public appreciation of the problems of air pollution. Without question, in the years after 1842 when the problems were widely discussed, the major emphasis had been placed upon the black smoke nuisance, reflecting the prevailing concern with steam engines and domestic hearths. However, as the chemical industry grew in size and importance, other perhaps even more pernicious problems were to share the concern of the sanitarians. The alkali trade amounted to more than two million pounds per year and employed over 19,000 men.[40] Without doubt the previous, optimistic notion that profit and purity went hand in hand had to be seriously questioned. A real dilemma faced those concerned with public health and with public wealth; the legislature had to be very circumspect when attempting to regulate so prosperous an industry as the burgeoning chemical production.[41] Yet the committee made some strong recommendations, following the suggestion of a former Manchester surgeon, P. H. Holland, that a nationwide system of inspectors be set up to enforce compelling legislation, and of Dr. Edward Frankland, former professor of chemistry at Owens College, Manchester, that such legislation "insist upon hydrochloric acid and sulphuretted hydrogen not being transmitted into the atmosphere," something "readily ascertainable by a chemical test."[42] The committee in its report recommended that certain effluvia (especially muriatic acid) be banned and that inspectors be given free access to any industrial establishment nationwide.[43]

The convening of the Noxious Vapours Commission was a response to the problems attendant upon the rapid growth of the chemical industry, and in particular of the alkali trade in the 1850s and early 1860s.[44] The so-called "alkali-trade" was a series of processes centering around Le Blanc's method for the production of soda alkali (sodium carbonate). The heart of the process was the treatment of salt ($NaCl$) with sulphuric acid, obtaining hydrogen chloride (muriatic acid gas) and sodium sulphate. The sodium sulphate or saltcake was then treated with limestone and coal and thereby converted into foul-smelling "black ash" containing sodium carbonate and calcium sulphide. The soda (sodium carbonate) was extricated from the black ash by a dissolution-evaporation process and thereafter used in the soap, glass, and textile industries and for gunpowder.[45] After the introduction of the Le Blanc process on a large scale by James Muspratt in the mid-1820s, the attendant difficulties of the process began to be felt, both by the manufacturers and by their neighbors. The alkali trade was faced with the disposal, for example, of large amounts of alkali waste, large heaps of which scarred the Lancashire countryside where the production was greatest and which offered visible evidence of the Faustian bargain which England had made in the Industrial Revolution. In addition, the first part of the process, the salt treatment to form saltcake, spewed forth huge quantities of malodorous and corrosive muriatic acid (hydrogen chloride); for every ton of soda produced, about three fourths of a ton of muriatic acid was emitted, destroying vegetation and corroding metalwork and doing injury to buildings.[46] However, at least for the muriatic acid problem, a technological solution already existed: William Gossage had shown in the 1830s that the muriatic acid would be absorbed efficiently if it were forced to pass through a water film in a tower filled with a porous material such as coke.[47] The Gossage towers became the technological basis for the Alkali Inspection Act. As Playfair asserted before the committee, the control of muriatic acid was practicable; the scenes of desolation reported to the committee, which were after all common knowledge, were improvable.

The evidence given before the committee concerning St. Helens was particularly dramatic. Within living memory, so the testimony went, St. Helens was a sort of paradise with luxuriant gardens and orchards. But with the coming of the Le Blanc soda manufactories and the copper works and silver-refining works, the countryside could now be described as "one scene of desolation — you might look around for a mile and not see a tree with any foliage whatever." As if in sympathy, the *Chemical News* reported: "The town wears a somber aspect. Every thing exposed to the vapours has become of a 'slaty-blue' colour."[48] The alkali makers saw trouble ahead and took collective action. The Alkali Manufacturer's Association was formed in 1862 and, led by Henry Deacon and John Hutchinson, the producers were able to avoid crippling legislation. The Alkali Act which was passed in July 1863 was to have force for a trial run of five years and provided only that ninety-five percent of the muriatic acid was to be condensed, something well within the technical limits of the Gossage towers and certainly no hardship on manufacturers. An inspector was appointed at a salary of £700 and four subinspectors at £400. The inspector selected was Dr. R. Angus Smith, who was expected to devote only part time to inspection, and who remained in Manchester for the tenure of the post.[49]

Smith was a logical choice for the position. He had already gained a national reputation as a sanitary chemist, and his studies on air pollution were published in the leading journals. Locally, in Manchester, he had risen rapidly to a position of scientific leadership, especially after the departure of Playfair. He was elected a member of the Literary and Philosophical Society in 1845, its secretary by 1852, and after 1859 served continuously as either vice-president or president of the group. In 1857 he was elected a fellow of the Royal Society. Moreover, as a Liebig student and as a friend and colleague of Playfair, who was doubtless consulted regarding the appointment, Smith would further enjoy a substantial edge on other suitable chemists.

Smith's first report, issued in 1865, gives a clear indication of his approach to inspection. Despite their input into the final legislation, many alkali manufacturers were reluctant and suspicious of the inspectors, who were able to come and go freely, and who performed chemical tests on their effluents with an eye to enforcing compliance. Smith, however, was sympathetic, and he won, early on, a reputation for tact and skill for which he received commendation many times: "Beginning as I did," Smith wrote in his first report, "with a strong desire to aid the public in the way most agreeable to the manufacturer, I could not forget that it was in his power to put many interruptions in the way." But few objections and fewer obstructions were forthcoming. Smith was pleased. He was pleased also by the excellent results. Whereas a limit of five percent escape was permissible, the effluent muriatic acid from the towers was easily reduced to only 1.28 percent. Smith was happy to report that "several letters . . . gave a flattering account of roses which grew where none had grown for years," though he made no claims for a reconstitution of the land.[50]

By the time of the appearance of the second report, interesting changes in the position of the inspectorate were already taking place. For most of the sixty-four alkali factories which they visited, Smith and his men were the only trained technical consultants available. The proprietors were quick to sense and to exploit the situation. Smith and his subinspectors rapidly became unpaid consulting chemists to the industry.[51] Given the limits of the law, moreover, Smith found himself in the role of defender of the trade against the public. While being called upon to show, many times, that an accused works was *not* in violation, "a strange result therefore takes place; we become defenders of the alkali works."[52]

Smith could not avoid, however, a sense of standing still. Whereas the percentage of escaping muriatic acid was decreasing, year by year the total tonnage of pollutant was increasing, owing to the increased number of works and to the increased amount of output at individual enterprises. He was troubled also by the unknown amount of muriatic acid leaking into the air—perhaps as much as two percent—which was not passed through the condensing towers.[53] The chief inspector suggested that "an addition to the present act" might be required.[54] It was becoming increasingly clear that the consulting role of the inspectors was growing; one manufacturer rather effusively stated "I hope you will not be long before you come again, and, if possible, often." The symbiotic relationship between inspector and inspected could not fail to strengthen Smith in his determination to aid the industry. He argued for an extension of the act to embrace other noxious gases, to protect not only the public but the manufacturer as well. "If the public ought to be protected against the manufacturer," Smith wrote in 1867, "we must remember that the latter is an important member of the public." In order to protect the producer against the vague charge of responsibility for a "noxious smell," *chemists* must be called upon to define a nuisance and inspect for it, or else "neither the public nor the manufacturer is protected."[55]

The original act was scheduled to lapse in 1868. Upon Smith's urging that it be continued and extended, a bill was introduced and passed which extended the act indefinitely. The inspectors took the action as a new mandate and began their attempt to account for leakage and to study the wider problems of the condition of the air. Smith began calling in his yearly reports for the founding of a new science, chemical climatology, "to express in distinct language the character of a climate, and certainly of the influence of cities on the atmosphere."[56] Smith had passed the boundaries of his original mission and looked far beyond it. His motives, however, were far from simple. There was a certain internal momentum generated by the technical problems of air pollution control. Just as surely, there were career considerations and the natural desire of the professional to advance in his field. But we cannot fail to see as well the mission of the Liebig school reasserting itself in a somewhat clearer form. All elements were summed up in Smith's own account:

In former years when I reported on the escape of muriatic acid from alkali works, being rather weary of the monotony as well as from the narrowness of the subject, I commenced to inquire into the amount of other gases. . . . I have thus extended my work from that of the inspection of alkali works to the examination of air generally for sanitary purposes, so far as chemical means are concerned; and I hold that it is of great importance that a chemical branch for such inquiry should exist.[57]

The fifth, sixth, seventh, and eighth reports of the alkali inspector, covering the years 1868–72, were marked, then, less by concern for the control of muriatic acid than by Smith's attempts to found a science of chemical climatology, and administratively to extend the act to include other noxious vapors. In the eighth report, published in 1872, he reported, for example, that "the greater part of the injury to vegetation and to the atmosphere by manufactures is not now done by muriatic acid, but by sulphur acid."[58]

The Public Health Act of 1872 appears to have, at least in part, rewarded Smith's work. His office was transferred to the Local Government Board and placed under the supervision of John Lambert; Smith was eager to view his responsibilities in a broader context, and it would seem that the transfer accomplished just that. Smith continued to push for a revision of the act. Chemical works were increasing, and the atmosphere was worsening. A percentage measure of offense was outmoded; a density measure was required. "The Alkali Act," Smith reported, "which was excellent for a time and has done some good, is becoming less valuable daily." His tenth report insisted that changes in the law be instituted "demanding a certain state of dilution for all escaping gases."[59] He was, in fact, heard. The president of the Local Government Board, George Sclater-Booth, entered a bill which was passed which set a volumetric standard for muriatic acid.[60] Smith looked upon this as a new lease on administrative life; he could now be more vigilant and more useful. If only others had faced the issue as squarely as he the larger problem of environmental quality would have surfaced more plainly: "We have not yet looked the matter with the greatest fullness in the face. If we are determined to keep towns to a certain standard of purity we must make the standard; if we keep only the works themselves then the number of works must not increase."[61]

Was Smith's faith in the ultimate reconciliation of purity and profit finally shaken? In any case, if he was willing to make only modest claims for environmental improvement, Smith could have taken some credit for the health of the industry. An ammonia-soda process for the production of sodium carbonate was patented by a Belgian, Ernest Solvay, in 1863; it was introduced into Britain by 1872 and was eventually to replace the LeBlanc process. It is remarkable, however, that the LeBlanc process persevered, reached its height by about 1880, and only then declined; final extinction did not come until after 1920.[62] It was efficient use of the muriatic acid effluent and the recovery of sulphur from alkali waste which kept the industry alive.[63] By the 1880s, for example, the production of bleaching powder from hitherto ejected muriatic acid reached 150,000 tons per year.[64] A supportable claim can be made that the example set by the inspectors, as well as the necessity to condense the muriatic acid, played an important role. Smith and his subinspectors created a demand in the chemical industry for trained chemists which in turn made possible technical innovation.[65] Smith himself noted these changes in his very revealing interim report of 1876. When the Alkali Act was introduced in 1863, he noted, few manufacturers had laboratories or trained chemists in their works. "Now things are entirely changed, and the frequent entrance of the inspector has caused him to be watched, imitated or criticised."[66]

We may grasp more clearly the elements of Smith's conception of his role as Alkali Act inspector, in the light of his background as Liebig student and situation as Manchester civic scientist, by comparing his view of the administration of the acts with Oliver MacDonagh's controversial model of what is termed the administrative "revolution" of Victorian Britain.[67]

According to MacDonagh's model, the first stage of governmental reform commenced with the exposure — often in dramatic form — of some social evil which leads to prohibitory legislation. The legislation itself is the result of a compromise between the demand for reform and resisting forces; nonetheless, responsibility was assumed. Soon it became increasingly clear that the legislation was insufficient to meet the evils. A second, stronger, act followed which designated officers to enforce the prohibitory legislation (stage two). Once appointed, these officers began to press for further legislation, the closing of loopholes, and the appointment of a larger number of officers (stage three), ending in fresh legislation and a superintending central body. MacDonagh's fourth stage involves a change in attitude on the part of the administrators, a change which substituted a dynamic concept of administration for a static one. They ceased to regard mere alterations in legislation or the naming of additional inspectors as the answer to basic problems. During the course of the fourth stage, as MacDonagh puts it, there is the "gradual crystallization of an *expertise*" of the field in question. Finally, during the fifth and last stage, the administrators demanded and often received discretionary legislation and began systematic and frequently experimental investigations; "they strove to get and to keep," MacDonagh writes, "in touch with the inventions, new techniques and foreign practices relevant to their field. . . . They even called directly upon medicine and engineering, and the infant professions of research chemistry and biology to find answers to intractable difficulties in composing and enforcing particular preventive measures."[68]

The conformity of Smith's experiences in Alkali Acts administration to the model outlined by MacDonagh is striking. By the beginning of the 1860s the failure of the public health measures of the 1840s and the Nuisance Removal Act of 1856 to deal adequately with the growing environmental devastation wreaked by the alkali trade was apparent. The Noxious Vapours Commission of 1862 heard considerable testimony, frequently of the melodramatic type, which pointed out the existence of what MacDonagh refers to as an "intolerable" evil. The compromise legislation, acceptable to the manufacturers and containing elements which provided hope for improvement, was passed and inspectors with Smith at their head were appointed. Almost at once, after the initial successes in obtaining conformity to the rather loose standards of the legislation were digested, Smith himself began to press for further legislation. By the time of his second report in 1866, he was disturbed by the insufficiencies of the enabling act, and as the five-year trial period was coming to a close Smith actively campaigned not only for extension but for improvement. After the passage of the continuation bill in 1868, Smith began developing a dynamic rather than static view of his inspectorate; he began to see his role as one of finding a scientific basis for legislation, in short, of founding a science of chemical climatology. He began to press for legislation which would enable the inspectors to control not merely a percentage of muriatic acid but the condition of the atmosphere. For once science could be brought into the service of law in a systematic and productive way. The outlines of MacDonagh's fourth stage, at least in the case of Smith, appear sharply and clearly. In his appearance before the Royal Commission on Noxious Vapours (1878), Smith testified that he had for a long time believed the first act defective and, although the second act had improved the situation, other noxious vapors required control. He told the commissioners that it was his opinion that the power of enforcing the act ought to be left to inspectors and that, in general, the Local Government Board should have the power to exert more pressure on the trade. Perhaps of prime importance is Smith's reply to Question 283, "Do you mean to vary the stringency of the Act?" to which Smith answered "Yes." Elaborating (elsewhere), Smith went on: "The Local Government Board should be empowered from time to time to fix by provisional order to be confirmed by Parliament, a standard of escape or to require the adoption of the best

practicable means for preventing escape." For Smith, the question revolved upon scientific and technical advance; enforcing administrators
required the discretionary powers and the flexibility to deal with progress.
"It is invention that is required," he replied to Question 115 of the Commission, "and not inspection merely."[69]

After pointing to the striking consonance of Smith's experiences with the
Alkali Act administration and MacDonagh's model, is there more to be
said? Can we, in fact, proceed to the more general attitudes of those who,
as Jennifer Hart implies, hold a "Tory interpretation of history"?[70] According to Hart, the "Tories" denigrate the role of men and ideas, especially
Benthamism, downplay the role of ideology, and support the notion of
governmental change by unpremeditated, pragmatic responses to situations arising in the course of administration after the initial responsibility
had been assumed. In short, was Smith's career as Alkali Act administrator
an example of such pragmatic development, or was there a lurking ideological commitment which shaped his evolution?

The answer, of course, may be impossible to secure, but considerable
advantage may be drawn from the examination of a fascinating paper
which Smith gave before the National Association for the Promotion of
Social Science and which was published in their transactions for 1857. The
paper, entitled "Science and Social Progress," gives us a glimpse into the
attitudes and presuppositions about the relationship between science and
society—between science and legislation—which Smith would carry with
him almost seven years later upon his entry into governmental administration.

"Science and Social Progress" is a call for the recognition and systematic
utilization of the scientist as expert in government and industry. Science,
for Smith as for so many others, was unquestionably the basis of progress
in industry and of all rational means of controlling the environment. The
laws of physical nature, he claimed, "are daily demanding to be made part
of a national code, and they are demanding for themselves a suitable
machinery by which they may be administered." To be sure, science has
come to the aid of society, and in ways which have been impressive. But,
Smith continued, "we find men seeking the assistance of science only when
insurmountable difficulties occur; we never find them looking to it as the
great source of their present knowledge and future progress." In short,
Smith is espousing the importance of research, of an on-going effort to
bring the fruits of science to bear on society, and not merely as a cure for
its most flagrant evils. He might have pursued the obvious analogy with the
preventive public health measures with which he had been associated (he
did not): what we require, he insisted, is not only remedial but positive
action as well. We require a corps of scientific experts in industry and
government. "It is proposed to make use of the knowledge of those who

have studied any particular subject up to the latest moment, and to make that knowledge generally and universally effective."[71]

One is forced to recall Liebig's *Familiar Letters on Chemistry,* published only a few years before, in which Liebig prophesied "a more vigorous generation will come forth, more powerful in understanding," who will "bring forth fruits of universal usefulness"[72] (see above, chapter 3). On the matter of sanitary chemistry, Smith was especially clear: "In sanitary matters the want of public officers is everywhere felt. In such a district as that in which I live there are continually arising cases of nuisance from manufactures. Many of these can be readily cured and the advice of a suitable officer would be taken thankfully by the offending party and accepted by the public, but no such officer exists."[73] In a nutshell, we have Smith's policy of cooperation with the manufacturers, "of leading instead of driving," which later won him by all accounts "golden opinions among all competent judges."[74]

Specifically, the members of Smith's scientific corps were not to be limited to minor technical duties but rather were to have range and freedom of action: "I propose as objects to be attained that Government should have in their employment a number of scientific men to give every department advice in the latest and soundest views, who shall not be occupied constantly with one particular duty but who shall have such freedom of action that they shall be able to give attention to all proposals and applications made to them."[75] It seems clear, then, from Smith's views as outlined in "Science and Social Progress" that the germ of the dynamic view of administration developed by Smith by the beginning of the 1870s (the fourth and fifth stages of MacDonagh's model) was already present by 1857. It was rooted in the complex of ideas about science and society which Smith brought back with him from Giessen—the sense of mission which one can detect in many of Liebig's students—and which was nurtured so beautifully by the Manchester environment. In a revealing letter to Edwin Chadwick on the subject of disinfection, Smith confessed that he saw "in it the germ of great things which I am longing to work out. . . . I expect to see many engineers get many thousands for working out these plans some day. . . . *We must push our thoughts, it is for this we live.*"[76] What was this complex of ideas, this ideological commitment? We need only to return for a moment to the "science of civic virtue," to that credo of public service and the ultimate reconciliation of the public good with the growth of industry (purity *and* profit) to see that Smith was in fact envisioning the realization of Liebig's dream. Indeed, Smith himself must have recognized it. It is no accident that the more general studies of the atmosphere initiated by Smith while he was alkali inspector and published by him in his inspector's reports were collected by him in 1872 as the basis of his book *Air and Rain: the Beginnings of a Chemical Climatology* and were dedicated by Smith to Liebig "not merely as a proof that I still admire him as I did when young and listening to his teaching, but also to remind my countrymen how much we owe to his genius and labours."[77]

As inspector, Smith could not rest content with meeting the letter of the law as if he and his trained men were mere gauges of muriatic acid condensation. Smith began his public career with much more ambitious notions of the scientist's role; he possessed at the very first a predisposition to what MacDonagh and others see as a dynamic conception. In Smith's case at least, the idea of the scientist-expert preceded the reality.

Chemistry and Industry

The movement of the civic scientists in the direction of public health–sanitary chemistry was obviously spurred by the urban crisis and encouraged by government's first efforts to stem what was increasingly seen as an intolerable evil. There was, however, another side to the notion of civic science, one which might easily be obscured by the early success of the sanitarians, like Smith, if it were not so fundamental to the economy of the region. It was, after all, industry which lured the small group of professionals to the Manchester area in the first place.

The "science of civic virtue" had as one of its two pillars the importance of science, especially chemistry, to the growth and expansion of the economy. Liebig himself had declaimed enthusiastically on the crucial character of the chemistry-industry nexus: the prosperity of empires depends upon it (see chapter 3). A clear idea of the attitudes which the Liebig students carried with them can be gained from Lyon Playfair's inaugural address to the School of Mines, delivered in 1851 but written in large measure some years earlier, possibly during his Manchester sojourn. In his address, entitled "The Study of Abstract Science Essential to the Progress of Industry," Playfair pointed out that should free trade come about, as the Manchester economists desired, "and the raw materials confined to one country become readily attainable by all at a slight difference in cost, then the competition in industry must become a competition in intellect." There never had been a time, Playfair continued, in the history of Great Britain when it was more necessary that "skill and science be united for the promotion of the industrial arts." Science, in its advance, has created hitherto undreamed of resources and has removed the disabling features of local geographic disabilities. "If England still continue in advance," he warned, "it will not be from the abundance of her coal and iron, but because, uniting science with practice she enables her discoveries in philosophy to keep pace with her aptitude in applying them." Quoting Humboldt to the effect that "nothing but serious occupation with Chemistry and natural and physical Science can defend a state from the consequences of competition," Playfair proceeded to make a case for the determined prosecution of scientific research; not only what could be termed applied science, but pure research was called for. Like his mentor, Liebig, Playfair believed that "it is but the overflowings of science that thus enter into and animate industry. . . . Science is too lofty for measurement by the yard of utility; —too inestimable for expression by a money standard." No easy balance sheet could be drawn for expenditure for science against commercial return. But science governs the principles behind manufacture and would without doubt continue to transform it. "Almost all the staple manufactures of this country are founded on chemical principles, a knowledge of which is absolutely indispensable for their economical application." The priorities must be set aright. If England is to meet the industrial challenge from abroad, "it must be by her sons of industry becoming humble disciples of Science."[78]

It was with this transcendent faith in the ultimate bond between pure science and industry and in the real power of science to succor the industrial arts that the Liebig students returned to Britain and attempted to find careers for themselves. Playfair, it will be recalled, found employment in printing-works of the forward looking James Thomson of Clitheroe. The others, including Crace-Calvert, Young, Smith, Schunck, and Gilbert, were all attracted to the Manchester area by the economic vitality of the region, by the reputation for the prosecution of chemistry (especially in industry) that the city possessed, and by the hope (initially) of industrial posts. Most were satisfied.

The cotton industry, the staple of Manchester, had long possessed a reputation for requiring chemists for its ancillary industries. The patently chemical processes of bleaching, dyeing, calico-printing (already by 1840 a very complex undertaking) and mordant production led many within and without the industry to look upon textile finishing as a branch of chemical technology. Edward Baines in 1835 called cotton mills and printing establishments each "a wonderful triumph of Science."[79] Love and Barton's guide, *Manchester As It Is,* of 1839 noted with some pride that "the arts of dyeing and calico printing have received great assistance from . . . men of science . . . and in most of the establishments in which these arts are wrought out, some one or more of the principals, are, practically, men of eminent scientific talent; in fact, it is almost indispensable that they should be, in order that the trades be rendered profitable."[80] Léon Faucher, in his remarkable *Manchester in 1844,* concurred, emphasizing that "Science, which is so often developed by the progress of industry has fixed itself in Lancashire. . . . Chemistry is held in honour." The Mancunian translator of Faucher's little essay added that "when Manchester ceases to honour science—science will cease to honour it."[81] With expressions of enthusiasm such as the foregoing, it is no surprise that trained chemists might seek to establish themselves, if anywhere, in Manchester.

These encomia to science and industry in Manchester might easily be dismissed as local "boosterism," were it not for the recent detailed researches of A. E. Musson and Eric Robinson in their valuable *Science and Technology in the Industrial Revolution.* Messrs. Musson and Robinson have documented the depth of the scientific interests of Manchester industrialists, especially in areas related to bleaching and dyeing.[82]

Thomas Henry, a leading figure in the early Literary and Philosophical Society, had already launched his career as a manufacturer of patent medicines by the mid-1780s, marketing Henry's Genuine Magnesia and Henry's Calcined Magnesia both in Manchester and London. Beginning in a small way at his Essex Street premises, Henry later took his son William as a partner and established works in East Street. By the beginning of the nineteenth century Henry also began relatively large-scale commercial production of artificially-carbonated beverages or soda water and established a plant off Deansgate which he also operated with his son, William. Spurred by competition from Schweppe's, the Henrys started a plant to produce their "Artificial Mineral Waters" in Birmingham, although by 1817 the soda water business had passed into other hands.[83]

Interesting though these early efforts are, far more important for the economy of the region and for the development of chemistry in Manchester were the efforts in textile-related chemistry, on the part of professional men like the Henrys and of cotton manufacturers, and bleachers, dyers, and printers.

Thomas Henry was himself a propagandist for a scientific approach to bleaching and dyeing and performed experiments in the 1780s which he explained to the Literary and Philosophical Society and published in their *Memoirs,* relying especially on the French chemists such as Macquer and Berthollet.[84] His work on the chemistry of mordants was particularly well-received and retained remarkable durability in the literature.[85] Henry was joined in the Literary and Philosophical Society by chemically minded cotton industrialists such as Theophilus Rupp, who delivered an important paper on chlorine bleaching before the society in 1798.[86] Henry, Charles Taylor, manufacturer, bleacher, and dyer, and Thomas Cooper, bleacher and dyer who later taught chemistry in America, were also all centrally concerned with the introduction of chlorine bleaching into the region; John Kennedy, the important spinner, credited Henry as "one of the first who carried it into practice" and who "gave to some of the principal bleachers in the country the first instruction they received respecting the new process."[87]

John Wilson, a dyer and printer, Rupp, Taylor, Cooper, and James Watt, junior, who worked with Manchester fustian manufacturers, all were connected with the Literary and Philosophical Society and had attempted to familiarize themselves with the latest developments in French theoretical and practical chemistry. By 1800 they had given the Manchester region a reputation for progressive views regarding the application of science to the industrial arts.[88] It was, therefore, not chemical manufacture but the textile finishing trades which spurred applied chemistry in Manchester in the early years of the nineteenth century. Even Dalton was drawn in, acting as a chemical consultant.[89] This situation continued naturally enough well into the century. It was, after all, James Thomson, a printer and dyer, who brought Playfair to the area, and it was presumably the young men of such works who staffed the meetings of Playfair's Whalley meetings.

To make this claim is not to ignore the importance of incipient chemical industry. In 1783 the vitriol works of Walker, Baker, and Singleton were founded at Pitsworth Moor, near Manchester, and in the 1840s Charles Macintosh had formed an association with H. H. Birley, a Manchester Gas Corporation director, to develop Macintosh's rubberizing cloth, using waste products of the Manchester gas works.[90] It was not, however, until the late 1830s that the markings of a new situation can be clearly discerned. In 1837, an offshoot of Charles Tennant and Company of St. Rollox (Tennant, Clow and Company) purchased E. P. Thomson's chemical works at Ardwick Bridge and began operations in the alkali trade. Thomson's works had been established in 1813 for the production of heavy acids and alkali. James Young was hired by Tennant's as works manager; William Hart and William Gossage, later to become very well known as an innovator in industrial chemistry, joined him as assistants.[91] It was at about the same time that Robert Rumney began his chemical works in Manchester, producing mordants for calico printers and dyers.[92] Subsequently there was considerable industrial activity in the Manchester area. By 1864 there were five registered alkali manufacturers in Manchester (Tennant's; John Metcalf and Sons; C. J. Schofield; Thomas Dentith; and William Howarth) and several nearby (William Barton of Leigh; H. Becker of Chadderton; and John Former of Gorton).[93]

All this activity was bound to have some effect upon the character and make-up of the Manchester community, although until the arrival of the civic scientists in the 1840s changes came slowly. Although John Dalton occasionally did consulting work for bleachers and performed water and gas analyses, and despite his close friendship with Thomas Hoyle, the calico-printer, he was, as Archie Clow maintains, not directly involved in industry.[94] He was, however, the teacher of one of the more important of Manchester's future young chemical industrialists — John Dale.

Dale was born in Birmingham in 1815; after preliminary schooling he was apprenticed to a druggist at Denbigh. In the period 1833–35 he moved to Manchester as an assistant to a Mr. Ansell, a Quaker acquaintance of

Dalton. Ansell is reported to have been, at the time, the only dealer in chemical apparatus in all of Lancashire and Cheshire.[95] Shortly thereafter Dale became Dalton's pupil; as Joule reported, Dalton had the gift of inspiring his students with a love of research, and it was not long before Dale, emulating Dalton, began a lecture tour of Lancashire, speaking on chemistry in all the major towns. Returning to Manchester, he made the acquaintance of Mr. Thomas Roberts, with whom he much later entered into partnership in a chemical firm, Roberts, Dale and Company. In the interim he was employed as manager in a calico-print works. Dale continued his work in chemistry and it was at about this time that he made his first discovery. Dale's own account of it is as follows: Naphtha from wood was employed at the time in hat factories to prepare a lacquer or varnish; attempts were made to substitute naphtha from coal tar but they failed owing to an unbearable smell which it gave off upon evaporation. Dale attempted to convert coal-tar naphtha into wood-naphtha. He processed it with sulphuric acid and nitre and obtained a liquid possessing the smell of bitter almonds. John Leigh, at the Manchester meeting of the British Association, pursued the matter further and is reported to have exhibited quantities of nitrobenzol and dinitrobenzol, proving that coal-tar contained considerable amounts of benzol.[96]

In 1852 the chemical firm of Roberts, Dale and Company began operation, producing coloring matters for textiles and for the increasingly popular wall-paper trade and manufacturing oxalic acid from Dale's own process, employing sawdust.[97] After the discovery of aniline dyes, Dale attempted to enter the artificial dyestuffs trade and met with remarkable success, with the help of his German chemists, Heinrich Caro and C. A. Martius.[98] Caro, trained in Berlin, came to Manchester in 1859 primarily through connections with his relatives, the Behrens. Supported by Magnus of Berlin and by Edward Schunck, the Manchester chemist, he obtained a position with Roberts, Dale and Company.[99] Martius joined the firm in 1863, after training at Munich and with Hofmann in London. The small investment in trained chemists paid off handsomely for Roberts, Dale; in 1864 a patent for a new dye, Manchester Yellow, was taken out jointly in the names of Dale, Caro, and Martius.[100] The group followed with Manchester or Bismarck brown, and thus Roberts, Dale and Company, with Caro (who became a partner) and Martius, launched Manchester into the era of artificial dyestuffs.[101]

Caro stayed in the city until 1866 when he returned to Germany. He appears to have taken little part in the scientific institutions of the city, apart from a paper delivered with William Dancer before the Literary and Philosophical Society in 1865, entitled "An Instance of the Injurious Action of Alkalies on Cotton Fibre.[102] Dale, however, joined the Literary and Philosophical Society in 1854 and remained a leading figure in it (although not as an officer) until his death in 1889.[103]

The course of Dale's career offers some clues regarding the manner in which young men were drawn to chemistry in and around Manchester. Peter Hart, almost a generation younger than Dale, provides further hints. Hart was born near Warrington in 1834, the son of William Hart, an assistant to James Young at the Tennant chemical works. The young boy served for a while in the works, returning to a position in the laboratory in 1849 after a brief career in a solicitor's office. He attended night classes at the Mechanics' Institution and pored over the books in the small library attached to the laboratory at Tennant's. When Young left the plant in 1852, William Hart succeeded him as manager and Peter became chemist to the works. Shortly thereafter he began publishing research papers, generally in analytical chemistry. When in 1866 his father was killed in a plant accident, Peter succeeded him as manager, by that time having established himself as a leading authority on sulphuric acid manufacture.[104] It was not, however, until 1862 that Hart joined the Literary and Philosophical Society; it appears that many industrial chemists who were not formally trained sought or were asked to join only after their local reputations were firmly established.

By the end of the 1860s Tennant, Clow and Company was the fourth largest chemical firm in northwest England; the third, behind Brunner, Mond and Company and United Alkali, was Peter Spence and Company, manufacturers of alum and other chemicals.[105] Spence's firm was the world's largest alum producer; over half the alum in England was produced by Spence or by his patentees.[106]

Peter Spence (1806–1883) was the son of a Scots handloom weaver who, after a short apprenticeship with a grocer, obtained employment at a gas works in Dundee in the early 1830s. While at Dundee, Spence's interests turned to chemistry; his familiarity with the manufacturing and purification operations of gas production provided him with the entree into chemical industry. In 1836 he took out a patent concerned with the utilization of refuse lime from gas manufacture and the production of Prussian blue and prussiate of potash. For about a decade, Spence struggled as a chemical manufacturer, until he perfected a revolutionary process for the manufacture of ammonium alum explained in 1845 and 1850 patents.[107] Alum, an important mordant, was essential to the printing and dyeing industries of Lancashire. Up until Spence introduced his new process, potassium alum was used; after Spence's firm went into production, it was rapidly replaced by the ammonium variety.[108] The old process was a laborious and time consuming process which involved the roasting of alum shale, lixiviation (water treatment), evaporation, precipitation of the alum by the addition of alkaline salts, washing and crystallization. It involves considerable labor and waste; an entire year was required for the conversion of the raw materials into alum for the market. Spence's process consisted in the calcination of shale obtained from Lancashire collieries; it is then treated with sulphuric acid and ammonia, mixed, drained, washed, steamed, and finally precipitated. The entire process is completed within a single month. By the late 1850s, Spence produced over 30 tons of alum per week, and by 1865 production had been stepped up to over 7000 tons per year.[109] Spence's success was recognized by the award of the medal for alum production at the Exhibition of 1862.[110]

The Pendleton Alum Works (near Manchester) had, however, to contend not only with its competitors but also with its neighbors. Increasingly during the 1850s, residents of the Pendleton area complained of annoying and noxious vapors emanating from the works. Spence took the complaints seriously and retained the professor of chemistry at Owens College, Edward Frankland, to advise him and commissioned inquiries by his colleagues in the Manchester Lit. & Phil.—John Leigh, the Manchester city chemist, Angus Smith, and Crace-Calvert—into methods of control. His caution came to no good end; in 1857 the complainants brought Spence to court on the charge of conducting an unacceptable nuisance at Pendleton. The case of Regina vs. Spence was tried before the *nisi prius* court, South Lancashire Assizes in August 1857. The voluminous six-count indictment included that of discharging "foul and noxious vapours," purveying important injury and "unendurable nuisance." At the time the works were producing eighty tons of alum per week and discharging as waste products hydrogen sulphide gas and sulphuric acid. The prosecuting attorney drew upon precedent: "Chemists of course made light of these matters and did not think anything should be complained of which was [not] palpably injurious to health or vegetation; but in the case of King v. Neale . . . [it was held that] it was not necessary that a public nuisance should be injurious to the health, but that if there were smells which were offensive to the senses it was sufficient, as a neighbourhood had a right to fresh and pure air." Edward Frankland, despite his service to Spence as consultant, was called as a witness for the prosecution. Frankland testified to the effect that certain species of hydrocarbons, carbonic acid, and sulphuretted hydrogen were given off from the plant, and that they could be noticed as far as a half-mile from the works. Frankland had customarily visited the plant, accompanied by his Owens students for five or six years, and appeared to be in a strong position to testify on the nuisance.[111]

The defense attorney, however, attacked not Frankland's expertise but his loyalty. "Dr. Frankland had," he began, "become a deserter, and after partaking of the bounty of Mr. Spence and taking advantage of his works for the instruction of his pupils he had been induced to join the ranks of the enemy."[112] The defense called, as opposing experts, Angus Smith and John Leigh. Leigh testified that if properly conducted there need be no nuisance; Smith disclosed that in 1854 he and Crace-Calvert investigated and found no sulphuretted hydrogen and only small amounts of sulphurous acid.[113]

The jury promptly delivered a guilty verdict regarding nuisance, but declared as not proven the charge that the nuisance was injurious to health. Spence's Pendleton Works were shut down, and he removed his business to Newton Heath, Manchester, where he retained the name Pendleton Alum Works until the 1880s.[114]

It is ironic that it was Spence who became the object of the famous lawsuit upholding the neighborhood's right to pure air. Atmospheric purity was a subject on which he himself was very concerned and remained so. In 1857 he published a pamphlet "Coal, Smoke and Sewage Scientifically and Practically Considered," in which he inveighed against the antismoke sanitarians, who concentrated their efforts on black smoke and soot and neglected the invisible evil; he called it "the sanitary smoke-consuming mania." Perfect consumption of smoke, he asserted, would only aggravate the production of the noxious invisible results of combustion. "I would at the same time say to the sanitary smoke consumer, that he had better not interfere further as every step he takes in enforcing the consumption of smoke will only tend to deteriorate the atmosphere, and that every cloud of visible smoke he is successful in dispelling is only making way for a more baneful though invisible agent."[115] Black smoke, he believed, actually was an antiseptic. He advocated, however, the connection of house flues to drains, which would then become gas as well as water conduits, the condensation of all possible gas wastes, and the erection of very high chimneys (600 feet) to diffuse the gases into the atmosphere.[116] In 1866, before the National Association to Promote the Social Sciences, he denied once again the noxious character of smoke and declared that the sanitarians "are following in *ignis fatuus*," and that they will at last "find that you have got into a mire." He warned that "black smoke will not be abated to any great extent till it can be done economically, and by bringing the law to your aid, you only compel the manufacturer to take up the first patent nostrum that is forced upon his distracted attention, and which as soon as your legal fangs are removed, he throws up in disgust."[117]

Crace-Calvert and Industrial Chemistry

Spence, like Dale and Hart, signifies the subtle changes occurring in the Manchester community: increasingly after 1840 men engaged in science-based industry would begin to make the effect of their presence felt. Not only did Spence play an active role in the Lit. & Phil. (serving on its council) and the Mechanics' Institution but as chemical manufacturer offered employment, at least on a consulting basis, to the chemists of the city. Manchester was becoming by steps a comfortable home for chemical professionals. It is in this regard as well that the career of Frederick Crace-Calvert is significant, for Calvert, even more clearly than Smith, demonstrated that a chemist could make his way in Manchester.

Immediately upon his arrival in Manchester Crace-Calvert began to exhibit his phenomenal energy and wide-ranging chemical interests. Well trained by Chevreul in the chemistry of bleaching and dyeing, Crace-Calvert at once proceeded to secure his fortune. Supporting himself, while honorary professor at the Royal Manchester Institution, by teaching chemical classes, by lecturing, and by consulting for industrialists and local government, he systematically propagandized for chemistry. Again and again he would assure his audiences that the chemistry of today was not the "theoretical chemistry of former times but was a science which enabled our manufacturers to improve their works and often even to make discoveries." Not only was chemistry important for the "staple trade of Manchester," but also in order for Britons to "maintain our position, the rising generation must study chemistry along with their mechanical trades." Indeed, "it was only by the diffusion of science in every shape and form that it could be rendered useful in promoting the wealth and general good of humanity."[118]

Like Playfair and Smith, Crace-Calvert would put his words into action. In his earliest days in Manchester, he turned his attention to what would naturally appear to be the most critical problems of the city's industry— those relating to cotton. His papers during the first years concern bleaching and dyeing, the chemistry of mordants, and the action of chemicals on cotton. In 1850 he suggested certain improvements in the bleaching of linen which would reduce the six to twelve week duration to a mere three weeks: "I trust," he stated, "that the slight improvements I have made in the bleaching of linen will still be . . . brought to perfection; thus rendering available to Ireland and Scotland those thousands of acres of land which are now uselessly covered and by enabling them to bleach in winter as well as in summer, allow the manufacturers to renew their capital more rapidly."[119]

Late in the 1840s, Crace-Calvert's attention turned to carbolic acid in connection with his work on mordants. He found that it was tannic acid which produced black coloring with the use of iron mordants, and he explored the reaction with the mordants of gallic acid. He discovered that impure carbolic acid added to the extracts of tanning matters prevented, for several months, the tannin from turning into gallic acid, thus retaining the tannin extracts as commercially useful.[120] He found the carbolic acid useful as well in producing picric acid, which when used as a dye produced pretty, bright yellow colors.[121] Crace-Calvert joined with McDougall and Smith in disinfection experiments, in the mid-1850s, using the acid, and he began investigating its medical uses—as a preservative of cadavers and for a variety of internal and external applications.[122] In short, Crace-Calvert was already very familiar with and interested in the variety of uses of phenol when M. Marnas of the Lyonnaise firm of Guinon, Marnas and

Bonnet visited Manchester in 1859 and asked him to supply the firm with a purer variety of the acid than had hitherto been available. By treating naphthas with weak alkaline solutions and after careful distillations, he was able to obtain a relatively pure product which was subsequently utilized by the French firm for the production of improved colors derived from phenol in order to compete with the brilliant aniline dyes. At this time Crace-Calvert established a small plant for the production of carbolic acid which he produced as white crystals melting at between twenty-six and twenty-seven degrees centigrade.[123] By the mid-1860s the demand for purer and purer carbolic acid was growing rapidly. Both Guinon, Marnas and Bonnet and medical men were interested in obtaining a purer product. The new company F. C. Calvert and Company was up to the task and continued to refine its product. By 1865 the company was greatly expanded and Crace-Calvert's position as chemical manufacturer was secure; he had given up taking pupils long before and devoted himself almost entirely to chemical industry.

During the 1860s, Crace-Calvert conscientiously attempted to expand the uses of his product. He proselytized among physicians, publicized the uses to which it was put by the surgeon Lister, advanced its employment as a disinfectant, as a preservative, and as a dyestuff. Owing to the company's purer product, the production of picric acid was much improved and made far less expensive.[124] About the same time Crace-Calvert took out a patent, with Charles Lowe and Samuel Clift, for new coloring matters, Emeraldine (green) and Azurine (blue) from aniline and its homologues.[125] Without question, Crace-Calvert's own career had demonstrated the intimate connection between chemistry and Manchester's "staple trade."

About the same time (1859) that his commercial success was beginning to blossom, other honors were arriving. He took his Ph.D. in that year,[126] and was elected a fellow of the Royal Society of London. Having made his way, he now considered relinquishing the honorary professorship which had served him so well as a launching post. This possibility caused him to enter into an interesting correspondence with another chemist, now struggling in London, William Crookes, a student of Hofmann at the Royal College of Chemistry, and founder-editor of the *Chemical News.*

Crookes' attempts to relocate in Manchester in the 1860s provide a fascinating glimpse into the professional life of a Manchester chemist. Crookes was a good friend of Peter Spence and Angus Smith, both of whom he presumably met at the British Association.

After a visit to Spence at Manchester in October 1864, Crookes wrote a revealing letter concerning a proposed removal to Manchester in hopes, as Crookes' biographer states, "of finding brighter prospects and more scope for his gifts in the industrial North." It is obvious that the chemical community in Manchester was beginning to attract national attention — as a community. The success of men like Crace-Calvert and Smith to make their own way professionally acted as a lure on struggling proto-professionals, like Crookes: "Since leaving your hospitable roof," Crookes wrote to Spence, "I have been constantly thinking of the proposed move. Inquiry amongst my friends and those able to advise has tended to confirm me in the opinion that such a change would be prudent." Crookes had written of his intentions to Angus Smith, who wrote a long, mixed letter, reviewing the Manchester careers of Dalton, Playfair, Stone, Roscoe, John Davies, Crace-Calvert, Frankland, James Allan, and himself. Smith concluded that "no man has ever in Manchester made a decent living, or indeed any living at all by *analytical* chemistry," although about five had made their way by consulting in addition. On the whole Smith was favorable to Crookes' coming, but reluctant to be overly encouraging. He indicated that he regularly passed up 5 shilling and 10 shilling analyses and could have earned considerably more had he been willing to testify in legal cases and to please lawyers. Crookes, however, remained undaunted: "The £200 or £300 a year which the promised copper assays will bring in will be quite enough to encourage me."[127]

Crookes was also in correspondence with Crace-Calvert and had made him an offer for the purchase of his laboratory and his equipment. He was, however, not overly optimistic about having his offer accepted. He wrote to Smith a long letter, dated 31 October 1864, which gives a clear picture of the London situation and the attractions which Manchester held for him. Its importance is such that it deserves quotation *in extenso*:

My dear Dr. Smith,

Your very friendly letter just received has set me thinking. I am grateful for the trouble you have taken in going so fully into the matter, and if I now proceed to argue upon the information you have given me, do not think that I wish to induce you to give me different advice, but only that I want you to look at the matter from my point of view, which may be different from yours. In the first place, I find "London" a *failure*. No doubt if I were to advertise constantly, and give puffing testimonials to tradesmen I could get a connection and make a decent living out of my Laboratory, but as for respectable work as consulting (in the wide sense you employ the term) or analytical chemist, I get next to none. I have made possibly £100 in six years at that work. My presence in a court of law, practically speaking, then my Laboratory and chemical education, are only of *money value* in so far as they enable me to exercise editorial supervision over the *Chemical News*. This being an amount of knowledge that any sharp person could get up in six months, nine-tenths of my "brain-force" is lying idle. This is the £ s. d. view of the case. There is another item. I was scientifically fortunate, but pecuniarily unfortunate some years ago, in discovering thallium. This has brought me in abundance of reputation and glory, but it has rendered it necessary for me to spend £100 or £200 a year on its scientific investigation. Of course, it brings more than the value of this, *ultimately,* in reputation and position—but "whilst the grass grows, the horse starves," and it is the utter despair that I feel of ever making anything out of my London Laboratory that makes me anxious to leave. It is not that I neglect taking steps to make my wants known. I attend societies and converse with all the leading men, but . . . and the public all go in one or two grooves — their work is sent as a matter of course to the College of Chemistry, or to some of the well-known chemical schools attached to hospitals. There is plenty of work to be done, but there are so many eminent professors twice my age and standing that they absorb it all. I am aware that persons who have not lived in London are of a different opinion, but mine is the general experience of beginners here. When a great prize is to be competed for there are one hundred applicants from my class of chemists. One gets it, and is held up as an instance of the advantage of cultivating science and living in London, but what becomes of the 99 unsuccessful ones? They starve in places of £150 a year, or are kept by their friends, or wait in the hopes of getting something better next time. . . . Within the last few years I have gone heart and soul into three competitions for a prize of £200 or £300 a year. I had a dozen rivals — everyone said I stood by far the best chance, but somehow or other I failed. I shall not try again.

The situation in Manchester appeared far brighter to him; Crookes reasoned that with diligence and a bit of good fortune he would be far better off in the north:

I have thus given my reasons for not caring to stay in London. Now for the attractions of Manchester. Supposing I have moved there and find the Laboratory brings in *nothing at all*. I still have my present income scarcely diminished. But it won't be so bad as that. You refuse the small analyses. I can't afford to do so, and besides I feel I could go through the routine. *I am promised by Mr. Spence's new copper smelting company sufficient copper assays at 10/- each to bring in £200 or £300 a year*, and I have other trifles also promised which will give me a certainty of the latter figure. *Now all that is clear gain to me.* Moreover, there must be taken into consideration the other small analyses, which I shall lay myself out for and organise the laboratory for. Then there are the chances of pupils. Calvert tells me there are frequent applications, but he always refuses. I am accustomed to teaching, and like it.

There are also to be considered the chances of profit from entering more into Manufacturing Chemistry or dabbling in patents. Money has been made by yourself and most of those you name by these means. I have not had much experience at manufacturing, but I fancy my *forte* lies somewhat in the line which would be found most useful at Manchester.

Your letter has shown me that Manchester is not the place where a chemist is likely to make a fortune, but he can make a tolerable living by his profession if aided by a little private income. That is all I want. A stationary income will not do with an increasing family, and domestic necessities are apt to make scientific men very mercenary.

I have told Calvert that if he will guarantee by his books that his laboratory has been bringing in £500 or £600 a year, and can secure the laboratory to me for a long term of years, and will give me all his working apparatus (except balance and air pump, &c.), I will give him during the next 3 years such a share in the profits as will make up the £100.

I have decided to come, and *wish especially* to have his Laboratory. Will you therefore kindly see the secretary and see what can be done, for in case Calvert rejects my offer I should like to be certain of having his Laboratory before he interests himself for anyone else? He told me he could hand it over as a matter of course. Possibly you would be able to find out better than I could the time Calvert talks of leaving. I am thinking of moving soon after Xmas.

Respecting the article on the Report of the Mining Commission, I did not get your letter till Friday night owing to a delay of the post. I should very much like to have such a review. Can you not manage to write it yourself? As an article on the subject it would be an admirable topic for the *Quarterly Journal of Science,* and no one could handle the matter so well as yourself. If you would rather not do it, I would see if I could undertake it, but I would rather it were by you. Where can I see a copy of the report?

Excuse this long letter — you have brought the infliction on yourself by the very kind manner in which you have answered my former one.[128]

After all, Crace-Calvert decided against relinquishing his post and laboratory. Crookes' formal application for the post of honorary professor at the Royal Manchester Institution came to nothing. Crookes' move to Manchester was postponed indefinitely.

The Impact of the Civic Scientists

The new professionals, the civic scientists, collectively had a dramatic effect upon the "ecology" of the scientific institutions which comprised the Manchester scientific community. On the macro-level, the small band of professionals provided a nucleus which drew others of similar mind and background to the city in search of a scientific career. Moreover, the group reinforced what scientific reputation Manchester already possessed by dint of Dalton's residence in the city and bolstered south Lancashire's reputation as a forward looking, progressive community. In short, the presence in the city of men like Playfair, Young, and afterwards Crace-Calvert and Smith, men with national reputations, enabled Manchester to present a face to the nation, and perhaps even to the world, as a "scientific" city.

The effect of the coming of the civic scientists was even more profound, however, internally. The character of the scientific community was forced by these men to undergo drastic changes. At once, upon their arrival, these men began to assume positions of leadership within the community. It was almost as if, by virtue of their special training and access to the most famous scientists of Britain and the continent, they possessed special rights. Playfair, and later Crace-Calvert, assumed a professorship created for him at the Royal Manchester Institution. The others, mindful of the importance of the proper institutional affiliations, sought teaching positions; Smith taught at the New College, Crace-Calvert and Allan taught at the Pine Street (later Royal) Medical School. Within the Literary and Philosophical Society they rapidly proceeded to join the ranks of the officers. Most of them were asked to join the society rather more quickly than their brief periods of residence would appear to have warranted, and most of them rather rapidly rose to positions of power and influence within the society. Together with distinguished devotee cultivators of science such as Joule and Binney, Angus Smith and H. E. Schunck were the most potent members of the Literary and Philosophical Society. Its *Memoirs* and the *Proceedings* became their organs, when they wished it, and the meetings their platform. During the years 1860-80 the presidency of the society rotated among these four men, the remaining men generally serving as vice-presidents.

Not only did the professionals rise quickly to the top of the community, as befit their qualifications, but they also altered the character of the major institutions themselves. Perhaps the most important shifts occurred within the Literary and Philosophical Society. By virtue of its preeminent place in the community as the medium of exchange for scientific research, it was the "natural home" of the professionally trained and research-oriented chemists. As we have already seen, the civic scientists came upon the scene at a critical time in the history of the society: the old guard was relinquishing its grasp upon the major positions within the society and the devotees were increasingly putting pressure on the less ardent members of the society. The influx of the new professionals served to sharpen the split between the dilettante element and the serious cultivators of science, a split which was broached openly in the Grindon affair. The increasing professionalization of the society was to continue throughout the rest of the century and in a curious way was even to touch the civic scientists themselves. Relative to the academic scientists and to the more intensively trained and more highly specialized chemists and physicists of the last quarter of the century, the "new professionals" of the 1840s became quaint and old-fashioned. Edward Schunck himself noted this trend several times. Before the British Association in 1887 he referred to himself as a dilettante and confessed the difficulty of keeping abreast with so rapidly advancing a field as chemical science: "The marvellously rapid progress of chemistry during the last twenty years has made it difficult for the most industrious cultivator of the science to keep abreast of the knowledge of the day, and for a *dilettante* like myself one may say it is next to impossible. I confess myself painfully conscious of my defects in this respect.[129] Ten years later, upon resigning the presidency of the Literary and Philosophical Society, Schunck outlined the changes in the society in the more than fifty years of his membership:

In one respect a change has taken place in the constitution of the Society which is still in progress. I mean the gradual effacement of what, without giving offence, may be called the *dilettante* element, of men who carried on science and literature not as a profession but as an intellectual diversion, and the substitution of men who cultivate science in a strictly professional spirit. This may be regretted — I regret it — but considering the great and ever increasing specialisation of science, and the difficulties attending its cultivation, this tendency must ever be on the increase.[130]

The Royal Manchester Institution, too, felt the impact of the coming of the civic scientists. The honorary professorship awarded Playfair provided him not only with enviable laboratory space, but with a springboard to civic and national attention. The lectureships at the Royal Institution were perfect propagandizing mechanisms; the title "professor" carried weight in building the size of his paying classes and of his consulting practice. The vigorous prosecution of the professional concerns of the chemists at the Royal Institution weighted the image of the institution towards that science; it was the effort made by Playfair and Crace-Calvert which brought about the fulfillment of G. W. Wood's ambition for the organization — to instruct the middle class in science, literature, and the arts — to fruition.

The immigration into the city of (relatively speaking) so many research oriented scientists facilitated an important change in the understanding of the scientific enterprise on the part of the community. The emphasis was beginning to shift from the diffusion of science to the advancement of science. In the nation at large the beginnings of this massive shift can already be seen in the 1830s in the founding of the British Association, and in the city of Manchester itself in the efforts of devotees like Joule and Binney. Research was increasingly coming to be seen as an activity to be fostered, that is, as something which is not merely the result of inspiration but as an enterprise.

It is no wonder then that the balance of institutions within the community was disrupted by the civic scientists and the reputations which they were to build in the years 1840-60. The importance of the Literary and Philosophical Society in the scientific community and in the city was strengthened. The most active and prestigious scientists were concerned with the health, welfare, and security of that group, which in fact prospered under professional attention. The Royal Manchester Institution, the sounding board for the locals Playfair and Crace-Calvert, and the locus for visiting luminaries to expound their views, rose dramatically in importance in the community during the tenures of Playfair and Crace-Calvert. After the latter's reputation and professional security were well established, however, his concern with the Royal Institution declined; he turned his attention increasingly to his chemical manufacturing company and to national affairs, and the honorary professorship withered.

Mabel Tylecote has described the process by which the Manchester Mechanics' Institution transformed itself in the years after 1830 and became increasingly less an organization for the dissemination and encouragement of scientific knowledge for the worker than a more broadly based one with respect to program and social composition.[131] Still, in the late 1830s and 1840s the institution drew to it bright young men eager for scientific training. From 1833 until 1841, when it collapsed entirely, the chemical class had a spotty history.[132] Before its demise, however, it possessed a certain vigor and even published the *Scientific Miscellany* in 1840 and 1841. Seventeen numbers appeared under the auspices of the "chemistry society of the Manchester Mechanics' Institution" and the motto "The object of all Science is to enrich human life with useful arts and invention." Aimed especially at the youth and ladies of the town, the *Miscellany* appears to have been edited by Daniel Stone, perhaps with the assistance of his brother-in-law, Henry Day.[133] It died an early death, largely owing to the same general lack of support for such enterprises which doomed the Royal Victoria Gallery.

The introduction of new scientific blood into the city in the 1840s, especially Sturgeon, Stone, and the Liebig students, gave temporary life to the scientific role of the Manchester Mechanics' Institution. But for the most part, with the exception of James Allan, who succeeded Stone at the Manchester Mechanics' Institution's chemical laboratory and class, the new professionals were not ardent in their support; they saw their audience not with working youths, clerks, or ladies, but rather elsewhere. Even Fairbairn, one of the founders and serious about science for the worker, asserted that the Mechanics' Institution and others like it had "failed so far as regarded the particular classes they were intended to serve. We were still in want of institutions and museums for scientific and industrial tuition."[134]

At best, after the coming of the civic scientists, the Mechanics' Institution, like the Royal Manchester Institution, became for aspiring professionals a way station on the road to recognition and security, the sooner passed the better. Even Stone, the managing director, left the institution in 1849, when he decided to make a go of it as a professional chemist (see chapter 3). Gone were the days when the leading scientific luminary, John Dalton, gave strong support and even contributed to the *Scientific Miscellany*.[135]

The Geological Society, owing to its reputation for the advancement and application of science, achieved some sort of stasis and success; the Natural History Society was forced to relinquish its last pretensions to scientific status and devoted itself entirely to its collections, to which the Geological Society gladly added theirs in 1850.[136]

There were, in addition, more subtle changes in the relationships between the scientific community and the city precipitated by the coming of the civic scientists. In the first place, the claims for the central role of *expertise* which science increasingly demanded throughout the last years of the nineteenth century were greatly advanced in Manchester during the period 1840-60. The scientists insisted that their profession was important, even essential, for the proper functioning of the city and the progressive management of key industries; through their efforts the case was significantly advanced by 1860. The selection of John Leigh as public health officer in 1868, discussed in chapter 2, signals the increasing acquiescence of the general public to the unrelenting claims for expanding the role of the scientific expert. The career of Angus Smith, after his appointment as inspector under the Alkali Act, likewise reinforced the image and status of the scientist in civic life.

It is important to note that these subtle changes reflect a continuing dialectic between the changing economic and social structure of Manchester and the conscious effort on the part of the scientists themselves to carve out careers for themselves and to fulfill—at least in the case of the students of Liebig—what they conceived to be their "mission." The needs of the city in public health and in commerce, the increasing industrial competition from abroad, especially from Germany and the United States, and the rise of science related and science based industries (gas, artificial dyestuffs, and later the electrical industries) were all consonant with the professional needs and ambitions of the scientists, who, as we have seen, spared no effort in propagandizing for the scientific cause, using the existing institutions and transforming them when necessary for their purposes.

The outline of the results of this dialectic is clear enough: there was in Manchester in this period a changing conception of the science and its institutions. Before 1840 the prevailing image of science pictured it as the collection of God's laws which can be uncovered, diffused to the productive and industrious classes of society, and applied for the general good. Research and discovery belonged to the realm of the isolated genius. The role of scientific institutions was essentially to spread the word of these remarkable men and perhaps, at least in principle if not in practice, make life a bit more comfortable for them. Institutions could, however, spread useful knowledge and make men better workers and more reasonable citizens. This effort provided, of course, the basis of a nationwide effort: the Society for the Diffusion of Useful Knowledge, the Mechanics' Institutions all over the country, and the Whig and utilitarian radical journals all contributed towards this end.[137]

In the president of the Manchester Mechanics' Institution, Benjamin

Heywood, this view found an articulate expositor. In the general discourses of James Joule, too, the union of natural theology and science found eloquent expression, as we saw in chapter 3. But with the coming of the professionals this view of science declines in importance and is replaced with new, more potent ones: science as *method,* and (institutionally) science as *enterprise.* The example and the message of Liebig and his students had been: science can remake the world. Learn the principles and methods of the sciences and the world can be won. The example set by Playfair, Young, Smith, Crace-Calvert, *et alia* in Manchester, and by Liebig, Graham, and Hofmann elsewhere, was unmistakable. Science was indeed reshaping the world—through quality control as well as through production.

From an institutional point of view the message was equally important. The scientific enterprise can not only remake society, its economic base and its conditions of life, it can produce *scientists* (through advanced training as at Giessen) and it can produce *science* (through concerted research effort). To be sure, the Ph.D. with which many of the new professionals came armed served as an impressive qualification and as a reservoir of status, but more importantly it provided living demonstration of the efficacy of science as enterprise.

P. W. Musgrave has outlined the changes in the definition of technical education which occurred around 1870, revealing a shift from the teaching of elementary science or training in industrial processes increasingly towards an emphasis upon the mastery of principles.[138] It is no surprise that this should be so. With the efforts of the civic experts during 1840-60, it was soon acknowledged and generally accepted that, as Playfair and others insisted, industry was science-based; that the economics of Great Britain and Manchester within it was science-based; and even that the quality of civic life was, as well, science-based. New processes, increased production, efficient management of resources, and, even more strikingly, quality control (the management of waste, the analysis of industrial processes and of urban effluents and so forth) became matters for the chemist.

As K. B. Smellie has indicated in his *History of Local Government,* local government in Great Britain has been shaped by the "creation of wholly new and subtle manners and customs whereby the thronging powers called into existence by the advance of science were made to serve the people."[139] Technical education rather subtly became education in applied science, and attention shifted from the improvement of worker efficiency to the training of scientifically sophisticated managers. Having few of the natural advantages of a great city, a correspondent to the *Guardian* insisted in 1867, Manchester must rely upon the "vigour, intelligence and training of her sons."[140] When Henry Enfield Roscoe, professor of chemistry at Owens

College, Manchester, testified before the Select Committee on Scientific Instruction (in fact established to investigate the state of technical education), he was able to express with confidence that "the district of which Manchester is the centre, has special need of a high class of science instruction."[141]

These changes, in part wrought by the coming of the civic scientists to Manchester and the ensuing development of the scientific community, were to play a large role in the birth and in the shaping of the character and development of Manchester's experiment in higher education, Owens College. The new professionals, the civic scientists, completed much of the spadework for the professionalization of the community, a development which was further continued and strengthened by the introduction into the scientific life of the city of the *academic* scientist. In 1851 the Owens College was founded, and an entirely new and sophisticated group of scientific practitioners entered upon the scene.

5
Academic Science:
Owens College Born
and Reborn

The Move for Higher Education in Manchester

At mid-century the star in the crown of Manchester culture was still missing. The eighteen-thirties and -forties had witnessed unparalleled growth in the number, quality and diversity of cultural institutions in the city, and yet it could not boast of appropriate comparison, let alone parity, with the metropolis. Manchester had neither university nor college for the liberal arts and sciences. At the inaugural meeting of the Manchester Athenaeum in January 1836, however, the subject was broached. The president of the Athenaeum, James Heywood, supported by his vice-presidents George William Wood and John Kennedy and his brother Benjamin, recalled the College of Arts and Sciences of Percival and Barnes. He saw that college as a noble failure: "Its own merits were not sufficient for the undertaking. The spirit of intellectual improvement required half a century more to elapse before it arrived at its full development. Now such a college is not only wanted, but actually called for in Manchester. . . . A general interest is already excited in favour of a college in Manchester."[1] The Heywoods, Wood, Fairbairn, Kennedy and others of the vigorous businessman-elite of the twenties and thirties saw the time as ripe for true higher education in Manchester.

After discussing the commercial success of the city, G. W. Wood insisted: "And while this preparation is going on for worldly affairs surely it is not too much to attempt to make some additional provision for the proper cultivation of our minds." There was, he claimed, a "strong temptation to the young men of this country to seek for advancement in Manchester." It was only proper that they be given every opportunity. Heywood concurred, and saw a brilliant future in "the strong vigorous and active minds of mercantile men."[2]

153

The "general interest" to which he referred was a reality. Heywood sponsored the project of Harry Longueville Jones (1806–1870) to establish a university in Manchester. Jones published a pamphlet at Heywood's expense, entitled *A Plan of a University of Manchester* (1836), which was read at the Statistical Society and which precipitated a general meeting of town dignitaries in November. Jones saw as major grievances the theological tests of the English universities, which barred dissenters, and the lack of any interest upon their part to weigh the demographic changes which the nineteenth century had wrought upon England. The restless intellectual energy of the Manchester region deserved another outlet beyond commerce. A university would provide for the profitable use of such energies; it would add to the cultural establishment of the city and elevate its avenues for happiness: "Nor one [William] Roscoe but many are wanted; not one Dalton standing like a patriarch of science at the head of the scientific part of the nation but a body of men who may appreciate and follow up his experiments or make similar advances in other branches of natural philosophy."[3]

In the details of Jones' *Plan* and the actual construction of the Owens College during the 1850s one may discern striking similarities, and the College's historians, Fiddes and Thompson, have already done so. "The similarities," Fiddes has written, "of Longueville Jones' plan and the scheme arranged by the Trustees are too great to be entirely accidental."[4]

Jones's *Plan* immediately precipitated action. In October 1836 a circular emanating from the Pine Street Medical School appeared announcing the intention of a group of Manchester gentlemen to meet and discuss the possibility of a college of general education in the city.

The meeting was held at the York Hotel on the 10th of November. Thomas Turner, surgeon of the Pine Street School, was the principal speaker; he urged the importance of establishing a college to be connected with the London University, and offered the assistance of the faculty of his school.[5] The Manchester *Guardian* reported that the facilities and premises of the Royal Institution were offered for the proposed "collegiate institution."[6] A general committee was appointed which included Turner and Jordan of the Pine Street School, Pendlebury of the rival Marsden Street Medical School, H. H. Birley, Richard Cobden, John Davies, William Fairbairn, who earlier in the year had advocated the erection of a University in the city,[7] the Heywoods, Eaton Hodgkinson, William Neild, George William Wood, Edmund Potter, R. H. Greg, Samuel Fletcher and others. As in Jones's *Plan* the committee urged the creation of six professorships (mathematics, chemistry, natural history, classics, English literature and history) as well as provision for instruction in design and modern European languages. A medical department was likewise to be provided for. The organization was designed as that of a joint stock company, with annual subscribers of three guineas and benefactors of £31.10s and more to be given votes towards the election of a twenty-four-member governing council.

Despite the initial optimism the plan faltered. A subcommittee proposed the election of a medical faculty but the two medical schools found it impossible to allay their rivalry and proceed with joint action. Few pupils applied for instruction, and the scheme withered.[8] John Davies, lecturer in chemistry, gave an obituary of the college at his inaugural lecture for the fall term of the Pine Street Medical School:

So anxious indeed had the members of the Pine Street School been to extend the means of a provincial education that they some time ago made arrangements for the formation of a college in Manchester and convened a meeting of their townsmen to co-operate in the undertaking. Difficulties . . . had however arisen to occasion delay; but the prospect had not been relinquished. The advantages of a college in Manchester would be too numerous . . . to specify. It would be not only of the greatest importance in respect to the liberal education of the medical student but it would afford to gentlemen of other professions and the public in general a source of instruction for which they were now obliged to resort to at a distance. It would raise the character of the town . . . it would be worthy of an enlightened, a wealthy and influential population. . . . The intellectual light which it would soon emit would not only shed its splendour over the community; but be seen reflected by all our other institutions.[9]

It is clear from all available evidence that the new college or university was indeed seen as the logical culmination of the move to establish a variety of cultural institutions in the City. It would be, as Davies stated, an ornament of a major sort, a brilliant ornament. It would serve as the emblem of "an enlightened, a wealthy and an influential population."

The first opportunity to realize these goals occurred upon the death of a wealthy Manchester merchant, John Owens, in 1846. Owens bequeathed a considerable portion of his estate for the establishment of an institution for the instruction of young men "in such branches of learning and science as now and may be hereafter taught in the English universities."[10] The trustees were left with a handsome sum, slightly under £100,000. Those named in the will included George Faulkner, a good and close friend of Owens, Samuel Alcock, William Neild, James Heywood, Alexander Kay, Samuel Fletcher, Richard Cobden, J. B. Smith, and John Frederick Foster, in addition to the mayor of the city, the Dean of Manchester, and the members of parliament for the borough.[11] It is important to note that more than half named in the will (Neild, Fletcher, Heywood, Cobden, and Foster) were members of the 1836 general committee and thus form an important personal link with the Heywood–Jones–Pine Street School enterprise.

The trustees, headed by their chairman George Faulkner, cut across all conceivable lines: Whig and Tory, protectionist and free-trader, Anglican and dissenter. Only two had attended a university. All were open to advice.

The models upon which they drew were, largely, the Scottish universities and the colleges of the metropolitan university with which they were to affiliate.[12] At the second meeting of the trustees for educational purposes it was decided to appoint professors in classics, mathematics and natural philosophy, mental and moral philosophy and English language and literature. The four posts were to be advertised in the newspapers of London, Edinburgh, Oxford, Cambridge, and Manchester. To provide premises for the fledgling college, Faulkner bought Richard Cobden's house in Quay Street and leased it back at £200 per annum.[13]

By May 1850 numerous applications had been received, and the trustees selected for consideration a small number of these. The leading candidates for the mathematics/natural philosophy post were Isaac Todhunter (B.A. St. John's, Cambridge) and George Lees of the Scottish Naval Academy, Edinburgh. For reasons which remain hidden, however, the appointment went to Archibald Sandeman, a relatively obscure mathematician. A. J. Scott was chosen for the chair in logic, mental and moral philosophy, and English Language and was in addition chosen principal of the college. His salary as principal was set at £200 and as professor £350 plus two-thirds of class fees. J. G. Greenwood was appointed to the chairs of classics and ancient and modern history at a salary of £350. Half-time appointments went, in chemistry, to Edward Frankland, Ph.D.; in botany, zoology, and geology, to W. C. Williamson (a surgeon who had been connected with the Manchester Natural History Society); in German, to T. Theodores; and in French, to A. Podevin. Frankland and Williamson received £150 plus fees; Theodores and Podevin a mere £50.[14]

There was particularly keen competition for the chemistry chair, as might have been expected. The local candidates included R. Angus Smith, Daniel Stone, and Frederick Crace-Calvert.[15] Other leading candidates were Dr. John Stenhouse (1809-1880) a Scottish chemist and product of Glasgow (T. Thomson), the Andersonian (Graham), and Giessen (Liebig) who subsequently was appointed lecturer at St. Bartholomew's (London); Dr. Frederick Penny (1816-69), another Liebig student and current professor of chemistry at the Andersonian; and Dr. Edward Frankland.[16] The successful candidate, Frankland (1825-99), had studied under Playfair, received his Ph.D. under Bunsen at Marburg, and subsequently had trained under Liebig at Giessen. In 1850 he was appointed to the Putney College for Civil Engineering. Frankland had a great deal in his favor. Although the youngest of the candidates, his record was superb and, beyond that, he was a Lancashire man. Along with considerable teaching experience, he had already demonstrated his research prowess. His list of testimonials was extraordinary: Graham, Gregory, Daubeny, Brande, Brodie, Bunsen, Fremy, Miller, Hunt, Hofmann, and Leibig all wrote glowing letters. Of Frankland, Liebig wrote: "I look upon Dr. Frankland as one of the most talented young chemists of Great Britain, who, through his already published researches, has opened up a new path in the science. I do not doubt, if outward circumstances present a favourable field for his talents, which for the interests of science I ardently wish may be the case, he will, as chemist and philosopher, be an ornament to his country."[17] The *Guardian* was delighted with the Trustees' selection. "It is of the utmost importance to Manchester," its leader ran, "to possess in the chemical chair of its local college a man of the highest scientific attainments," and the newspaper deplored "the inadequacy of the remuneration offered to such a man. "£150 a year for talent and ability like that we have named is indeed a poor allowance."[18] The professor-designate, however, reported that he was overjoyed at his success.[19]

If the trustees saw the chemical chair as a part-time post, at least at first, they did take seriously the furnishings necessary for the prosecution of teaching and research. Frankland was authorized to buy apparatus and chemicals for fitting up what was optimistically labeled "the Laboratory." The stables behind the Quay Street house were to be inspected regarding their suitability as a temporary site, and a local architectural firm, Travis and Mangnall, were contacted to prepare plans for a laboratory and lecture hall. In the interim, a house on John Street nearby was rented for six months for Frankland's use, and the trustees began raising nearly £3,000 for the renovation.[20]

Shortly after the chemical chair was filled, the appointment of W. C. Williamson, surgeon, to the chair of natural history was announced. The disciplines of zoology, botany, geology, and physiology were included in his duties — all at the bargain rate of £150 plus a portion of the fees. The *Guardian* was similarly pleased with Williamson's appointment: "He is a good geologist; has more than ordinary acquaintance with botany . . . ; [and] has pursued the study and manipulations of physiology with considerable success." Williamson was, of course, no stranger to the Manchester scene. He had been curator for the Natural History Society, had been connected with the Mechanics' Institution, and was well known to scientifically inclined Mancunians.[21]

The Opening of the Owens College

The college officially opened on the 13th and 14th of March 1851 for a truncated session. Williamson had the honor of opening the session for the science professors with his inaugural address on the fourteenth. In effect it was a personal and professional manifesto; in it he declared the radical separation between his past and future scientific careers in the city.

[I] aim not to impregnate the students with a taste for scientific dilettantism but to make them in the highest sense of the term naturalist. Manchester had long been celebrated for names illustrious in the annals of natural science. . . . Several living witnesses proved that the demands of commercial pursuits could be satisfied without [devoting one's entire life to them]; for we have amongst us those who had attained a high scientific position but whose energy and skill as men of business was undoubted. A normal condition could only be secured by bringing all the faculties into healthful operation; and by the action and reaction which would be thus produced they would be mutually strengthened.[22]

It was only through systematic study, then, that the great benefits of science would accrue to active men. Williamson's primary idea "would be to lay before the students those great elements of scientific truth which exhibited the connection existing between the conformation of an animal and its habits." Some, he continued, postpone the pleasures of science for their maturer years, after business success had been secured. This was, he claimed, nothing but baneful, for any period of inaction saps the intellectual faculties. "This could not be a state of things intended by Providence," for no pursuit could be more easily studied than natural history.[23]

For the scientific community, however, the major event occurred a week later, at Frankland's inaugural address. In attendance were the younger, vigorous elements of the community: E. W. Binney, James Joule, John Leigh, Henry Day, Peter Spence and Daniel Stone. Curiously absent were Frankland's unsuccessful rivals for the chair, F. Crace-Calvert, and R. Angus Smith.[24]

In an address entitled "On the Educational and Commercial Utility of Chemistry," Frankland told his audience what they had assembled to hear. After a brief review of the history of his discipline, he stressed the importance of chemistry as an instrument of education:

[It] can scarcely be over-rated; as an analytical science it is not greatly excelled by mathematics, whilst the close and minute observations, and the careful analogies which it requires, form an invaluable discipline for the faculties of youth that can scarcely fail to give a high degree of tone and energy to the mind, and to exercise a most important and salutary influence in future life even should the student not be destined for any profession in which the knowledge acquired can be rendered practically useful.[25]

Moreover,

[Required] for the full enjoyment of the science [is] a deep religious feeling, which will lead the young student to see God in all his works, and to admire and contemplate with delight the wonderful adaptation of the most minute parts of the universe to each other, and the whole to the happiness and enjoyment of His creatures. Such contemplations cannot fail to fill the mind with the highest satisfaction and will prove an effectual barrier against the possibility of the Christian degenerating into the mere man of science.

I am far from coinciding with those persons who urge upon you the study of chemistry merely on the ground of its numerous applications to the arts and manufactures. I would take much higher ground than this, and recommend the science for its own intrinsic excellence, for the intellectual delight which every student must find in its pursuit and for the bright glimpses of the Deity which it discloses at every step.

But of course there are more tangible benefits of the pursuit of chemistry as well.

But I shall not fail constantly to point out to you . . . the numerous important and useful applications of this branch of knowledge which have been made or may yet be accomplished; in fact destined as most of you no doubt are for the active pursuit of business and manufactures, I should fail in a most important part of my duties did I not take every opportunity of bringing these practical applications prominently before you.[26]

Chemistry's utility was, he thought, almost too obvious to enter upon in detail, but concerned as all should know, medicine, public health, dyeing, printing, supplying towns with gas, water and fuel, metallurgy, foods, and agriculture. It was, however, to the needs peculiar to Manchester that Frankland especially addressed himself:

The advantages of chemistry to the chemical manufacturer, the dyer and the calico printer are almost too obvious to require comment. . . . It is now an acknowledged fact that these processes cannot be carried on without some knowledge of our science, yet with the exception of some few firms who have not the aid and co-operation of their distinguished chemists, this knowledge is too often only superficial, sufficient to prevent egregious blunders and ruinous losses, but inadequate to seize upon and turn to advantage the numerous hints which are almost sure to be constantly furnished in all manufacturing processes. It is well known how many valuable discoveries of the highest practical importance have been made by the acute observation of a single minute phenomenon exhibiting itself during a manufacturing process and which would perhaps never have come to the cognizance of any one if the intelligent and scientific conductor of the process had not at once comprehended the reaction and chronicled the fact.[27]

Calico printers especially, Frankland continued, are now aware that the brilliancy and tone of their colors are comprehensible through rational chemical study. However, despite objections that researches occupy too much time, it must become clear that the work of Manchester's Schunck on madder "have the double satisfaction of perfecting art and advancing science." The science of gas-making, too, advanced by Manchester's Henry and Leigh, "shows how much room there is yet for discovery and improvement," as does metallurgy, medical science (especially toxicology) and agriculture.[28]

Finally, turning to his own plans, Frankland announced that at the beginning of the fall session the laboratory will be open for students to join the analytical and manipulative classes "conducted on the plans which have been found to work so well in the laboratories of London, Giessen and Marburg. Each student will have his own working table and set of apparatus, and will himself conduct his own analytical experiments."[29]

By August the new laboratory was completed, and it was one of the most capacious, modern, and best-designed in Britain, Frankland himself having had a serious input. The old stables behind Cobden's Quay Street house had given way to a new two-story structure comprising a laboratory, a lecture hall, and more. The floor plan measured eighty-one feet by almost thirty-seven. The lecture room (34'6" x 26'4") was to accommodate one-hundred-fifty students. The laboratory (51' x 21') reached thirty feet towards the skylight, encompassing the second story and the handsome open timbered roof and carved braces. It possessed separate tables for forty-two students and in addition a cloak room, balance and weighing room, and an apparatus store-room. Frankland had, for himself, a private room, a private laboratory, and an additional room cryptically designated "for experimental purposes." From below there was access to the roof, on which had been built a platform (12' x 12') for "experiments with gases, etc. which are deleterious in a confined space."[30]

Surely Frankland had reason to be optimistic and cause to be gratified with his chair. To be sure, his salary was indeed hardly commensurate with his talents, but it was not a derisory sum. The trend of prices during the previous three decades had been downward, and the inflation of 1854–55 was yet to come.[31] In the late 1860s, when prices had risen considerably, a skilled artisan could earn approximately £100: Frankland's salary, lecture and laboratory fees an consulting income would have, even then, placed him at the upper end of the lower middle class or perhaps even safely within the confines of the "middle" middle class.[32] Frankland himself later reported that his enthusiasm for Owens College was so great in his early years there "that nothing would have induced me to leave the college."[33] His facilities were excellent; trustee Samuel Fletcher termed them "second to none in the kingdom," and there is every reason to believe that he was correct.[34] A comparison with the University College (London) laboratory completed several years earlier shows the Owens facility as far more capacious, at least as well equipped, and more intriguing in design.[35] Frankland had cause to be optimistic about his reception. The Manchester *Guardian,* representing important segments of city opinion, was very enthusiastic; his principal, A. J. Scott, seemed receptive; the trustees had outfitted the college well on behalf of the discipline; and there was, as we have seen, the potential for a flourishing chemical culture.

At the close of the first session in July of 1862, J. F. Foster, the chairman of the board of trustees for Owens, declared that "We have every reason to rejoice and to congratulate ourselves upon the present state of the college." Frankland's classes in systematic chemistry (lecture) and practical chemistry (laboratory) were attended by eighteen and by seventeen students respectively, or, in all, twenty-five young men. This number was a considerable portion of the sixty-two–student enrollment at the college. Frankland reported upon his progress "with great pleasure."[36]

During the next academic year (1852–53) the question of the Dalton testimonial was raised once again. At a public meeting Alderman Neild proposed the abandonment of the old scheme for a Dalton professorship and the establishment instead of Dalton scholarships. William Fairbairn added his powerful support and seconded the motion. Principal Scott applauded the move. "Every course of lectures in chemistry in the present day [is]," he claimed, "a monument to the memory of John Dalton." Manchester ought to have a great school of chemistry, he added. He confessed to his ambitions along these lines: "[I do] not see why Manchester should not become what comparatively small places like Giessen [have] become." Indeed, "chemistry is the only study within Owens [which is] directly and practically applicable to the probable practical pursuits of this place," and although the trustees did not believe that vocational training was the proper goal of the college, they believed in expansion of the chemical curriculum. "They value practical pursuits; they value the relation between science and practice."[37]

A letter soon appeared in the *Guardian* complaining that the scientific men of the city were not invited to the testimonial, evidence perhaps of the city's continuing perception of the scientific enterprise as a cultural ornament. Where were Binney, Stone, Crace-Calvert, Joule, Smith, and even Frankland? The *Guardian* retorted the "Invitations were very naturally and as we think properly sent to those who were most likely to set an example to others by heading a good subscription list."[38] The fund amounted to over £5300, part of which was appropriated for a Dalton monument at Ardwick cemetery, and part to a bronze statue in front of the Royal Infirmary, but the bulk went to Dalton scholarships in chemistry and mathematics and a Dalton prize in natural history. In his annual report to the trustees, Scott evinced his hope that the chemical scholarships would give to experimental science its rightful place as a mental discipline and not merely by its practical application.[39]

By the end of his second academic year, however, Frankland's well-warranted optimism had been dampened. Although the College's enrollment had swelled to 99 (up 37 from the previous year), Frankland's own enrollments had declined, particularly in the laboratory class. He reported that he had "mixed feelings of pleasure and disappointment." A few students had distinguished themselves, but others, he claimed, had been "remiss in their exertions, especially those belonging to the laboratory class." He proceeded nonetheless to institute a *third* class, in technological chemistry, during the 1853–54 academic year. Despite the obvious consonance with the needs and aspirations of important segments of the community and despite Frankland's own belief that the class "stood alone in this country," the class attracted only seven students. The systematic class declined commensurately, although Owens had now grown (largely owing to the introduction of evening classes) to a total of 144 students.[40]

Frankland's teaching ambitions for 1854–55 were considerable. He planned, in addition to his day-school classes, to offer in the library a ten-week course "On the Powers of Matter in their relations to chemical force," a twenty-week course on the nonmetallic elements, and an evening laboratory course as well. Nothing came of these admirable plans. Enrollment was insufficient to proceed with the evening courses, and, worse still, the technological course was canceled as well. Scott reported to the trustees that "the Technological Chemistry and Logic classes were not being held for want of students. This was thought to be caused in part by the stiffer entrance examination which is held," but he defended Frankland by noting that half the regular students at Owens attended the chemistry and natural history classes.[41] At the end of the session, it was reported that no competition for the Dalton scholarship was held, for "the standard of intellectual labour in Manchester is low." Frankland remained, at least publicly, stoically optimistic. Although only one session of chemical classes was held, "The attendance and interest shown had even been more exemplary than since the opening of the college."[42] On balance, by all exterior criteria, Frankland's courses were neither blazing successes nor dismal failures.

From the very beginning Frankland had attempted to demonstrate the relevance and importance of chemistry to the community. His inaugural address, which stressed the special relationship between Manchester and chemical science, was not mere rhetoric; Frankland had every intention of using it as the basis of a chemical program. His very first course treated of the manufacture of vitriol and nitric acid; the chemical principles of bleaching and calico printing; disinfection; the extraction of metals from ores; and the analysis of adulterated foods and of manures. His course, "Chemistry and Its Applications to the Arts" given the following year examined the production and use of coal gas, chlorine bleaching, the industrial uses of sulphuric acid; the chemistry of water supply; and the composition of the atmosphere.[43] His first-prizeman for 1852, Edmund Atkinson, went on to take his doctorate at Heidelberg and hold professorships at Cheltenham College and Sandhurst.[44] His students were not only given the benefit of lecture and laboratory but were led by Frankland into the field to visit chemical manufactories such as Spence's Pendleton Alum Works (see chapter 4).

The scientific community was furthermore enriched by the succession of interesting teaching assistants whom Frankland attracted to Owens. The first, William James Russell, was a student of Graham and Williamson, and after two years left Manchester to take his Ph.D. under Bunsen at Heidelberg. Russell later taught at St. Bartholomew's Hospital and served as president of the Chemical Society.[45] He was succeeded for a year by C. J. Tufnell and subsequently by Dr. B. Wilhelm Gerland who came with his doctorate in hand, and left Owens for the Alderly Edge Copper Works, where he did his well-known work on metavanadic acid.[46] Frederic Guthrie, too, already possessed his doctorate from Marburg under Kolbe when he came to Owens in 1855. He remained to serve under Roscoe until 1859.[47]

Frankland went beyond the confines of the college, however, to demonstrate his good "chemical" citizenship. He demonstrated his interest in local industry and agriculture and supplemented his income at once by serving as consultant to such firms as that of Spence — although, as we saw in chapter 4, with unfortunate consequences for the firm. He took out a patent (no. 938, 25 April 1855) concerning the production of ammonium and potassium sulphates which "can be used in the manufacture of artificial manures." He lectured at the Royal Institution of London on lighting and concerned himself with the analysis of Stephen White's hydrocarbon gas (water gas) at the time a serious competitor for coal gas. Using facilities at the Pollard Street gas works of Messrs. George Clark and Company, Frankland produced a report before the Chemical Society outlining the advantages of hydrocarbon gas which the Manchester *Guardian* "regarded as the most interesting and instructive document on the chemistry of gases which has ever appeared."[48] The classroom interest Frankland evinced for public health (air and water pollution, disinfection) carried over into public duty and he attended meetings of the Manchester and Salford Sanitary Association in which he participated both as citizen and as expert.[49] He served as well on the council of the Manchester Photographic Society, an organization to which many Manchester chemists belonged.[50]

The scientific members of the Owens faculty — Williamson, Sandeman and Frankland — were elected members of the Literary and Philosophical Society at once upon their arrival to the city in their new posts. The nonscientific faculty, with the exception of Principal Scott, did not join. The participation of the Owens men in the society, at least during the 1850s, was not of major significance. Williamson published two papers in volume 9 (1851) of the society's *Memoirs,* and Frankland published his "Contributions to the Knowledge of the Manufacture of Gas," in volume 10 as well as in Liebig's *Annalen* and elsewhere. Beyond this initial burst of energy, however, Owens' faculty preferred the larger stage. It was in Manchester that Frankland performed important work on organo-metals, and the paper before the Royal Society in which he outlined his views on "atomicity" or valency was completed during his first full academic year at

Owens. But the significant work was published in national British journals or important continental ones, and Williamson attempted the same route.

It is clear then that Frankland and the other Owens faculty were not at all aloof from the city. They participated in its organizations, did consulting work of great value to the community, and performed experiments on matters of major civic concern. Frankland, for one, saw Owens not as a more conveniently located imitator of Oxford or Cambridge, but rather as a fine opportunity for teaching and research in chemistry and for demonstrating to Manchester the important role of chemical science in the service of man.

Yet it is also clear that the Owens faculty in the 1850s participated in the *scientific* community in a secondary way. Unlike the great devotees such as Joule, Hodgkinson, Fairbairn, and Leigh, and those new professionals of the 1840s who remained in the city—Smith, Schunck, and Crace-Calvert—the academics eschewed what they saw as narrow parochialism and sought the larger stage. They published their important work exclusively in the journals which would capture international attention; while they did not ignore local duties, *Science* made upon them more compelling demands.

It was, in fact, a sense of distance from the scientific frontier as well as the frustrations of pioneering which caused Frankland to leave Manchester. Writing to Bunsen on 3 March 1856 Frankland complained: "It must be in the highest degree satisfactory to you to be surrounded by such a number of students so many of whom are engaged in original research. Unfortunately this is far from being the case here and there is rarely any further desire for knowledge than the testing of "Soda-ash" and "Bleaching Powder.""[51] Owens College was in serious trouble.

A Serious Decline

Optimism about the College had flowered at the summer convocation in 1854. A *Guardian* editorial of July first reported that "The middle classes are beginning to discover that the advantages arising from a college education are not to be regarded as a waste of time and energy. . . . We trust therefore that Owens College may still prosper and that its benefits may be extended so that Manchester may become renowned for the high tone of its education as at present it is for its commercial prosperity and importance."[52] Owens began with only 25 day students. By the 1852–53 academic year the numbers had swelled to 72 day students, supplemented by 28 evening pupils. When the summer 1854 convocation took place there were 71 day students, and 73 evening students. Frankland's three chemistry classes mustered 41, and Williamson's evening natural history sessions enrolled 41 gentlemen, "many of them merchants."[53] Hopes were high for the fall semester. Frankland planned to offer both day and evening courses, as did Williamson. When the fall convocation opened the term, respectable Manchester businessmen such as Schwabe, Behrens, and Garlick took the time and trouble to attend.[54] But 1854 turned out to be a critical year. The technological chemistry and logic classes were canceled for want of students, and the number of day students dropped precipitously from 71 to 48. Scott bravely insisted that the cause of this decline was to be attributed to "the parents having found that their sons are not sufficiently prepared for entrance to the college."[55] While the chemistry and natural history classes generally held up well, the rest of the college suffered badly.

Scott blamed the city. He remarked on its "distaste for thorough and systematic teaching." The chemistry classes had held up well because they were the least abstract and the most immediately practical. What was to be done? Scott's reply: ignore public opinion. "It has proceeded," he said, "from the wildest ambition to make a College a perpetually flowing fountain of all the information required by manufacturers, engineers and tradesmen."[56]

The decline continued. For the 1856-57 session only 33 day students enrolled. Eleven new students entered, eight of which dropped out by the following fall and the rest by 1858. What had seemed doldrums two years previously had now become disaster.

The issue quickly turned public. At the 1855 summer convocation Scott disregarded the warning signals: "There seems little cause to look beyond the ordinary causes of fluctuation to explain the differences in number."[57] By 1856 he turned bitter:

Names of greater weight than mine might easily be cited for the uncertainty of the experiment made in offering to Manchester not popular lectures, not an education of the school or of the workshop but a College education; an education not professional but general. If there is disappointment, it is no more than just to inquire whether Manchester had cause to be disappointed in Owens College or Owens College in Manchester. . . . The demand has yet to be created for the kind of instruction which it is our duty to furnish.

Frankland, softer in tone, was encouraged by the interest and attendance, but complained nonetheless of the "want of sufficient preparatory school preparation."[58] In private, all the professors of Owens maintained strong opinions. Letters were solicited in May 1856 by the trustees from each regarding the causes of and remedies for the present despair.

Williamson submitted a list of fifteen causes, of which the first was the fact that "high education is not deemed requisite for the acquisition of wealth, whilst the social position of the wealthy is but slightly affected by their want of education," and numbers thirteen and fourteen deplored the lack of a professor of natural philosophy and experimental science, and the want of a degree.

Frankland's views were particularly pointed. "It is impossible," he said, "to reside in Manchester for even a much shorter time than five years without becoming painfully conscious of the very slight appreciation which the Manchester people in general possess for a higher education." However, he believed it within the powers of the college materially to alter the situation. Frankland opposed what he saw as "the almost exclusively classical character which has been impressed upon the institution." He proposed the following:[59]

1. the establishment of a junior school (which was done);
2. greater emphasis on applied sciences ("In a commercial and manufacturing community like Manchester the department of 'Applied Sciences' ought to occupy a very prominent position in the Collegiate education offered to the public.");
3. the establishment of a chair in experimental physics. ("No institution having any pretensions to the character of a College for the higher departments of education is now without such a chair.");
4. incorporation of the medical schools within the college.

In April 1857 the college meeting voted to establish a preparatory school in connection with the college, to suspend the entrance examinations, and to open up single-course attendance by dropping the entrance fee. After a faculty meeting at Frankland's house produced a few organizational proposals clearly aimed at Scott, the principal resigned "owing to ill health" on 28 May. J. G. Greenwood, Professor of Classics, was named to succeed him. Frankland, in a letter from Zurich, resigned in August.[60]

During Greenwood's first full year as principal, Owens' fortunes continued to decline. The number of day students remained abysmally low at 34, and the number of evening students was cut from 121 for 1856–57 to only 59. The Manchester *Guardian* in an often-quoted leader, charged that "Explain it as we may, the fact is certain that this college, which eight years ago it was hoped would form the nucleus of a Manchester university, is a mortifying failure." The cause, it seemed to the *Guardian,* was the college's attempts to rival Oxford and Cambridge rather than tailor its efforts to the needs of the community. However, "A Former Student" replied in a letter on 12 July that "if the College is the mortifying failure you make it out to be, the fault is in the parents of Manchester for not justly valuing education for their sons." A. J. Scott retorted in the same vein: "That love of information which leads to a 'marvellous development of a cheap press' is not identical with a taste for systematic study." The *Guardian* continued to dwell on differences in educational philosophy: "Mr. Scott thinks and writes of Owens College as a rival of Oxford and Cambridge; we speak of it as an institution for supplying the middle classes of Manchester with suitable education. . . . We want Mr. Scott to take society as he finds it."[61]

From this low point in 1858 the college began to lift itself, based partly upon the suggestions of Frankland and Williamson, partly upon external developments and partly again upon the exertions of the new professor of chemistry, Henry Enfield Roscoe, who took seriously the needs of middle-class Manchester. Indeed, the trustees' first priority had been to replace Frankland, who taught Owens' most popular courses, and who resigned on such short notice in August 1857.

A New Era

Almost immediately the trustees inaugurated a search for a new chemistry professor. The post was advertised and fifteen candidates for the position presented themselves. The leading candidates were the locals R. Angus Smith and Frederick Crace-Calvert, and a young student of Bunsen, Henry Enfield Roscoe. Roscoe had a great deal in his favor: in addition to presenting testimonials from Bunsen, Liebig, Graham, and Williamson, Roscoe came from a distinguished Lancashire family, and this latter fact was surely helpful to him. The choice, it appeared, was between Roscoe and Smith.[62]

The Manchester *Courier* reported at the end of August that the post was "likely to be filled by Dr. Angus Smith." Rather too confidently, the *Courier* insisted that "There is little doubt of his election," an allegation quickly denied by the Owens trustees.[63] In fact it was Roscoe who was chosen. Without any fanfare, not even that approaching the attention which Frankland had won in 1851, Roscoe began his duties. His salary was slightly better than Frankland's; he was to receive an annual salary of £150 plus £10 per lecture weekly and in addition *all* class fees. At first these were not immense, but by 1885 Roscoe had built his Owens income into something over £2000 per annum.[64]

Roscoe (1833–1915), the son of a Liverpool barrister, had been the student of A. W. Williamson at London and subsequently had trained with R. W. Bunsen at Heidelberg. After returning to London, Ph.D. in hand, Roscoe assisted Williamson and scratched out a professional existence as part-time teacher, and private and governmental consultant. He was assisted in his work by W. Dittmar who continued as his private assistant in Manchester.[65]

It was mid-October when the new twenty-four-year-old professor of chemistry gave his inaugural lecture. In it, he defended experimental physical science as mental discipline. The student of experimental science gained, he claimed, practical experience which others won more slowly in the world. The experimentalist acquired what has been termed "common sense" but what were actually "the most valuable habits of thought." The mode of instruction best adapted to an institution such as Owens was a middle way between the professorial system of the German universities and the tutorial system of Oxford and Cambridge; Roscoe believed he saw such a model in the colleges of the United States with their oral examinations and recapitulations.[66]

The inaugural lecture was a very clever and revealing piece. Roscoe was able to hammer home important points concerning his views on scientific education as well as to flatter his audience. It is clear that Roscoe saw that success for Owens College lay in the possiblity of convincing Manchester commercial and industrial society that their values were consistent with and reinforced by science at Owens. The identification of experimental, systematic training in the sciences with "common sense" was a flag certain to be noticed by his audience, as was his reference to the United States. If Roscoe was to have his way, Owens College would, in the words of the *Guardian,* "take society as [it] finds it" and provide "the middle classes of Manchester with suitable education."

Roscoe did, in fact, have his way. All accounts of Owens College in the period after his arrival credit Roscoe with providing momentum and leadership, especially with regard to the science curriculum.[67] Within two years after the coming of Roscoe, Owens College began its sharp ascent toward stability.

The college reached its nadir during Roscoe's first year. Morale of both students and faculty had ebbed badly. Only 19 new students entered for 1857. The number of day students was only 34, and the evening school enrollment dropped from 121 to 59. But matters improved dramatically thereafter. In 1859 thirty-seven new day students entered the college, supplemented by 77 evening students, despite a 50 percent increase in evening class fees the year before.[68]

By 1861, after the introduction of Robert Clifton's physics courses, the increase in student numbers was more dramatic still: for the 1861–62 session there were 88 day students and 235 evening class participants. For the 1864–65 terms the enrollment climbed to 127 and 312 respectively. At the commencement in June 1861, the College was able to claim that the "usefulness of the Institution was much improved"; in 1863 it was tendered that it had "every reason to hope that this progress was not of a transient character but that it would be permanent."[69]

By all accounts the morale of the students and of the faculty had markedly improved. "We had," a former student later wrote, "a kind of feeling that we were the fore-fathers of a great race to be." "Corporate feeling," another student of the sixties recalled, had "begun to declare itself."[70]

The reasons for the resuscitation and ultimate success of the college are extremely complex. No doubt, Professor Charlton is correct in suggesting the importance of Roscoe, A. W. Ward, and the German research ideal, but that is only part of the story.[71] The marvelous coming to life of the college after such dark days and times of trouble was only partially an internal matter.

The years 1857 and 1858 were economically difficult ones for Britain. Owens College had been founded during a period of vigorous industrial growth and of a new confidence in the future of British capitalism. The year 1857 spelled crisis and the end of easy optimism; 1858 brough economic depression. A new and in many ways untested college, already hit by falling enrollments by 1856, would naturally continue to suffer during times of economic stress.[72] When the recovery of 1859 began, Owens already had a new look about it, and had begun to increase enrollments. It was prepared for the dramatic changes which were to affect all colleges affiliated with the University of London.

The key to Owens' revival was perhaps more to be found in London than in Manchester. Beginning in the mid 1850s and gaining momentum thereafter, British society demanded qualification for careers. It was a period of increasing emphasis upon degrees, courses, and examinations. Owens College, it ought to be remembered, was not alone in its period of difficulty. The 1850s put University College, London, in serious jeopardy.[73] The number of students dropped precipitously at King's College, London, and at the University of Durham where in 1860 enrollment was less than half its size in 1850. Queen's of Birmingham was on the edge of bankruptcy.[74]

For Owens College the picture began to brighten with the decision to hold London matriculation examinations in Manchester, at Owens, beginning in July 1859. The matriculation was highly sought after, informally establishing itself after the 1850s as a scool-leaving certificate.[75] In October of the year the Bachelor of Arts examinations were held in Manchester, for the first time obviating the need for candidates to travel to the metropolis. Principal Greenwood had agitated for these changes as critical for Owens.[76] It was, however, the new charter awarded the University of London in 1858 which altered the situation dramatically. The new charter contained several elements. On the negative side, the university opened its examinations to all, collegiate preparation no longer remaining a prerequisite. On the positive side, the university introduced new science degrees designed to be more attractive to the sons of the industrial middle class.

In July of 1857 a letter was presented to the Senate of the University urging its consideration of new degrees in science: "We . . . beg leave to submit to your consideration that the present system of Degrees in Arts does not afford sufficient encouragement to the study of Experimental Science nor sufficient recognition of proficiency in that important branch of knowledge. . . . We find that many of our most distinguished students who devote themselves especially to Experimental Science either take no Degree at all or go to Germany for that of Ph.D."[77]

A committee was appointed including Michael Faraday, Grote, Brande, Sir James Clark, and others.[78] It met in April 1858 and soon received a second appeal. This second letter underscored the desirability of progress toward professionalization of the various disciplines of science.

The attainment of proficiency in any one of these sciences [electricity, heat, magnetism, etc.] is acknowledged to be the worthy object of a life's labour; and society, appreciating the value of their fruits in alleviating the wants of man, practically regards the pursuit of these sciences as Professions. . . . The Academic bodies, on the other hand, continue to ignore Science as a separate Profession.[79]

The testimony before the committee is particularly revealing, especially the testimony of those scientists, like Alfred S. Taylor, FRS, lecturer on chemistry at Guy's Hospital:

There are many persons engaged in the practise of Chemistry as applied to Arts and Manufactures [Taylor began] who have no Degree and no certificate of qualification from any British University. Secondly, . . . some of these have procured the Foreign Degree of Ph.D. and have assumed the title of Doctor in this country, thus acquiring unjust advantage over men equally or more competent than themselves. Thirdly, the Degree of Ph.D. furnishes no test whatever of a man's knowledge of Science. It is, or was, procurable by purchase without residence or examination. (p. 109)

There are two important elements in Taylor's testimony which struck responsive chords among many British scientists. First, Taylor is claiming that there exists in Britain a desire and a need for science qualifications — for degrees equal in dignity to the Bachelor's and Doctor's degrees. Secondly, Taylor raises the point that Britain suffers from foreign competition in providing science certification. Was the problem of certification legitimate? The question was immediately raised and Taylor was prepared for it: "Questions are constantly arising in Courts of Law in civil suits on Chemistry and Physics," and indeed the whole question of sanitary legislation emphasized the increasing role of science in local and national government (pp. 110-11).

Edward Frankland concurred and pointed to yet additional benefits that the introduction of science degrees would confer: "I think that a distinct Degree recognising more moderate attainments would have the effect of inducing a considerable number of students to undergo scientific training; a result which I should consider very desirable. . . . I should think it [the Bachelor of Science Degree] would be [sought after] very considerably" (p. 119). The testimony of W. B. Carpenter, Registrar of the University and well-known biologist, agreed with that of Frankland on the matter of the latent desire among young men for science degrees. "I know," he maintained, "that there are a number of young men who are preparing for various departments of business who would gladly pass through such a course of education" (p. 124). A large proportion of drop-outs, he went on, would in fact continue for a science degree which, as opposed to an arts degree, is useful in business (pp. 131-32).

Carpenter had cities such as Manchester very much in mind. The future of the University and its daughter-constituents in the provinces seemed to lie with the business class and its wants: "They are a class who are rising into a most prominent position in most of our provincial towns, and who direct the course of intellectual and mental cultivation in those towns; and it seems to me of great importance that they should be led to feel that there is something beyond their mere business" (pp. 131-32). Michael Faraday, the distinguished chemist and physicist of the Royal Institution was only too happy to ask a helpful question: "Is it not the case that in modern times the development of the arts of life is far more entrusted to men of science than it used to be; that is, that capitalists and persons concerned in foundries and manufactories refer more than they used to do to men of science for their skill and knowledge?" (p. 133).

The new degrees were in fact instituted with the new charter of 1858. The B.Sc. degree required knowledge of mathematics, chemistry, biology, physics, logic and ethics. The D.Sc. required the first degree, further training in a principal and subsidiary subject, and research or further examination.[80]

On the whole, the science alternatives appear to have been well-received. The *Lancet,* an important journal for the medical community, heartily approved, pointing to increased social demand for scientific guidance in commerce and industry. "These demands," it maintained, "have created a new profession whose vocation is specially to apply the advanced knowledge of the day in the Physico-Chemical Sciences." The *Lancet* perceptively saw the coming of the age of the expert. Its view was becoming more broadly diffused as the 1860s dawned. "We have no doubt," the *Lancet* editorialized late in 1859, "that the engineer, the miner, the architect of the next generation will, as well as the chemist, recommend themselves to their clients by the guarantee of training and acquirements which the initials B.Sc. and D.Sc. will afford."[81] The importance of science training in modern Britain was seen not only by science-conscious physicians, but by the lay public as well. Charles Dickens' *All the Year Round* likewise approved of the instituting of the new degrees while claiming that "the foremost place of modern science in the knowledge of today, and the immense extent of it, has to be recognised in any university that shall endeavour to be truly national."[82]

Were the scientists correct in their assertion that there existed a huge demand for the science degree merely waiting to be filled? The statistics show a significant although not massive response. The first B.Sc. degrees were awarded in 1861. There were five London bachelor's degrees in science that year, compared with fifty-five B.A.'s, and twenty bachelor of medicine degrees. In 1862 the number rose to thirteen (compared to fifty-five in arts and eighteen in medicine). For the rest of the decade the numbers ease slightly to 10 and 15 percent of the number of arts degrees.[83]

In short, the absolute numbers are initially small, but they are comparable to the number of other degrees offered. Moreover the number of degrees completed only partially reflected public interest. Officials of the University seemed pleased with the results. The chemist A. W. Williamson, dean of the Faculty of Science at University College, felt able to claim in 1870 that "the demand for instruction in Science and the recognition of Science in education are facts beyond question. . . . The importance of Science as a preparation for industrial pursuits is now generally acknowledged."[84]

Just as important for Owens College as the introduction of science degrees were the changes wrought in the early 1860s in preparatory medical education. Graduates of the University of London who took the degrees of bachelor or doctor of medicine were entitled to practice medicine by the University of London Medical Graduates Act of 1854. The University College calendar for 1857-58 reports that candidates for the degree of Bachelor of Medicine were required to pass two examinations and have attended a course in practical chemistry, especially dealing with the problems of adulteration. In 1859 the Educational Committee of the General Council of Medical Education and Registration recommended stiff changes in the requirements for beginning professional studies, including the passing of the matriculation examination and the attaining of a degree of the university.[85] The council accepted the substance of these recommendations and the University College Calendar for 1861-62 announced that candidates for the B.M., after matriculating, must pass a preliminary scientific examination as well as two B.M. examinations. The preliminary scientific examination included natural philosophy, mechanics, inorganic and organic chemistry, botany and zoology. After September 1861 every student had to pass the matriculation examination to continue in the sequence. This test included mathematics, natural philosophy and chemistry. Thirty-two students passed the preliminary scientific in 1864, and five the first B.Sc., the main difference between the two being the inclusion of practical chemistry in the former and mathematics in the latter. In the following year, the numbers leaped to 48 and 17 respectively, and remained near or at these levels into the 1870s.[86]

The effect of the new regulations was to ease the fears of the provincial colleges and University College itself regarding the external degrees introduced after the new charter of 1858, and actually to swell the enrollments of their science classes. The London matriculation examination was after all a preferred method of qualifying either for further training towards a London medical degree or for hospital training. It should be remembered, however, that these changes in the medical preparation were only part of what was termed a "mania" for examinations. Written tests for the domestic and Indian civil services, for the Indian Engineer Establishment of the Department of Public Works, for the Pharmaceutical Society, and later for the Institute of British Architects and the Institute of Actuaries all contributed to the increasingly widespread view that collegiate training was becoming indispensable, particularly in the sciences.[87]

In 1857 one commentator claimed before the National Association for the Promotion of Social Science that "no phenomena in the educational horizon at all approach in importance the rapid extension of a method of examination hitherto almost exclusively confined to the students of the Universities. . . . On the whole, these movements are among the most hopeful signs of the progress we are making. They originate, doubtless, in an increased perception of the value and importance of knowledge, and in an increased desire to apply to the knowledge imparted in our various places of education a thorough and searching test."[88] The Civil Service reforms instituted in the mid-1850s with the creation of the Civil Service Commission and the new system of examinations reinforced the need for systematic and quality preparation for middle-class careers. The Indian Civil Service examination, first held by the Civil Service Commissioners in 1858 quickly became what one historian has called the "honours examination of the public service." Sixty-seven candidates stood for the first examination in 1858; 21 were successful. By 1865 there were 458 successful candidates, the overwhelming majority of whom were trained at universities and university colleges. The Civil Service examinations as a whole, both Indian and domestic, drew 46,523 candidates in the years between 1855 and 1868.[89]

The Indian Civil Service examinations included sections on mathematics, and the natural and physical sciences. Of a total 7375 marks in the 1859 examinations, 1750 were for science and mathematics.[90] The science examiners were distinguished educators, including Spottiswoode, Maitland, Liveing, Challis, Carpenter, Stoney, and Story-Maskelyne.[91]

These new developments in higher education and in public examination outside Manchester had a marked effect within it. Principal Greenwood noted before the trustees of Owens College that the students, increasing in number, were likewise showing a new "earnestness in preparation for the London University degree."[92] At the opening of the 1860 autumn term, a college trustee, William Neild, acknowledged the importance of the public examinations and the changing professional requirements of the medical and the engineering professions. At this time, and in connection with these new demands placed upon the College, Neild announced the appointment of a professor of natural philosophy, Robert Bellamy Clifton. The appointment gave the trustees "considerable anxiety," Neild reported, for the proposed post had stirred some controversy for several years. Until Clifton's arrival mathematics and natural philosophy were taught by Archibald Sandeman, a mathematician insensitive to the needs of aspiring engineers and physicians, and a man who was notorious as a poor teacher. In his letter of 1856 suggesting reform of the college, Edward Frankland had insisted that a prime desideratum was a new man to teach technical classes in mathematics and physics, and his call was seconded by Principal Scott.[93] Upon arriving at Owens, Henry Roscoe continued the call for a professor of experimental physics, and it was he who was named to head a subcommittee charged with investigating the appointment of a professor of natural philosophy for the fall of 1860.

The Trustees were spurred to action by the professors who had most to lose from Owens' decline and who had the most to gain from a strengthening of the institution. The professors voted "almost unanimously" (presumably Sandeman dissented) in favor of initiating a course of lectures in experimental physics. Their resolution stressed the new position of physics in higher education: for the matriculation examination, for the bachelor's degrees, and for premedical education.[94]

Roscoe's subcommittee solicited advice from the mathematician Augustus de Morgan in London and the eminent physicist George Gabriel Stokes at Cambridge. De Morgan's letter of 10 February 1860 stressed the need for hiring a sound mathematician for the post; Stokes's letter was probably of more use to the group:

I think it a great defect in our system here that our students have so little opportunity for attending or encouragement to attend lectures of this kind [in experimental physics]. The study of Natural Philosophy for its own sake & not merely as a field for the exercise of Mathematics is, I think, too much neglected among us. . . . I think that in any establishment for higher teaching there is ample room for a chair of Natural Philosophy, distinct from one of Mathematics.[95]

Overriding Sandeman's objections, the Trustees voted to establish the chair of Natural Philosophy at Owens and shortly afterwards appointed a Cambridge man, Robert Bellamy Clifton (1836–1921) to it, at an annual salary of £200 plus two-thirds of the fees.[96]

Clifton, the son of a wealthy Lincolnshire landowner, took his degree in 1859, finishing as sixth Wrangler and second Smith's Prizeman. He was subsequently elected a fellow of St. John's before accepting the new chair at Owens. He seemed a good choice for it; his forte was teaching. Both at Owens and later at Oxford for which he left in 1865, he did very little research, but established reputations for superior teaching ability.[97]

Clifton's classes at Owens began dramatically enough. His was the largest of the regular classes during the academic year 1860-61, and his debut was termed a "great success" at the June commencement. Clifton was nothing if not energetic. He instituted an evening course in natural philosophy, and during his second year he began classes in experimental mechanics and mathematical natural philosophy. By the commencement exercises of June 1864 Clifton had drawn 44 students to his experimental mechanics course, 28 to what was now termed experimental physics, and

10 to his mathematical physics courses, out of a pool of only 110 full-time students.[98]

However, it would be Henry Roscoe's attempts to establish a school of chemistry at Owens which would command the attention of the city and of the larger scientific world.

Building a Chemical School

Roscoe capitalized upon his advantages of situation, training, and the momentum built by his predecessor Edward Frankland. Manchester and the south Lancashire area was, after all, one of Britain's leading centers of chemical industry and the city had a long history of interest in chemical research. Roscoe's training at the research frontier with Robert Bunsen was a second reservoir upon which he could draw in creating a center for chemical teaching and research.

Like Liebig, Bunsen was able to build in his students an esprit de corps which combined a commitment to *Wissenschaft* and a "craft" consciousness which could only be developed in a laboratory of enthusiastic co-workers. "To work with Bunsen," Roscoe later wrote, "was a real pleasure. Entirely devoted to his students, as they were to him, he spent all day in the laboratory, showing them with his own hands, how best to carry out the various operations in which they were engaged." He was always, to Roscoe, "a soldier of science" and his master.[99]

Bunsen's example was always before his eyes. But it was Roscoe's own combination of energy, enthusiasm, and political skill which thrust him to the forefront of Owens' professors in the reviving college. To be sure, Frankland had made a good beginning; his chemistry classes were by 1857 the single bright spot. But it was Roscoe who spearheaded the move to bring Clifton to Owens over Sandeman's objections, and it was Roscoe who saw most clearly the new opportunities opened by the changes of the late 1850s.

Roscoe knew, however, that his effectiveness within the college depended in large measure upon his effectiveness in the community. If in the city, Roscoe made himself available to industrialists and to local governments as chemical consultant. For industrialists he advised upon the recovery of waste products, the treatment of cloth, air pollution, and the properties of cloth. He performed numerous water quality analyses for local Boards of Health and was active in the Manchester and Salford Sanitary Association before which he often lectured and of which he later became vice-president.[100]

His efforts were intended to reach all levels of society. Along with other members of the Owens staff, Roscoe was active in establishing a branch of the Working Man's College in Manchester. A People's College had been founded Sheffield in 1842, and upon its example in the early 1850s Frederick Denison Maurice initiated the Working Man's College Movement in London. There were three such institutions in the Manchester area—first at Ancoats, then at Manchester itself and later at Salford. Roscoe taught physical geography at the Manchester college which was later absorbed into the Owens College evening school.[101]

Almost immediately Roscoe was elected to the Literary and Philosophical Society (January 1858), and within two years he had become an officer of the society. He served the society as secretary, vice-president, and ultimately as president. He assumed a major role in the British Association meeting at Manchester in 1861, contributing to a session important for local businessmen, "On the Recent Progress and Present Condition of Manufacturing Chemistry in the South Lancashire District," along with R. Angus Smith and Edward Schunck.[102]

The year 1862 marked severe distress in south Lancashire, owing primarily to the cotton famine. The cotton industry's supply of raw materials was severely disrupted by the Union blockade; Manchester's Confederate cotton was ended. Large numbers of men and women were placed out of work. In Mancunian fashion a committee was formed, of which Roscoe was secretary, to provide evening recreation for those displaced. Using shut-down mills, the committee presented over 100 recreational evenings for audiences of more than 4,000 weekly. Among the programs were science lectures by such local luminaries as Clifton, Roscoe himself, and Crace-Calvert.[103] Encouraged by his success, Roscoe planned "penny" science lectures after a model previously tested in Birmingham. The lectures, begun in 1866, were presented in a large hall in a poorer section of Manchester; the published version sold for a penny. They were discontinued after the 1866–67 academic year but were resumed again in 1870. The entire eleven series included lectures by such notables as Carpenter, Lockyer, Huxley, Tyndall, Huggins, and Spottiswoode, as well as Owens faculty.[104]

His efforts to build local constituency paid handsome dividends. As early as 1861 William Fairbairn recommended Roscoe for fellowship in the Royal Society; Roscoe was eventually elected in 1863.[105] In the long run, this constituency was the mark of his success: "To make a school of chemistry worthy of the great manufacturing district of South Lancashire," he wrote in his memoirs, "was my ambition, and after thirty years of work I think it must be admitted that this was to some extent at least realised, for there were, I believe, few engaged in that district in any large way of business in which chemistry plays a part who did not show their appreciation of the value of scientific education by sending their sons or their managers to learn chemistry at Owens College."[106]

Roscoe's success depended heavily upon convincing Manchester's merchants and industrialists of the necessity for scientific education in training their sons for managerial positions in industry and commerce. During the period in which Roscoe was building his chemical school, the college's income of approximately £6,000 drew about half from the original endowment, a sixth from income from gifts and bequests, and a substantial one-third from student fees.

The average age of the day students was about 18; the youngest entered at 14. The average fee paid (as of 1868) for all students was £15, but the laboratory classes demanded a much higher price (£21 for six days a week). About half the students in the day school were drawn from the sons of manufacturers and merchants, one-fourth from the children of the professional class, one-eighth from tradesmen and shopkeepers, and the rest unknown. The evening students were generally young men employed in warehouses and factories, some teachers, a small number of artisans, and a few factory foremen eager for advancement.[107]

Those students preparing for a London degree followed a complete three-year sequence, and paid fees considerably larger than the average. During the first year both the B.A. and B.Sc. students prepared for the matriculation examination and took Latin, Greek, mathematics, natural philosophy, history, inorganic chemistry, and either French or German; for these subjects they paid 16 guineas. During the second year the science students took mathematics, natural philosophy, the junior chemistry class, chemical laboratory (two days per week), anatomy, physiology, and French or German; for the second year courses a student paid approximately 23 guineas. Usually at this point such students took the first B.Sc. examination. During the third year the student prepared for the final B.Sc. examination and took a sequence of courses including logic, mental and moral philosophy, mathematics, natural philosophy, organic chemistry, chemical laboratory two days per week, geology and botany; for these courses a student paid approximately 23 guineas, as compared with about £36 for similar courses at the Royal College of Chemistry.[108] The training itself must be judged, relatively speaking, a quality one. In 1867 of the 17 successful London first B.Sc. candidates almost a quarter were Owens men, and of these three were in the first division; of the 10 successful second B.Sc. candidates three were from Owens.

In his 1887 retrospect of his years as director of the chemical laboratory at Owens, Roscoe recalled that for years he had been convinced a prime *industrial* need was the establishment of the intimate connection between science and practice. In bringing about a marriage between British science and British practice Roscoe attempted to prove "to the practical man that the youths trained in the Chemical School of Owens College were able not only to take a more intelligent part in the operations of the various manufactures than those who had not had such advantages, but that this education had given insight into these processes such that those thus trained were able to effect improvements or even to make discoveries of importance."[109]

The point was, looking back, the training of chemists for superior operation, for development of processes and, extraordinarily, even for research.

Essentially the same points had been made by Roscoe almost twenty
years earlier in his testimony before the Samuelson Committee: "This
district of which Manchester is the centre has special need of a high class of
science instruction, inasmuch as a great portion of the national industry
there carried on is founded on scientific principles and laws. A knowledge
of those laws and principles is in the first instance necessary in order that
such industry should flourish and grow." Chemists are needed in industry,"
he continued, for "determining the strength and value of each chemical
product," for product improvement, for quality control, and for the
economic recovery of waste products. A chemist, recently graduated,
might receive an annual salary of up to £100 with possibilities for advancement. The training of such men was a primary goal of the Owens curriculum.[110]

A second immediate purpose was the training of science teachers; "I
think at the present time that is the great work which we . . . have to do,"
Roscoe reported. Oxford and Cambridge were incapable of meeting the
nation's needs in this regard, and the demand for well-trained teachers of
scientific subjects would increase each year.[111]

The students came to Owens virtually untrained, and often with strong
external pressure to learn the chemical "trade" quickly. With stories
strongly reminiscent of those of Liebig, Roscoe insisted that Owens was not
a trade school but a school of science, the principles of which were the basis
of industry: "I deprecate altogether the idea of teaching . . . the arts or
manufactures themselves. It not infrequently happens that the fathers of
intending students come to me and say, 'I wish you to teach my boy the
principles of calico printing,' . . . and I always answer that I can teach
them chemistry upon which their art or manufacture is founded."[112]

During the 1858–59 academic year Roscoe taught three day school
courses: a course of chemical lectures, a laboratory course and a course of
chemical physics, the need for which was obviated by the coming of Clifton
in 1860. By 1864–65 the demand had increased so much that Roscoe felt
it appropriate to offer both a junior and a senior course of lectures, a
technological chemistry class, laboratory, and a class in chemical calcula-
tions. One hundred students attended classes and 49 the laboratory, drawn
from a day-school population of only 123. The accommodations for
chemistry were already severely strained.[113]

Like Bunsen, Roscoe began the working day with lecturing. After some
attending to correspondence and administrative matters, and after permit-
ting the students to begin their laboratory work, Roscoe would make his
rounds, offering counsel, giving directions, and assessing their progress.
Roscoe saw this latter activity as the key to the program:

The personal and individual attention of the professor is the true secret of success; it is absolutely essential that he should know and take an interest in the work of every man in his laboratory, whether beginning or finishing his course. The Professor who merely condescends to walk through his laboratory once a day, but who does not give his time to showing each man in his turn how to manipulate, how to overcome some difficulty, or where he has made a mistake, but leaves all this to be done by the Demonstrator, is unfit for his office, and will assuredly not build up a school.[114]

The program for students began with the elementary laboratory which was intended to inculcate in them the principles of method and accuracy in both practical and theoretical work. "The student must be put on a sound track, and made to understand what he is doing and why he does it."[115] This first course, which stressed qualitative analysis, was succeeded the following year by a more rigorous course of inorganic quantitative analysis. "On this firm foundation of a competent theoretical knowledge of Inorganic Chemistry, and of a thorough practical acquaintance with Qualitative and Quantitative Inorganic Analysis, and on this alone, can, I have always been convinced, the proper and higher education of the Chemist, whether for purely scientific or for technical purposes be based."[116]

The final stage of development involved the student's introduction to original research. Time and space were made available to the students for such work, and incentives were provided by the Dalton prizes for original research by students, which were awarded only once before Roscoe's tenure but virtually annually after 1860. The results were published, and Dalton Scholars (1860-72) included T. H. Sims, William Dancer, W. M. Watts, Arthur McDougall, T. E. Thorpe, Harry Grimshaw, and Thomas Carnelley. "The stimulus for Original Work," Roscoe later wrote, "must be given by the teacher, and it is he only whose head, hand and heart are thus occupied, who can induce others to follow the same difficult though delightful path. The spirit of research must be felt in the atmosphere of the laboratory."[117]

It would be a mistake, however, to overrate the emphasis upon original research by students. Roscoe was perhaps being more candid in his presidential address before section B of the B.A.A.S. in 1884:

I feel I am doing the best for the young men who, wishing to become either scientific or industrial chemists, are placed under my charge, in giving them as sound and extensive a foundation in the theory and practice of chemical science as their time and abilities will allow, rather than forcing them prematurely into the preparation of a new series of homologous compounds, or the investigation of some special reaction, or of some possible new colouring matter, though such work might doubtless lead to publication. My aim has been to prepare a young man by a careful and complete general training, to fill with intelligence and success a post either as teacher or industrial chemist, rather than to turn out mere specialists who, placed under other conditions than those to which they have been accustomed, are unable to get out of the narrow groove in which they have been trained.[118]

The wisdom of this course was substantiated by the results.

In his testimony before the Devonshire Commission Roscoe provided data which, when properly analyzed, may be of use in creating a portrait of his chemistry school. Roscoe lists sixty students in his laboratory (1870–71), of which the youngest was 14 years of age and four were over 25. Of the sixty, 48 listed their object of study. By far the largest number (32) intended careers in manufacturing, industry, or business; eight sought academic, professional scientist, or "science degree" qualifications; three were pre-medical; and five, "general education."[119] The numbers of students had grown significantly during Roscoe's Owens period. In 1857–58 only 11 students registered for the lecture classes, and but 15 for the laboratory. By 1864–65 the numbers had grown to 100 and 49 respectively. After a drop in 1865–66, owing probably to economic conditions, the enrollees climbed dramatically to 142 and 47 in 1868–69 and to 195 and 60 in 1870–71.

Owens students did well in their London examinations, with a few, like W. Marshall Watts, carrying off honors year after year; a surprising number of students of the 1860s, such as Watts, Bottomley, C.R.A. Wright, and Thomas Carnelley, went on to take their London D.Sc. Many of the best students, like T. E. Thorpe and J. B. Cohen, were sent off to Germany for the Ph.D. By the end of the 1880s the Owens school of chemistry was turning out graduates who populated the newer university colleges: F. W. Babbington (Toronto), G. H. Bailey and H. Baker (Owens), P. P. Bedson (Newcastle-on-Tyne), C. A. Burghardt (Victoria University), Carnelley (Dundee), J. B. Cohen (Owens), G. Dyson (South Kensington), D. E. Jones (Aberystwith), C. A. Kohn (Liverpool), T. M. Morgan (Victoria, Jersey), G. S. Newth (South Kensington), L. T. O'Shea (Firth, Sheffield), W. Ray (Yorkshire, Leeds), S. Shaw (Newcastle), Watson Smith (Owens), Arthur Smithells (Yorkshire, Leeds), W. C. Williams (Firth, Sheffield), C. R. A. Wright (St. Mary's Hospital, London), Sydney Young (Bristol), as well as M. M. Pattison Muir (Cambridge).[120] This was a record unmatched anywhere in Britain, and one of which Roscoe was justly proud.

The clues to the success of Roscoe's chemical school lie buried among these statistics. First, and perhaps most important of all, was the congruence of Roscoe's conception of the relation of science and industry and the needs of Owens' essentially upper-middle-class clientele. Roscoe, and others at Owens as well, perceived and seized upon the new relations between science and industry which were only beginning to emerge at midcentury. Fairbairn had glimpsed it in lamenting the failure of the Mechanics' Institution (see chapter 4); the Owens College faculty and trustees proclaimed it. Britain needed not scientifically minded workmen but scientifically trained managers. "Technical education" was to be transformed from the education of workers and artisans in the elements of science or in the particulars of manufacturing into the training of the officer-staff of modern industry.

Roscoe had often stressed the intimate connection of industry and the principles of science; in a period which was marveling at the new artificial dyestuffs — the fruit of chemical research — the point began to have more thrust. More difficult to bring home was the point that the day of the older view of technical education — of Brougham and of the early mechanics institution movement, of the Society for the Diffusion of Useful Knowledge — had passed. Roscoe told the Samuelson Committee, "I think that the attempt to teach science completely to the working class is a mistake, but I think we should give an opportunity for the best men in the working class to rise."[121] Hence, Roscoe took great interest in systematic evening instruction for the best of the workers and artisans; he took little interest in the Mechanics' Institution. It was to the foremen and to the managers that he looked for the future of scientific and technical education. The foremen were in his eyes an "overlooked class" despite the important fact that they were "persons who from their positions hold the lives of their fellows in their hands."[122] These men, he insisted, ought to have scientific training.

A letter of the Chairman of the Owens trustees Alfred Neild to the *Economist* in 1868 puts the matter plainly. Neild was writing with reference to an article in the journal on technical education. The article exclusively discussed technical education in terms of artisan training. Neild, reflecting the opinion of the Owens leadership, wished to redefine the terms of the discussion:

That a want exists in this country in the matter of technical education will probably be hardly disputed; but what kind of education is needed, and for what classes it should be provided, is another question.

I am not disposed to underrate the importance of a high level of intelligence among the masses of our workmen; but I think that the urgent need is for greater facilities for a thoroughly scientific education of those who have to direct them. The effect of a large expansion and systematising of manufactures is, undoubtedly, to lessen more and more the sphere of ingenuity of the worker, and to confine him to doing one or two things uniformly and accurately. On the other hand there is more and more needed on the part of the directing head; nor merely the experience which can only be gained in the workshop, but a knowledge of scientific principles and methods. . . . The scientific intelligence, which it is of the first importance to secure, is that of the *chiefs* of industry; and that their science should be thorough, and not merely popular.[123]

It is this systematic training in the principles and methods of science for the future chiefs of British industry which Owens College aimed at providing, and had made great strides in securing.[124]

Engineering and the Coming of Reynolds

While Roscoe was building a nationally famous school of chemistry, the courses on natural philosophy and mathematics, separated from chemistry since Clifton's arrival, failed to gain special eminence. Sandeman, both as a pedagogue and as a mathematician, was generally considered a detriment to Owens' reputation. There are too many Sandeman stories ridiculing his effectiveness as a teacher to disregard them.[125] Under great pressure from the trustees in 1865, he resigned his position and was succeeded by Thomas Barker, who published no more than Sandeman and likewise possessed a reputation for being abstruse; he was, according to one recollection "unintelligible to all but the elect (probably to all but F. T. Swanwick and J. H. Poynting)."[126] Clifton, however, as mentioned above, had a reputation as a fine teacher; physics enrollments were considerable during his time at Owens and remained high. He resigned his post in November 1865 to assume a professorship in experimental physics at Oxford, a post which he gained with the enthusiastic support of Roscoe and Joule, among many others.[127] His successor was a Scot, William Jack of Glasgow. Jack (1843–1924) was a graduate of the University of Glasgow who proceeded to Peterhouse, Cambridge, where he distinguished himself in mathematics. He finished as fourth Wrangler in 1859 but won first place in the competition for the coveted Smith's prize, a greater distinction than placing first in the mathematical tripos, in the view of many mathematicians. After serving several years as an inspector of schools in Scotland, Jack was named in 1866 professor of natural philosophy at Owens. He remained at the post until 1870, when he resigned to assume the editorship of the *Glasgow Herald*.[128] Armed with hindsight, it is neither ungenerous nor unfair to view Jack's appointment at Owens as a stop-gap. Far more significant for the growth of science at Owens was the creation of an engineering chair and the filling of it by Osborne Reynolds.

There had long been sentiment in Manchester for the creation of such a chair. Ample precedent existed: Glasgow, London, and Belfast already possessed academic posts devoted to engineering. Manchester was, after all, one of the truly significant centers of engineering and had long been such. By the early nineteenth-century Manchester and Salford boasted iron-founding and machine-making firms which not only catered to the textile industries but also to the construction of steam engines, railway components, and all other civil and mechanical engineering requirements.[129] Early scientific engineers such as Roberts, Hodgkinson, Nasmyth, and Fairbairn were joined by Whitworth, Beyer, and Peacock.

The mid- and late 1860s marked the beginning of an upsurge in organizational activity in engineering. Emerging from industrial practice, engineering had begun the inevitable process of differentiation and specialization, and had demonstrated its close links with mathematics and physical science — in short, engineering was becoming both profession and discipline. Recognition of these facts contributed to the establishment of chairs for such distinguished practitioners as W. J. M. Rankine, James Thomson, and Fleeming Jenkin, and to the rapidly increasing organizational opportunities for engineers of whatever kind. The Institution of Civil Engineers had been founded in 1818, requiring its members to demonstrate evidence of a respectable apprenticeship. The civil engineers were joined in 1847 by the Institution of Mechanical Engineers, originally composed of prominent engineers, factory owners, and managers. To these can be added those new societies resulting from the knowledge explosion: the Institution of Gas Engineers (1866), the Institution of Electrical Engineers (1871), and the Institution of Municipal Engineers (1873).[130]

Manchester was by no means immune from these developments. In addition to taking an active part in the national organizations, Manchester engineers created wholly new societies including the Manchester Association of Employers, Foremen and Draughtsmen of the Mechanical Trades of Great Britain (later shortened to the Manchester Association of Engineers, was formed in 1856 "for the purpose of bringing together those engaged in the direction and superintendence of Mechanical Works." It was founded by foremen at Galloway's, Ormerod and Sons, and Sharp, Stewart and Company, and it flourished under the paternal protection of leading engineers. Sir William Mather's firm, for example, paid the rent until the society was on its feet.[131] The Manchester Institution of Engineers first met in November 1867, with George Peel of Williams and Peel, Engineers, as president. The M. I. E. was a more "aristocratic" engineering society, appealing for its membership from the highest ranks of engineering companies. The Scientific and Mechanical Society was established in May 1870, to promote the advancement of engineering, the arts, manufactures and industry "by the propagation and diffusion of industrial science."[132]

In addition to this local activity, Manchester engineers remained important members of such national organizations as the Institution of Civil Engineers and, more prominently, the Institution of Mechanical Engineers. Fairbairn and Whitworth were presidents of that society. By 1868 C. F. Beyer, John Robinson, and C. P. Stewart had joined them as officers or members of council. In short, the professional character of engineering was strongly asserted in Manchester and much in evidence by the mid- and late 1860s.

The development of engineering as profession and discipline, coupled with the vitality of engineering industries in the Manchester area, hardly went unnoticed at Owens College. The college had only too recently emerged from its steep decline to permit any chance for adding real strength to pass. In April 1865 Clifton presented a report to the college on a proposed course of study in civil engineering. The "great progress of Practical Science and especially of Practical Mechanics in the last few years," he wrote "tends to render desirable some corresponding modification in the education of persons intended for such pursuits." He noted the formation of engineering courses in London, Dublin "and other places." "As Owens College is in a great center of Engineering Science" he continued," it would be desirable that some effort should be made to supply such practical instruction, thereby meeting a want much felt by local engineers, and at the same time drawing local sympathy towards the College." The report was adopted.[133]

Subsequently, according to Thompson, some of the leading engineers of the region met at Town Hall on 11 December 1866 and drafted a resolution to the effect that "it is expedient to establish a professorship of civil and mechanical engineering, together with a special library, a museum of models, a drawing class, etc. in connection with and under the management of the trustees . . . of Owens College."[134] They aimed at gathering £10,000. John Robinson served as secretary of the fund; Joseph Whitworth was the treasurer. By 19 November 1867 the fund had collected subscriptions totalling £8,750 including £3,000 from Beyer, Peacock and Company, and £1,000 apiece from Whitworth, C. P. Stewart, and John Robinson. The final sum turned over to the college was £9,505.[135]

The committee of engineers (including Whitworth, Fairbairn, Beyer, and Robinson) which had raised the money continued to act for the college; it was they who solicited candidates for the new chair at £250 by advertisement. It was this committee, too, who found the original candidates wanting. Charles F. Beyer offered to supplement the post's salary for a time to bring it to £500, and new candidates were sought. In March of 1868 Osborne Reynolds was chosen.[136]

Reynolds' appointment reflects very clearly the dual constituency of the new engineering chair. On the one hand, the local engineers who raised the money and solicited candidates saw the new professor as serving the engineering community. On the other, he was seen by Owens as an academic. He was to be not only the source of training for a new generation of scientific engineers but was also to provide a tighter link between community and college — a source of new ideas and inspiration. Roscoe's example must have encouraged both groups. In this light the appointment of the recent 26-year-old Cambridge graduate does not seem quite so unusual.

Reynolds, born in Belfast of an Anglican clerical family, had already combined an academic career with practical training; his background must have seemed alluring to the committee. At age 19 Reynolds had been apprenticed to a mechanical engineer in order to become a working mechanic before entering Cambridge in October 1863. At the university Reynolds had excelled at mathematics, graduating B.A. as seventh Wrangler in 1867. He continued to occupy two worlds: elected a fellow of Queens' College, he embarked upon a civil engineering career in London before his decision to seek the Manchester chair.[137] Undoubtedly, Reynolds' letter of application appealed to the engineers. He appeared to them as a man of the workshops, and yet impeccably trained in mathematics and science:

From my earliest recollection I have had an irresistible liking for mechanics, and the physical laws on which mechanics as a science are based. In my boyhood I had the advantage of the constant guidance of my father, also a lover of mechanics, and a man of no mean achievements in mathematics and their applications to physics. . . . Having now sufficiently mastered the details of the workshops, and my attention at the same time being drawn to various mechanical phenomena, for the explanation of which I discovered that knowledge of mathematics was essential, I entered at Queens' College, Cambridge, for the purpose of going through the University course, previously to going into the office of a civil engineer.[138]

Reynolds was, for his own part, sensitive to his double constituency, and like Roscoe worked constantly to link city and academy through his science. His inaugural lecture, "The Progress of Engineering considered with respect to the Social Conditions of this Country," was the start of his Manchester career. In it, Reynolds confirmed the efforts of his predecessors, the civic scientists, and his colleagues such as Roscoe, who began the attempt to convince the city and the nation of their need for science and scientists: "Though another mile of railway should never be made, there will still be room for all the engineering skill the country can find. As the ladder of science gets higher it requires better training and greater dexterity to reach the top. The more perfect our mechanism becomes the more knowledge and labour it requires to improve it."[139] Here, succinctly put, is the argument which was to gain momentum and strength during the next century: enterprise and expertise were now inextricably tied together. As Lyon Playfair was later to express it: "The competition of the world has become a competition of intellect. In the future of the world the greatest industrial nation will be the best educated nation; it may not be so today, but it certainly will be so to-morrow."[140]

Despite his sensitivity to the situation and the initial enthusiasm of the community to the creation of the chair, it must be stressed that Reynolds got off to a poor start. His sustained efforts to win the respect of the local scientific and engineering communities, efforts which ultimately were to bear fruit, would take time; his tremendously important engineering and scientific contributions were far in the future. What people saw was a shy, somewhat awkward young man who had difficulty communicating his views. J. J. Thomson's recollections give more than a hint of the situation: "He was one of the most original and independent of men, and never did anything or expressed himself like anybody else. The result was that it was very difficult to take notes at his lectures, so that we had to trust mainly to Rankine's textbooks. Occasionally in the higher classes he forgot all about having to lecture and, after waiting ten minutes or so, we sent the janitor to tell him that the class was waiting." When he did, in fact, come to class, Reynolds' performance was, for most of the students, impossible to manage. "He wrote on the board with his back to us, talking to himself, and every now and then rubbed it all out and said that was wrong. He would then start afresh on a new line, and so on. Generally, towards the end of the lecture, he would finish one which he did not rub out, and say that this proved that Rankine was right after all." For a student of genius, like J. J. Thomson, it "showed the working of a very acute mind grappling with a new problem"; for even the serious engineering students it was an unmitigated disaster.[141]

Enrollments began to drop; the matter became a cause for concern. The Principal reported to the trustees in July, 1870 on "the non-success of the Engineering classes." He concluded: "The Professor has a genuine enthusiasm for his subject & is, I am sure, sincerely devoted to the interest of his students. But it would be an affectation to conceal that our anticipations as to these classes have not been realised & that the great falling off in numbers in the second session is a cause of disappointment, &, I had almost said, of mortification."[142] To make matters worse, except for two short papers, "On an Oblique Propeller" and "On the Suspension of a Ball by a Jet of Water," his interests were turning towards what was, for a while, termed "cosmic physics." Articles began to appear under his name on comets, coronas, the sun's induction and terrestrial magnetism, meteors, clouds, and so on.[143] His colleagues found him a difficult, cynical, paradoxical, and often perverse man; fortunately, for the most part they saw past it. Reynolds' ability and enthusiasm for his discipline, for the college, and for the scientific community of Manchester surfaced, flowered, and survived.

One of the ways in which Reynolds managed to surmount his difficulties came through his active participation in the scientific and technical institutions of the city. The Scientific and Mechanical Society of 1870 was a key group bringing together the academic scientists of Owens and the practical men of affairs of the city. As was the case for so many other Manchester organizations, its demiurge was William Fairbairn. His address to the meeting of 21 December 1870 gives some clue regarding the temper of the society. The issue was Britain's relative decline in the industrial arts after its apparent success at the exposition of 1851. Fairbairn "did not think it would be doubted that from the want of sound and first-class education amongst the better class of mechanics and artisans we were getting behind." Through instruction in the first principles of mathematics and chemistry "there would be raised for the public service a much superior class of men than we possessed at the present time." Reynolds was an enthusiastic member of the society and often spoke before it. As vice-president he addressed the society both on special matters ("On Structural Mechanics") and on more basic themes. In his vice-presidential address "Future Progress" Reynolds laid out a defense of what must have seemed his own esoteric interests. "We have reached a point," he claimed, "from which it is almost impossible to advance [technologically] except by the utmost aid that science can give. . . . It is clear that experimental and scientific is of the highest importance; it is, in fact, of more importance than the immediate advance to which it leads."[144]

This theme was pursued by Fairbairn in his last address to the society, which had to be read for him owing to his failing health. In October 1873 he called Britain to scientific and technological arms: "I am anxious to impress upon the society the necessity for exertion in every scientific pursuit if we are to maintain our position and cope with the natives of other countries who have equal opportunities and are better educated than ourselves." Referring specifically to the academics of Owens College, he emphatically insisted that the "age of the rule of thumb [is] at an end."[145] Reynolds himself succeeded to the presidency of the society for 1876–77, and addressed the organization on the subject "Engineers as a Profession," in which he continued to urge the union of theory and practice.[146]

Reynolds also made himself accessible to the foremen and managers of local industry as well. He pleased this constituency with narrowly focused talks upon such subjects as "Elasticity and Fracture" (1871); "The Use of High Pressure Steam" (1872); "High Pressure Steam" (1873); and "Some Properties of Steel" (1874).[147]

The new professor of engineering likewise lavished his attention upon the Literary and Philosophical Society, which had traditionally welcomed Owens professors at once into its ranks. Reynolds was elected a member of the society in November 1869, was almost immediately chosen a member of the Council, and soon succeeded Roscoe as a secretary of the society.[148] Reynolds later served as vice-president and ultimately was chosen to lead the society in 1888. Unlike Frankland before him and even Roscoe, Reynolds chose to publish his early work in the *Memoirs* and the *Proceedings* of the society. Perhaps Reynolds chose to publish locally rather than nationally because of his sensitivity to his problems among local constituencies; or perhaps, unlike his more aggressive colleagues, he was less ambitious for national and international arenas.

The year 1870 appears to have marked the ebb-tide of Reynolds' teaching career. During the 1869–70 session only 22 students completed his three courses, aside from the 16 in the mechanical drawing class, and, as we have seen, Principal Greenwood was highly disturbed. By 1872 his classes had increased in number from three to four, and in enrollment to 34, plus 27 in the drawing class.[149] A milestone was passed; Reynolds' teaching position was far more secure. It should be noted that personal tragedy stalked him during these years. His wife, a bride of only a year, died in 1869 after bearing him a son. The young man managed this blow only with difficulty. He remarried in 1881 — two years after the death of his son.[150]

Reynolds' conception of his teaching role centered upon his insistence upon teaching engineering as a science; that is, while he acknowledged the importance of practical experience, it was not his aim to train men for special engineering tasks. Like Roscoe's chemists, his students were to be taught a scientific approach to engineering problems. His model was Rankine, and his goal the application of scientific principles to engineering requirements. In 1884, before the Royal Commission on Technical Instruction, Reynolds outlined his aims.

Reynolds' engineering course was designed as a three-year program and included (in addition to engineering) drawing, conducted by Reynolds' assistant, J. B. Miller, practical surveying, and the basic sciences: physics, geology, mathematics, and chemistry. Roughly half the student's time was spent on subjects other than engineering. In the engineering classes the first year was devoted to surveying, measuring, and descriptive engineering (earthwork, masonry, timber); the second year comprised applied mechanics (including the study of the strength of materials); the third year was devoted to the dynamics of machinery, hydraulics, and the steam engine. "All through," Reynolds told the commission, "the effort is to give the students a thorough hold of what is brought before them, rather than to give them an imperfect acquaintance with a great deal." The classwork was supplemented by experimental demonstrations, site visits to local works, and examination of the college's own works.[151]

The core of the curriculum, as Reynolds saw it, was applied mechanics, of which "the educational value . . . with the necessary substratum of mathematical and scientific knowledge, is as high, if not higher, than that of any scientific subject."[152] Other subjects of special local interest were introduced, such as mineralogy under Dr. C. A. Burghardt in 1871, which was seen as adding strength to the third year courses. The course was originally intended for Roscoe's student, T. E. Thorpe, but upon Thorpe's removal to Glasgow, Burghardt was asked to assume to the post as lecturer which he held until 1898.[153]

In Reynolds' opinion, engineering education had suffered after the founding of chairs in the 1850s and 1860s from the arbitrary character of the curriculum — a difficulty cruising from the novelty of the subject. Using the books written by Rankine and drawing upon the experience of French engineering schools, engineering education had lately become "clear and distinct." New support, mainly from the engineering profession itself, was mounting. The reason was clear: As he informed the British Association: "It cannot for one moment be doubted that this movement has been brought about by the conviction of the necessity of an education which, in its subjects and methods of teaching, is much more closely related than was the older system of the Universities, to the actual work which the students may eventually be called upon to undertake."[154]

The next stage of development in the curriculum was the development of engineering laboratories, pioneered in Great Britain by Professor Kennedy of University College, London; these laboratories would keep "alive in engineering schools a real scientific interest in the practical work which is going on around them." In these laboratories students could familiarize themselves with the actual subjects of classroom attention; furthermore, as on the continent, scientific research in engineering required them. The construction of such laboratories promised a new era for engineering. What the Cavendish Laboratory had done for physics, the engineering laboratory would do for their discipline: "What for the time may appear to be a visible end or practical limit will turn out but a bend in the road."[155]

Even as Reynolds spoke before the British Association in 1887, Owens College was preparing to open its new Whitworth Laboratory. Twenty years had passed since Reynolds' chair in engineering had been endowed. Throughout the following two decades, engineering interests had continued to support the subject. Important gifts had come from C. F. Beyer, of Beyer, Peacock and Company, who bequeathed over £100,000 in 1876 for endowed professorships in science and engineering, and from Charles Clifton of the U.S.A., who left over £21,000 for the mechanical arts and engineering. A committee was convened to consider the disposition of this relatively large sum; it reported to the council in April 1877, recommending the extension of science and engineering teaching. In 1881 a new professorship was created in applied mathematics; the first incumbent was a former Owens student, Arthur Schuster.[156]

Matters had moved slowly for Reynolds, however. In April 1884 he had proposed the construction of a new engineering laboratory. The council acknowledged the importance of such a facility but claimed that the college was financially unable to proceed with the matter. A committee was formed, including the engineers John Ramsbottom, and John Robinson, Thomas Ashton, and Reynolds.[157] An appeal for funds fell far short of needs. The council, reminding itself of the huge bequests intended for engineering, agreed to proceed, with the proviso that not more than £6,000 be expended for buildings and equipment. This sum was later met by the legatees of Sir Joseph Whitworth. Reynolds aided in the planning of the building, which was constructed under the supervision of architect Alfred Waterhouse. The plans included testing machines, an experimental compound steam engine, experimental boiler and furnaces, an Otto gas engine, dynamos, electric motors, pumps, precision instruments, and workshops for carpenters, smiths and fitters.[158]

The building was opened for use for the fall term of 1887. The subsequent scientific importance of Whitworth Laboratory is witnessed by the fruitful work of Reynolds and his students, which served to justify the initial hope that the laboratory would serve both as an important engineering research and testing facility as well as a vitally significant teaching instrument.[159]

Owens College Extended

It is clear from the foregoing that Owens College rebounded in amazing fashion from the doldrums of the late 1850s and built itself into an important "third force" in English higher education by 1870. In purely physical terms, it quickly outgrew its confines; Cobden's renovated old house and its additions were totally inadequate once enrollments began their steep incline. In an intellectual sense, too, the testament of John Owens and the old college idea under the careful stewardship of Scott had been made obsolescent by events both internal and external. With the aggressive Roscoe and the popular Clifton, the natural and physical sciences came to dominate the other chairs. In an era in Britain which required higher training as well as education, in a period which was beginning to lay great emphasis upon examinations and qualifications, the Owens College idea was pressed from within and without to change as well. These changes would require new financing; they would demand, it seemed, a second foundation.

The pressure for change began innocently enough with dissatisfaction with facilities. In January 1865 the trustees appointed a committee to consider new buildings in connection with the proposal that the college assume the Natural History Society's museum. The professors and Dr. Greenwood were requested to offer their own suggestions.[160]

Professor Clifton presented a letter in early February which declared the overcrowded conditions of his classroom and laboratory to be a health hazard; after a lecture the carbonic acid content of the air (as measured by Professor Roscoe) was said to be about 0.32 percent. The maximum consistent with good health was thought to be only about 0.07 percent. "These facts will show," he wrote, "that my lecture room gets into a state, not only destructive of efficient study, but most injurious to health." Clifton's contentions were supported by Prof. Williamson and by Roscoe, who reported that the condition of his rooms evidenced "from twice to nearly four times the amount of carbonic acid which is injurious to health."[161]

These charges of a health hazard at Owens were understandably suppressed in the official report of the principal and professors which was printed later that month.[162] The report instead laid stress on the obvious fact that the college was bursting at the seams. The College had grown from 34 day students in 1857-58 to 127 in 1864-65. The number of separate courses offered had more than doubled since the college had taken Cobden's house. There simply was not room for a single additional student (pp. 1-2). There were hints in the report of a grander plan. Implicit was the ambition to command an important place in English education: "While we can never hope [the report stated] to rival the ancient Universities in the study of Literature and Mathematics, we see no reason why Owens College should not aspire to be the first school of applied and experimental science in the country. But for the establishment of such a school more space is the first requisite" (p.4). In effect what was proposed was a revolution, liberating Owens College from the shackles of its foundation, and recognizing *de jure* what *de facto* it had made great strides towards becoming — a school of applied and experimental science.

The present site offered little hope. While it provided a convenient central location, it was hampered by poor drainage, a deteriorating neighborhood, especially to the north and east, and no easy access to a "leading thoroughfare" (p. 5). The faculty urged a suburban location, taking advantage of lower land costs, the light and air of a close-in suburban setting, and accessibility of open space for various college purposes.

The die was cast, and a new campus was projected. For such a major capital venture, however, large new financing was required. Further reports were forthcoming, suggesting a medical school, chairs in civil and mechanical engineering, astronomy and meteorology, and applied geology and mining. A figure of £100,000 was mentioned.[163] On 1 February 1867 a meeting of the principal, faculty, trustees and friends of Owens met at the Town Hall to organize the drive for extension. Accounts of the meeting make it absolutely clear that the purpose of recasting Owens was nothing less than to enable it to become the equivalent of a northern university specializing in the pure and applied sciences. The advertisement taken by the committee formed at Town Hall declared the aim of "rendering it on an extended basis, in effect the local University of Lancashire and the neighboring counties."[164] The group included leading businessmen, industrialists and scientists: Thomas Ashton, Solomon Behrens, C. F. Beyer, Fairbairn, John Hopkinson, Oliver Heywood, James P. Joule, H. E. Schunck, R. A. Smith, S. Schuster and Joseph Whitworth. Their goal, in part, was to raise not less than £100,000 and as much as £150,000.

They were supported by an editorial in the *Guardian* which pointed to the fact that, in science degrees, "to which special attention [is] paid in Manchester," Owens has sent up more students than University College or King's College. The *Guardian* warmly supported the proposal that Owens should become preeminent in "the natural sciences as applied to what are called practical purposes." It is natural," the paper went on, "that it should be sedulously cultivated among a section of the people into whose daily occupations the knowledge which it yields daily enters."[165]

The *Guardian* carried an extended account of the proceedings under the heading "A North of England University." Principal Greenwood admitted that "our special line should be the study of experimental and applied science. . . . It [is] not an unreasonable aspiration that Manchester, the city of John Dalton and Dr. Joule, should have chemistry laboratories which need not fear comparison with Germany."[166] An executive committee with Thomas Ashton as chairman was formed; it included Greenwood, Roscoe, and R. C. Christie. Almost £25,000 was pledged at once, some of it earmarked for the engineering chair. Many of those attending the February 1 meeting met again a few days later to form the Manchester Reform Club, and heard Goldwin Smith of Oxford admit that "many of the colleges instead of promoting, actually retarded education."[167]

Owens was to be a science-oriented, modern, liberal rival to the ancient universities. It was to be, as the *Spectator* enthusiastically exclaimed at the time "the University of the Busy."[168] For these goals, the choice of Thomas Ashton of Ford Bank, Didsbury, as chairman was especially fortunate. Ashton (1837–1898) was born in Hyde of a cotton manufacturing family, the son of the owner of a print works. He was educated at Mr. Voelke's school in Liverpool, a particular favorite of wealthy Lancashire manufacturing and merchant families such as the Rathbones, the Heywoods and the Fieldens. At age 18 Ashton attended Heidelberg but soon returned to assume an active public and commercial life. A Unitarian, he attended the Cross Street Chapel, along with Fairbairn and others of the dissenting elite. Like Fairbairn, he welcomed the challenge of participating in the creation and execution of civic and national groups; he was an early and energetic member of the Anti-Corn Law League and was a founder of the Manchester Reform Club. He became a leader of the Liberal Party in Lancashire and twice declined offers to stand for the party for parliament. His friend Henry Roscoe ultimately represented the South Lancashire constituency which Ashton had declined in 1852.[169]

During the cotton famine Ashton was a popular and responsible member of the Central Relief Committee. It was perhaps then that he became acquainted with Henry Roscoe. Roscoe claims the credit for attracting Ashton into the Owens extension orbit. He records in his memoirs "I walked up to Ashton's house in Didsbury one Sunday and said to him: 'You must help us in placing this college of ours on a footing worthy of the city,' " and he notes that it was mainly owing to Ashton that the reform of Owens was made possible.[170]

By October 1868 a site along Oxford Road for the new campus had been selected and almost £77,000 had been subscribed to the general fund. In December Alfred Waterhouse was selected as the architect for the new buildings. In May of 1868 a new subcommittee was appointed to draft a new constitution; the subcommittee engaged James Bryce to write it.[171] (Bryce later became professor of law at Owens and Regius Professor of Civil Law at Oxford.) It was this constitution which was the basis of the Owens College Extension Acts which were shepherded through Parliament in 1870 and 1871 with the help of Ashton, though not without difficulty.[172] The Owens Extension College was constituted and immediately merged with the Owens College to form a new entity, known by the latter name. By the new constitution, the college was freed from the legal authority of the old trustees and placed under the stewardship of a court of governors, a council, and a senate. Other changes, such as the admission of women and the raising of age of entrance, hitherto prohibited because inconsistent with John Owens' will, were now permissible.

The constitution divided authority between the faculty, the external governors, and the alumni. The court was in principle the chief authority, meeting twice a year, primarily to ratify the actions and recommendations of the council. The court was primarily composed of laymen, headed by a ceremonial president. The Duke of Devonshire was elected to the post in 1870. Life governors included important contributors and active members of the extension committee like Ahston, C. F. Beyer, Murray Gladstone, Joseph Whitworth, S. J. Stern, Richard Johnson, and later Hugh Mason, James Joule, and John Ramsbottom. Other members of the court were nominated by the president (e.g., William Fairbairn), by the City of Manchester and by Salford, by the members of Parliament from four counties, by the Privy Council, by the alumni, and by the senate. The council was far more important for the regulation of the college; for all intents and purposes it was the executive committee of the court and controlled finance. In 1870 leading members of the council were Ashton, S. J. Stern, Christie, Roscoe, Ward, Alfred Neild (chairman), and Murray Gladstone.[173] The senate, as before, was constituted of the professors and the principal.[174]

All told, the Extension fund-raising effort raised over £211,000, including the Medical School Fund, the Engineering Instruction Fund, and the Chemical Laboratory Fund. To this amount was soon added almost £122,000 from the bequests of Charles Clifton and C. F. Beyer. The money certainly enabled the college to realize its immediate goals and to make a brilliant beginning towards becoming the Manchester university. The sum did not, however, relieve Owens from financial concern. The college remained hard pressed for funds during the remainder of the century; the expansion, coupled with general economic difficulty during the 1870s and after, left Owens, if not distressed, certainly unembarrassed by riches.

One of the early aims of the extension movement was the establishment of a medical school. As early as 1866 older suggestions of a union between the College and the Manchester Royal School of Medicine (the old Pine Street School) were revived.[175] Most of the professors favored amalgamation; Williamson, the professor of natural history, objected on several grounds. Not least among them was his fear that the older medical students would exert a baneful effect on the regular Owens students, particularly as the former came from an "inferior social class, especially in Provincial schools." Furthermore, scientific teaching at the existing schools was of lower quality than that found at Owens.[176] At any rate, Williamson's objections were not heeded, and amalgamation took place in 1872, aided by Miss Brackenbury's gift of £10,000.[177]

One condition of the merger was the amalgamation of staff except in those subjects already taught at Owens. For the first sessions, however, Daniel Stone, formerly of the Mechanics' Institution, and Leo Grindon, the amateur natural historian, taught chemistry and botany respectively. By 1874, however, Roscoe and Williamson were teaching in the medical department, Arthur Gamgee, M.D., F.R.S., was teaching physiology, and Julius Dreschfeld had been added to the pathology staff.[178]

The biggest changes at Owens, however, came in the teaching of natural history. In March 1871 geology was separated from Williamson's already swollen duties. W. Boyd Dawkins was appointed lecturer in geology at a salary of £140 plus two-thirds of the student fees.[179] Dawkins had been curator of the Natural History museum, and his appointment was ostensibly occasioned by the gift of the Natural History and Geological Societies museums to Owens, but it is likely that natural history was viewed as an anachronism, and specialization was deemed necessary by powerful members of the Senate such as Roscoe. While teaching geology, botany, comparative anatomy, and palaeontology, Williamson had continued to hold a position as surgeon to the Manchester Ear Institute. When pressed, Williamson had resigned his medical post, and when Dawkins was appointed for the 1872-73 sessions, Williamson later granted, "it afforded . . . much needed relief."[180]

The next assault upon Williamson's chair came later in the decade. It was reported to the council in 1877 that new second B.Sc. regulations for the London degree required examinations in animal physiology, botany, and zoology. The new regulations demanded a large extension in the practical or laboratory portions of Williamson's remaining subjects. It was suggested by a committee that instead of merely hiring a demonstrator, the chair itself should be divided. The council informed Williamson that he was "invited to surrender the chair of Zoology" and retain that of Botany at £250 plus his usual share of the fees. Marcus M. Hartog was invited as demonstrator, and A. Milnes Marshall assumed the professorship of zoology.[181]

The physical sciences experienced a similar differentiation and specialization. The disciplines involved had become too complex for a single individual. Owens recognized very early the need for a broader base if the institution were to become what it had promised: a science university. In 1870 Balfour Stewart and T. H. Core were appointed professors of natural philosophy; Francis Kingdon was named demonstrator, to be joined in 1873 by an Owens student, Arthur Schuster. Originally, James Thomson Bottomley, a graduate of Queen's and Trinity, Dublin, was offered Core's post as junior professor but, it appears, declined it in order to join his uncle William Thomson (later Lord Kelvin) at Glasgow.[182]

Similarly, in 1874, the chemistry chair was split, and Roscoe gladly relinquished part of his duties to his assistant Carl Schorlemmer.[183] German-born Schorlemmer (1834–1892), was the son of a Darmstadt carpenter who along with his boyhood friend William Dittmar developed early his chemical interests. In 1854 he was apprenticed to an apothecary, and eventually he became a pharmaceutical assistant at Heidelberg, where he took the opportunity of attending R. W. Bunsen's lectures. His friend Dittmar, meanwhile had studied with Bunsen and had become Henry Roscoe's private assistant in London and later demonstrator at Owens College. Schorlemmer followed in his friend's footsteps. After a brief time with Will and Kopp at Giessen, Schorlemmer traveled to Manchester, where he first became Robert Angus Smith's assistant and then Roscoe's private assistant when Dittmar accepted an official position at Owens in 1859.[184] The pattern continued. When Dittmar left Owens to join Playfair in 1861, Schorlemmer became Demonstrator and began a fruitful career in research in organic chemistry which was only ended by his death in 1892.

In 1861 John Barrow of Gorton sent samples of cannel coal-tar oils to Owens, and Schorlemmer began their analysis. It had been generally believed certain hydrocarbons (the normal paraffins or alkanes) were capable of existing as two isomers — the alcohol radicals and their hydrides. Schorlemmer showed by analysis of naturally occurring paraffins and synthetically produced ones that the supposed double (the dimethyl radical CH_3-CH_3 and ethyl hydride C_2H_5-H) were in fact one, thus contributing to the Kekulé theory that carbon is tetravalent and that each of the valencies is equivalent. During the decade which followed the publication of his first important paper in 1862, Schorlemmer worked steadily on the investigation and classification of hydrocarbons, publishing over twenty-five papers by 1871. He was elected to the Royal Society of London on the first petition in 1871.

Schorlemmer's scientific reputation quickly outgrew his position at Owens. Roscoe, who was little interested and little skilled in organic chemistry, supported the creation of Britain's first chair of Organic Chemistry for Schorlemmer in 1874, and the two engaged in fruitful collaboration until Roscoe's resignation from the Owens chair in 1885. Schorlemmer collaborated with Roscoe on the unfinished but nevertheless impressive *Treatise on Chemistry,* and published by himself books including his influential *Manual of Organic Chemistry* and *The Rise and Development of Organic Chemistry* (1874). In 1888 Schorlemmer was honored by an LL.D. by the University of Glasgow.

As a teacher Schorlemmer has received mixed notices. Roscoe noted that "as a laboratory teacher Schorlemmer was excelled by few, merely as a lecturer by many." Arthur Schuster confirmed that he "was not a brilliant lecturer but he was an eminently successful laboratory teacher."[185] Schorlemmer's lack of brilliance as a public performer may be traceable to his difficulties with the English language (his *Sprachschwierigkeiten*) and perhaps to his retiring personality.[186] These difficulties did not prevent him from becoming an active member of the Literary and Philosophical Society, and serving as a member of its Council. He was at home, however, only among his communist friends, the closest of whom were Friedrich Engels, Samuel Moore, the translator of *Kapital,* and the Marx family, for whom he was the affable "Jollymeyer." He visited Marx regularly in London and planned with Engels a "Spritztour" of America in 1888.[187]

Together with Roscoe, Schorlemmer participated in the constructing of Britain's foremost chemical school; they continued in harness until Roscoe's election to parliament in 1885. The two continued as friends until Schorlemmer's death.[188] Roscoe's resignation marked the close of an era at Owens. His great work in effect was completed. Upon receiving the resignation in December 1885, the council expressed its very deep regret and noted that it was owing to Roscoe's efforts and attainments that Owens' chemistry department "has long held a position second to no other academic institution in the United Kingdom."[189] The council, if it had wished, could have claimed much more.

Owens College, Industrial Chemistry, and the German Threat

Henry Roscoe's testimony before the Samuelson committee in 1868, it may be recalled, referred to the region's need for well-trained chemists who could "carry on the improvements which are daily being made, and especially for determining the strength and value of each chemical product which is sent into the market; the quality of each article has to be guaranteed."[190] The addition of artificial dyestuffs added yet another requirement: chemists with laboratory ingenuity would be needed for the discovery, development, and production of wholly novel products; *research* had begun to peek into the marketplace. The ordinary cooperation between academic scientists and industry, which involved consulting work, analysis and the training of competent workers, was about to be transformed. Within a relatively few years the need for scientists for the production of novelty, would be recognized, if not yet sufficiently met.

Much attention was paid Owens because it was one of the few institutions recognized as meeting what came to be seen as a national threat. It was during the period of Owens' rise, and especially after 1867, that foreign competition became a critical national concern. And, of all foreign competition, the brilliant political and economic success of Germany appeared most ominous. On many fronts, Germany's success was interpreted as the result of its skill, energy, and scientific expertise. Could Britain, the pioneer of the Industrial Revolution, maintain itself against its rivals? If it were to do so, it would require intellectual as well as economic momentum.

The changing of the economic game rules may be seen most clearly for the Manchester scene in the chemical industry, particularly in the production of new dyes during the late 1850s. Marvelous new dyeing substances were produced from what hitherto had been only a nuisance: coal tar.

Coal tar is an oily fluid formed by the destructive distillation of coal; the oil is obtained as a by-product in the manufacture of coal gas. At the Manchester meetings of the British Association in 1842 John Leigh displayed considerable quantities of what would later be called benzene, nitro-benzene, and dinitrobenzene prepared from coal tar; Leigh was not, however, capable of pursuing his find. It was Hofmann and his students at the Royal College of Chemistry who would pursue the chemistry of coal tar.[191] Hofmann put his students Charles Mansfield and William Henry Perkin to work examining tar hydrocarbons. Mansfield subsequently isolated benzene and its homologues. Perkin was set upon anthracene, a project which yielded no immediate results, but which eventually bore fruit.[192]

In his home laboratory Perkin tried, early in 1856, to synthesize quinine from toluidine by oxidation. The attempt was a failure, and he repeated the experiment with aniline. The resulting precipitate dissolved in alcohol to form a solution which dyed silk a magnificent purple color. Perkin patented the new dye which was called "aniline purple" or, later, "mauve" in August of 1856. In the following year Perkin established his factory at Greenford Green. There Perkin developed important methods of securing sufficient aniline from coal-tar benzene by nitration into nitrobenzene and reduction to aniline.[193]

Within two years the beautiful but fugitive mauve became an international sensation. *Punch* referred to the fad as "The Mauve Measles": "Dr. Punch is of the opinion that it is not so much a mania as a species of measles. . . . One of the first symptoms by which the malady declares itself consists in the eruption of a measly rash of ribbons about the head and neck. . . . The eruption which is of a *mauve* colour, soon spreads, until in many cases the sufferer becomes completely covered with it. . . . Married ladies have been cured by amputation of their pin money."[194] In 1859 Verguin in France began the commercial development of "aniline red" or magenta and thereafter numerous aniline dyes appeared on the market. After the work of Peter Griess, another student of Hofmann, on the action of nitrous acid on aniline, important new dyes known as "azo-dyes" — *Manchester yellow* and *Manchester brown* — were discovered by Caro and Martius in Manchester and produced by Simpson, Maule and Nicholson, and by Roberts, Dale and Company.[195]

The appearance of the aniline dyes drastically altered the dyeing business. The natural dyes previously available were generally dull, expensive, and not color-fast. The new aniline dyes were simpler to use and cheaper; furthermore, unlike their natural counterparts, they were easy to mix. The aniline colors quickly grew in demand. Supply rapidly followed: in 1863 twenty patents for artificial dyestuffs were taken out by British firms, a figure to remain unmatched for at least half a century.[196]

The years 1860–65 were the golden years of the British dyestuffs industry. By 1862 the value of the annual production of coal-tar dyes was already estimated at over £400,000. The first dyes were produced in England, Britain was the world's largest producer of coal tar, and on the scientific side with Hofmann and his students at the Royal College, and Caro, Martius, Griess, and Schunck in or near Manchester, Britain seemed unchallenged in its preeminence.[197]

New firms joined the pioneers. Ivan Levinstein (1845–1916) founded a firm in Manchester in 1864 which was ultimately to rank among Britain's foremost. It was joined in the region in 1876 by Clayton Aniline in which Swiss interests were heavily involved. Roberts, Dale and Co., Levinstein, Clayton, and Read Holliday of Huddersfield ensured Lancashire's stake in the future of British dyestuffs. Heinrich Caro of Roberts, Dale expressed the exhilaration felt at the time: "The beauty, the fastness, and the brilliant success of the first aniline colours acted like sparks on tinder. A new world was disclosed full of magic promise, and all joined eagerly in the search, the manufacturer and the professor, the business man and the adventurer; for the one a new goldmine, the other new opportunities for fruitful investigation."[198]

The research and development of the artificial dyestuffs in this early period has been justly described as semi-empirical.[199] Certainly exact knowledge of the structure of the organic compounds involved was very limited; after mid-decade and the work of Kekulé, the situation would be radically altered. The manufacturers initially depended upon relatively simple processes, the factories were small, and the dyes merely supplemented and did not replace natural colorings. In the race for commercial advantage, the rule of thumb was still close at hand.

The case of alizarine both illustrates the point and introduces the next stage of development. Alizarine is the active coloring agent in the madder root. The color obtained from madder, known as "Turkey Red" was at the time one of the most popular of all natural colors. Vast areas of the continent, especially in France, were devoted to the growing of madder. As late as 1868 the world's production was estimated at about 70,000 metric tons.[200]

Alizarine had been extracted from madder some years before by Schunck and, given its economic importance, it is not surprising that attempts were made to synthesize it from coal tar. Its formula was imperfectly known, and it was long wrongly understood to be a derivative of napthalene. In 1868 Carl Liebermann (1842-1914), a former student of Bunsen, and Carl Graebe (1841-1927), a former Bunsen assistant, both working in Adolf von Baeyer's laboratory, drew upon Kekulé's theory and Baeyer's zinc dust method of synthesis to produce anthracene from alizarine. They demonstrated the converse as well: they oxidized anthracene to anthraquinone, brominated the product, and then suppled alkali to it to yield alizarine.[201]

The synthesis of alizarine ushered in a new period, guided by what the dyestuff pioneer Otto N. Witt (1853-1915) has termed "the principle of goal-directed synthesis."[202] Armed with Kekule's new organic chemistry, chemists trained at the research frontiers would rapidly replace vocationally trained foremen as the key personnel in the dyestuffs industry.

The synthesis of alizarine was stunning primarily because for the first time a natural dye was synthesized in the laboratory, and the artificial product was capable of replacing nature's own gift. Nature in a sense was beaten at her own game. Caro, Graebe, and Liebermann took out a patent, dated 25 June 1869, involving Caro's fuming sulphuric acid treatment of the anthraquinone; the very next day Perkin registered the same process. Perkin's Greenford Green works began at once the production of the artificial alizarine, and Caro, who had left Manchester for the Badische Anilin u. Soda Fabrik, continued investigations on alizarine, preparing for Germany's large entry into the field several years later.[203] The day of rule of thumb was near an end.

For the new scientific era—that is, for the era in which the laboratory application of theory was to dominate—Germany was far better prepared than Britain. Hofmann, who had trained many of the British pioneers, left for Germany in 1865, Caro in 1866, and Perkin left the industry in 1873. More serious for the industry and for the profession in Britain, however, was the paucity of facilities for training research chemists. Until 1874, when Schorlemmer assumed the Owens College chair, there was in fact *no* chair in organic chemistry; the Manchester professorship remained unique for many years afterwards.

Testifying before the Devonshire Commission on Scientific Instruction in 1871, Edward Frankland decried Britain's inferior position with regard to research and maintained:

In my opinion the cause of this slow progress of original research here depends in the first place upon the want of suitable buildings for conducting the necessary experiments connected with research; secondly, upon the want of funds to defray the expenses of those inquiries, these expenses being sometimes very considerable; but thirdly and chiefly, I believe that the cause lies in the entire non-recognition of original research by any of our Universities.[204]

Professor Adolf von Baeyer, before the Munich Academy of Sciences in 1878, repeated what was by then a common refrain among chemists in Britain:

Germany, which in comparison with England and France possesses such great disadvantages in reference to natural resources, has succeeded by means of her intellectual activity in wresting from both countries a source of national wealth. . . . It is one of the most singular phenomena in the domain of industrial chemistry that the chief industrial nation and the most practical people in the world has been beaten in the endeavour to turn to profitable account the coal tar which it possesses.[205]

In addition to the relative lack of will to mobilize her academic resources Britain fell behind also because of the skill of Germany's industrialists who made clever and efficient use of these resources. John J. Beer has outlined the situation clearly. Chemists were brought quickly into the management of firms. They sent samples to university chemists, who would assign their students and assistants to research projects involving them. Upon graduation these students would be encouraged to continue their investigations within industrial firms. These firms furthermore donated chemicals and competed for close ties with academic institutions. "The academic chemist," Beer writes, "felt himself to be a prospector who turned any new claim over to the industrial chemist, who then undertook the actual exploitation. The factory chemist did just the reverse; if he suspected a rich vein, he usually called in his prospecting colleague for definite verification."[206]

No such symbiotic relationship existed in Great Britain; Owens College came closest, if at all, on its still modest scale. With the coming of the University colleges of the 1870s and 1880s the situation only slowly improved. Percy Frankland's presidential address to section B of the British

Association in 1901 still regarded pioneer Owens as "the largest and best-equipped school of Chemistry in the British Islands."[207]

The concern for this new German challenge which became so apparent during the 1880s was anticipated among scientists as early as 1867.[208] The occasion was the International Exhibition at Paris during the spring of that year. Lyon Playfair, one of the British jurors, returned to Britain determined to initiate a crusade. In a widely circulated letter to Lord Taunton of the Schools Inquiry Commission, Playfair charged that the consensus at Paris was "that our country had shown little inventiveness and made little progress in the peaceful arts of industry since 1862." The causes, he continued, were apparent: there was, in Britain, a lack of "industrial education for the masters and managers of factories," and a "want of cordiality between the employers of labour and workmen, engendered by the numerous strikes." Lord Granville, in the *Times* for 29 May 1867, concurred and reminded his countrymen of the "lessons which the late war in Germany and the present Exhibition at Paris afforded us, if we wish to hold our own with other nations in the arts of peace and war."[209]

Numerous scientists joined the debate. John Tyndall testified that "the facilities for scientific education are far greater on the continent than in England. . . . England is sure to fall behind as regards those industries into which the scientific element enters. . . . England must one day — and that no distant one — find herself outstripped by those nations both in the arts of peace and war. As sure as knowledge is power this must be the result."[210]

Britain's supremacy seemed directly threatened by the central European scientific colossus. From the Royal College of Chemistry, Edward Frankland gave his analysis of the causes:

This want of progress in the manufactures of this country [is due] chiefly to the almost utter lack of a good preparatory education for those destined to take part in industrial pursuits. . . . [The masters and managers] have but rarely had any opportunity of making themselves acquainted with the fundamental laws and principles of physics and chemistry; they therefore find themselves engaged in pursuits for which their previous education has afforded them no preparation, and hence their inability to originate inventions and improvements. . . . [The] laws and applications of the great natural forces . . . must always form the basis of every intelligent and progressive industry.[211]

Scientist-industrialist James Young added that "Unless we add skilled instruction to manual labour England cannot expect to maintain her position in the industrial race."[212]

The *Edinburgh Review* rather gloomily concluded, with Tyndall, that we "can no longer hope to forestall other nations and to fill half the markets of the world with products of our looms and factories unless we can advance beyond our former selves. The inquiry now is not what we have been, but what we relatively are; not whether we have gained a high place and reputation, but whether we can retain them? To rest upon our oars is to drift backwards; and not to be the first is soon, in this age of activity, to be the last."[213]

The chemical industry continued to be the most dramatic example of this crisis of expertise. Not only was it the industry which unveiled British deficiencies in scientific and technical education, but it was also in dye-stuffs that early British supremacy was suddenly lost to German ingenuity. Graebe and Liebermann's production of alizarine in Germany, compared with that of Perkin and Sons' Greenford Green works, tells part of the tale. In 1870 Perkin produced 40 tons to the Germans' none; by 1873 the German firm was producing two and one-half times that of the English. By the end of the 1870s German companies held half the world market of all dyestuffs. By the turn of the century they controlled around 90 percent.[214]

Manchester and the Scientific Industry Question

Manchester was drawn almost at once into the flurry of concern over the foreign challenge. Since Manchester was a locus of considerable scientific-industrial activity, it was not unexpected that the city would be addressed as a potential spearhead to put right what many already saw as a deterio-rating situation. Lord Granville, whose letter to the *Times* had drawn considerable notice, spoke publicly in the city:

Many competent persons [he told an audience assembled for the certificates of award for the Oxford Local Examinations] who had been to the Paris Exhibition, both Englishmen and foreigners, concurred that in the application of science to industry other European nations were decidedly making greater advances than this country was making. . . . The men of Manchester [ought] to come forward and . . . push this question of science as far as they possibly could.[215]

The very next day a correspondent, probably William Jack of Owens College, added that if Manchester "is to keep her position at the head of the natural industries, she must trust to the vigour, the intelligence and the training of her sons."[216] Owens College, alert to the obvious connection with their plans for reform, attempted to elicit support from the Society of Arts and from the government. At a meeting of the society called for the purpose, Jack urged the establishment of science colleges in the major industrial centers and the support of Owens as one such existing center.[217]

The extension of Owens, and its plans for a science university fit pre-cisely into the plan. By the early 1870s, steps were taken by others in the community to initiate additional institutions which would be charged with promoting scientific industry. In 1870, as we have seen, Fairbairn and a number of engineers established the Manchester Scientific and Mechanical Society, which drew together those with engineering interests. In June 1872 Frank Spence urged a similar Association of Chemical Manufacturers. In a letter to the *Chemical News,* Spence denounced what he saw as "the extraordinary isolation" of British chemical manufacturers."[218]

Spence, the son of Peter Spence the alum manufacturer, continued to lobby for a new organization, to little effect. In October Spence wrote to the *Manchester Guardian* that the existing institutions were not suitable for the large task before them: the constitution of the Manchester Literary and Philosophical Society, he pointed out, "precludes it from discharging the functions of a technological society. The new Society will on the other hand, deal only with science in its applications. . . . Lancashire, if she is to hold her own during the next ten years will now have to put forth a supreme effort to foster and stimulate what is after all her only real source of material prosperity — industrial science."[219]

By May 1873 a Society for the Promotion of Scientific Industry was underway, and its inaugural meeting was held on 16 January 1874 at the Manchester Town Hall. The Earl of Derby presented the opening address, stating that "Our object is a purely national one — to do our part towards increasing and further developing the industrial resources of this country. . . . I think it is clear that scientific associations having a practical end in view are playing a considerable and perhaps an increasing part in our social organization, and I venture to add that if there is in England, or in the world any place which is a natural and suitable centre for such a society, it is this city of Manchester."[220]

The society, with the Earl as its president, rented a suite of rooms at number 9 Mount Street as its headquarters, and established a council which included as its chairman Hugh Mason (1820-66), the president of the Chamber of Commerce, as well as Frank Spence and Joseph Thompson. The society was organized around a plan submitted by W. G. Larkins, editor of the *Journal of the Society of the Arts.*[221]

The society was unable immediately to implement part of its program, especially with regard to speakers and to the establishment of sections, but it proceeded at once with the planning of an Exhibition of Appliances for the Economical Consumption of Fuel, which opened at Peel Park, Salford. It attracted over 50,000 visitors and was, by all accounts, a success.[222]

Still, the society faced rough going. William Mather, one of the chief patrons of the group and a vice-president, complained of the "coldness" of the other scientific institutions of the town. Hugh Mason was concerned about the size of the membership. Although membership grew from 238 in 1874 to 316 in 1875, it fell far short of expectations. "I think it reflects a little unfavourably," Mason said rather bitterly, "not only upon the public spirit of Manchester but upon its regard for art, science and industry."[223]

The society pushed on, establishing sections on patent laws, chemistry and sanitation, textiles and engineering. Another exhibition, this time on the economy of labour, was scheduled, and plans were inaugurated to look into an amalgamation with the Scientific and Mechanical Society. The new exhibition, held in May 1875 at Cheetham Hill, was a financial disaster, the losses being made up by gifts from Derby, Mason, Mather, and Peacock. The Scientific and Mechanical Society refused the offer to merge and the scientific industry group faded from existence after January 1876.[224]

Perhaps if Spence's original idea had been carried through, the society would have had a more vigorous existence. In broadening the scope of the group, its leadership may have fatally weakened it. While many local business leaders were at least nominally associated with the society, most of the city's scientific leadership, including Roscoe, seem to have ignored it. The Society of Chemical Industry which followed in 1881 was a very different sort of organization. The scientific and the commercial were intimately intertwined in the S.C.I.; for all important purposes of the society they were indistinguishable.

The S.C.I. grew, not from the Promotion of Scientific Industry group, but rather from specialized chemical societies of the north of England. Among the earliest such organizations were the Newcastle Chemical Society, founded 1868, the Tyne Chemical Society and the Faraday Club formed in 1875. The Faraday Club drew its membership largely from chemists in Widnes and St. Helens.[225]

A meeting was held on 21 November 1879, attended by members of the established chemical groups. Subsequent meetings were held in Liverpool and Manchester with E. K. Muspratt, Ludwig Mond and Roscoe in attendance. Slowly the aims of the new society altered. Mond, especially, urged a national organization and ultimately the same "Society of Chemical Industry" was chosen over some opposition which preferred "Society of Chemical Engineers." In April 1880 at Manchester a committee was formed with Roscoe as chairman and George E. Davis as secretary. The committee included (among others) Schunck, Smith, T. E. Thorpe, John Hargreaves, Frank Spence, Peter Spence, Mond, and Muspratt.[226]

In April 1881 a general meeting was held in London at Burlington House where the election of officers proceeded. Henry Roscoe was selected as the society's first president; vice-presidents included F. A. Abel, A. W. Williamson, I. Lowthian Bell, Perkin, Siemens, R. Angus Smith, James Young, E. K. Muspratt, and Walter Weldon; the committee included William Crookes and Peter Griess.[227]

It is clear that the society, dedicated to the promotion of research, development and application in the chemical industries, was envisioned as a union of the nation's leading academic and industrial chemists and *entrepreneurs*. The central position held by academics indicated general recognition of the critical importance of the union of pure and applied science.

It is clear too that the new group saw itself not merely as a regional nor even a national group. Sections were soon formed in the major cities of England and Scotland and also in the United States, Canada, and Australia.

Owens College was crucial to the formation of the society; the first annual meeting of the whole society was held at Owens College on July 5 and 6, 1882. Roscoe gave the opening address, in which he recounted the organizational successes of the previous year. Not least was the appearance of a journal, edited and published in Manchester, which aimed at nothing less than "a complete epitome of the progress of Chemical Industry." The editor was Watson Smith (1845–1920), who was at the time a demonstrator at Owens College and who became in 1885 lecturer on technological chemistry. Roscoe once again drew attention to British-German rivalry and attempted to illustrate something of which his audience was already convinced: "that in certain branches, at least, of Chemical Industry, it is only the highest and most complete scientific training that can insure commercial success."[228]

Late in 1883 the Manchester section was formed and the group boasted a considerable number of distinguished chemists and industrialists. The chairman was Henry Roscoe. George E. Davis served as the section's vice-chairman. The committee included industrial chemists such as Ivan Levinstein and Peter Hart, and consulting chemists such as Angus Smith and Charles Estcourt.[229] By the end of 1884 the committee was emended to include Harry Grimshaw, Carl Schorlemmer, and Watson Smith.[230]

The key figure in the progress and development of the Manchester section was Ivan Levinstein (1845–1916) who succeeded Roscoe as chairman in 1883. Levinstein was born at Charlottenburg, Prussia, and was educated at the Technische Hochschule and at Berlin University. He came to Manchester as a young man and in 1864, at age 19, founded an aniline dye factory at Blackley, which was ultimately to be the most important in Britain. Levinstein was a tireless publicist for reform in the chemical industry, and the S.C.I. was his natural base. For a period of over thirty years he campaigned for scientific education and for patent law reform, the keys, as he saw it, to Britain's chemical rivalry with his homeland.[231]

From 1883, when he delivered his early addresses from the chair of the Manchester section, to the end of the century, Levinstein continued to stress the same themes, which can be summed up in brief remarks to the group in 1897: "It was undoubtedly the fact that the position which Germany held today in the chemical industries was due to their great masters in organic chemistry and to their patent laws; whilst we, on the one hand, did not possess the highly trained staff of men, and on the other hand, stifled our inventive genius and handicapped our industry by absurd laws."[232]

What Britain required, he contined to stress, were more training grounds for organic chemists like the laboratories of Owens College, and an end to the unfairness of the patent laws. Towards these ends he not only lobbied for reform from the forum which the Society of Chemical Industry supplied but also served Owens College as a governor of the court, and helped found the Manchester School of Technology along with J. H. Reynolds and James Hoy.[233]

The principal defect in the British patent law concerned the lack of a requirement for the owner of the patent to work the patent in Britain; German chemical firms could hold patents in Britain and prevent any British firm from operating it, and yet restrict production to Germany. An example of the absurd conditions to which this legal situation led was presented to the society in 1897 by Levinstein himself. He recounted the tale of the invention of a new organic acid by a German firm which had not the means or the will to work the patents granted in Germany and in England. The German patent was soon upset, and other companies entered manufacture; but the original patent was held in Britain, and the acid was never produced there.[234] This situation obtained virtually until Lloyd George's Act of 1907, of which Levinstein was widely held to be the practical orginator.[235]

Levinstein also actively opposed interference with industrial activity on the part of local and national government. Indeed the Manchester section of the S.C.I. worked effectively as a laissez-faire lobby against such regulation. The situation offered paradoxes for the scientific community. It may be recalled that the professional scientists in the city spearheaded environmental concern; sanitarians like Robert Angus Smith who saw no conflict with efficient and profitable industry were able to live comfortably in both worlds. By the 1880s, however, the view that industry could practicably control its wastes and remain competitive was not seriously entertained. As one member of the Manchester section complained: "Chemicals could not be made in drawing rooms or greenhouses."[236] The industrialists actively opposed air and water pollution control measures, and most scientists, fully aware of the intimate connection between science and industry, supported them without reservation.

The first issue joined in these matters was the Rivers Pollution Bill of 1885–86. The previous river pollution control act of 1876 distinguished between pollution caused by sewage and that caused by mining and manufacturing. The Hastings-Birkbeck Bill of 1885 made no such distinction. It immediately aroused the ire of the mining and industrial community, inviting counterattack. In February 1885 the Manchester section unanimously passed a resolution declaring the bill "imperfectly drawn and utopian in character, and could not in its present form be carried into operation without serious injury to the manufacturing and commercial community." Ivan Levinstein would assert that "the liquid discharges from chemical works will, if anything, in some instances lessen the evils caused by sewage pollution."[237]

Henry Roscoe, by now a member of parliament, led the move for the rejection of the bill. Before the House of Commons in March 1886, Roscoe declared the bill to be "complicated, impracticable, and dangerous" and claimed that it would imperil "the very existence of many of our important industries, and that at a period of unexampled depression." "Science," he said, "was now in a position to deal satisfactorily with sewage. . . . But as regarded manufacturing pollution, the same could not be said." When Mr. Chamberlain agreed, and said that the bill "would interfere with, if not entirely destroy some of the most important industries of the North of England," the bill was easily defeated.[238]

The Manchester section of the S.C.I. likewise reached considerable agreement upon the smoke question. That the air of Manchester and its environs was a disgrace and a nuisance all admitted. But, as Harry Grimshaw, a former Owens demonstrator in chemistry, urged them, "if they were not indifferent to their own interests," they must not engage in "harassing any particular trade almost to the verge of extinction."[239]

Levinstein, perhaps less blunt, claimed that "no chemical manufacturer endowed with ordinary common sense was desirous of creating a nuisance. . . . On the other hand no Act for suppression, mitigation or prevention of any alleged evil . . . could be enforced without serious injury unless practical means were available for prevention." G. H. Bailey of Owens College made a study of the town air of Hulme and attempted to show that it was as impure on Sundays as on weekdays, proving, he claimed, that domestic smoke was far greater an evil than factory smoke. Grimshaw estimated the factory contribution as only 15 percent. H. E. Schunck insisted that "because the atmosphere of the place was injurious to plants it does not follow that it was injurious to health."[240] Unless it were proven to his satisfaction that smoke and suspended particles in the air were health hazards it would be merely injurious to industry to proceed against them. In a chairman's address he insisted that "at a time when manufacturers are being harassed, on the one hand by the exorbitant demands of workmen and on the other hand by the increasingly hostile character of foreign tariffs it is a question whether it is wise and politic to increase taxation or lay on fresh burdens except for purposes absolutely required by the public welfare."[241]

Knowledge was power, but power at the service of industry and commerce. Not all citizens of Manchester accepted these arguments. Dr. Simpson, representing the city council, protested what he saw as "an elaborate apology for the production of smoke and the perpetuation of a nuisance." He had "not heard one word or one suggestion for the diminution of the notorious evils which the production of thick black smoke certainly brought in its train."[242]

Whereas Charles Dreyfus (1848–1935), the chemical manufacturer, opined that "some consideration should be extended to [manufacturers] if only because a large capital was employed in the trade and a great number of workpeople depended upon the success of industry for their livelihood," Fred Scott of the Manchester and Salford Sanitary Association insisted that much could be done if manufacturers were willing to go to the necessary expense. He "objected to the public health always being treated as a secondary matter and commercial interests put in first place."[243]

One of Levinstein's most dramatic successes as chairman of the section was the prominent place offered the chemical industry in the Royal Jubilee Exhibition of 1887. The exhibition was planned during the summer of 1886, begun by a meeting called by the deputy mayor of Manchester. William Mather, the engineer, called for an exhibition "widely representative of the arts, science and industry of Lancashire," and the Manchester section of the Society for Chemical Industry threw itself into its part of the project with great gusto. At the December meeting of the section, Levinstein looked beyond the exhibition itself and foresaw conversion of the exhibition buildings "into a great technical institute" which would be "supported by a first-class teaching staff, and guided by a chief of real attainments, experience and high position in science."[244]

Levinstein himself was chosen as chairman of section 3, the chemical industries, and he was assisted by Watson Smith. The committee to oversee this part of the exhibition included Schunck, J. C. Bell, Charles Estcourt, and F. Ermen. The official catalogue of the exhibition declared its object as the illustration of progress during the "Victorian era." Regarding the chemical section itself, its aim was "an illustration of the practical value of Science which ought to satisfy the sternest advocate of utilitarianism. It was the boast of Socrates that he had brought Philosophy down from heaven to dwell among men; and here we have a proof that Science has left the closet and laboratory to do good and useful work in the factory and mill."[245]

In his report on section 3, Watson Smith termed it "probably the most complete exhibition of objects illustrative of the progress, advance and present position of the chemical industries that has ever been brought together." The exhibit included relics of past success: "classical" apparatus from Dalton, Joule, and Davy and manuscripts such as Dalton's *Principles of Chemistry* and Faraday's laboratory notebooks. It included also stalls illustrating present-day research. Henry Roscoe presented a collection of specimens of pure vanadium and tungsten to illustrate his own work in inorganic chemistry; H. E. Schunck exhibited natural coloring matters; Peter Griess' stall laid out the history of the discovery of diazo and azo compounds; Perkin's booth had specimens of his early aniline dyes; Carl Schorlemmer and R. S. Dale exhibited their work on aurin and aurin derivatives. Levinstein had his own little pavilion which furnished specimens of a wide-range of coal-tar products. There were seven groups in all, drawing exhibitors from all over Britain and the continent, but primarily from Lancashire and London.[246]

The Exhibition was a financial success; the surplus, which amounted to over £43,000 was channeled into technical education via the Manchester School of Technology. This money was added to a gift of land worth £47,000 by the Whitworth legatees (Lady Whitworth, Richard Copley Christie, and R. D. Darbishire) which was intended for an institute of art, a museum, and a technical school.[247] The old Mechanics' Institution had become a technical school in 1883 in their facility at David (now Princess) Street. The new School of Technology was merged into the Whitworth scheme in 1887, the legatees adding a grant of £1,000 annually for several years to their gift of land. Levinstein's dream, in late 1886, of using the Royal Jubilee Exhibition to found a "great technical institute" was well on its way to fruition.[248]

Another of Levinstein's goals, that of mobilizing the business-industrial community for more active efforts on behalf of the chemical industry, was reached through his successful campaign to establish a chemical section of the Manchester Chamber of Commerce. The Chamber of Commerce had up until this time almost exclusively concerned itself with questions directly related to the cotton industry. A chemical section was, for Levinstein, the "means of breaking down the barrier of exclusiveness and extending the scope of [the Chamber's] usefulness." On 18 January 1889, at a meeting of representatives of the chemical industries of the Manchester region held at the Victoria Hotel, the following resolution was passed: "That it is desirable that a committee be attached to the Manchester Chamber of Commerce charged with the duty of representing the interests of the Chemical and allied industries." A deputation was sent to the Chamber, and on November 27 the Chamber sanctioned the establishment of such a section. Levinstein was jubilant: "It will be in the power of this committee to render considerable assistance in the development of the Chemical and allied industries in the district, and in making Manchester not only the most important centre in England, but in the whole world."[249]

The Chemical and Allied Trades Committee was established early in 1890 and included besides Levinstein himself Charles Dreyfus, W. H. Bailey, G. E. Davis, Harry Grimshaw, David Spence and T. Jackson. Levinstein was unanimously elected chairman, and Davis served as its secretary. The concerns of the Committee were much the same as those of the Manchester section of the S.C.I. Patent reform was perhaps the most widely discussed issue of the decade of the 90s, and much attention was paid the smoke nuisance, river pollution, tariffs, railway charges, new forms of transportation such as the tramway and motor carriage. In short, the Committee served as a well-placed special interest group which attempted to lobby for its constituency with both the local and national governments.[250]

The activities outside Owens College had their effect within its walls. While remaining true to his view that the starting point for all technical chemistry must be solid and thorough training in chemical theory and in the methods of qualitative and quantitative analysis, Roscoe had introduced during the course of the 1870s a substantial curriculum in the applied aspects of the subject. The 1871-72 Owens College Calendar describes the technological chemistry course as one providing an introduction to the "Chemical Principles of Chemical Manufacture," including in the curriculum heat (heat of combustion of coal); light (coal gas, distillation of coal); water and air (sanitary and technological); alkali manufacture; dyeing and calico printing; the manufacture of acids; and the manufacture of glass and porcelain. By 1875 the major part of the course was devoted to the artificial dyestuffs industry and much of the remainder was spent on hydrocarbon fuels. Similarly Schorlemmer's organic chemistry course, as evidenced by the examinations, tended in the same direction. Roscoe added a metallurgical laboratory course in 1876 in which he was assisted by James Taylor, a graduate of Owens who later made a career in Australia as a government and consulting metallurgist.[251]

Despite Roscoe's efforts, however, the situation regarding technical chemistry at Owens remained inadequate for the new needs of both the industry and the students. Roscoe became increasingly, during the late 1870s and 1880s, a statesman of science and, eventually, a member of parliament. Attention naturally focussed upon Schorlemmer who simply was not, as a teacher, up to the task. Julius B. Cohen recalled his experiences as a student in the years 1878-1880: "During those two years I never made or saw a single organic preparation. All knowledge of the practical side of the subject was derived from the text-book and from some rather dingy-looking specimens (shown in the lecture room). It was only when I was subsequently in the employ of the Clayton Aniline Co. that I was brought into contact with the practical side of the subject."[252]

The beginnings of renewed life in the technical chemistry program may be said to date from 1880 when Watson Smith was appointed demonstrator and assistant lecturer. He was joined in 1882 by Harry Baker, later chief chemist for Castner-Kellner Ltd. Smith (1845-1920) began as a student at Owens in 1862 where he studied under Roscoe. He completed his chemical education at Heidelberg and at Zurich where he studied under Lunge. After a number of years as an industrial chemist he returned to

Owens, teaching technical chemistry and serving as first editor of the *Journal of the Society of Chemical Industry*.[253]

Owens, in its role as a university college of the new Victoria University, was afforded far more latitude in the training of industrial chemists than was possible hitherto. The curriculum reform, which was fully implemented by 1885, followed upon Smith's promotion the previous year to the post of lecturer of technological chemistry. The program bore strong resemblance to that of the Zurich Polytechnic, at which Smith himself studied, but was tailored to the B.Sc. offered by the university. The course was conceived as a four-year sequence, the first year serving as preparatory both for the technological chemistry course and the matriculation for the bachelor's degree. After three years' further study the successful candidate would receive a B.Sc. with honors in Chemical Technology and a certificate, analogous to the Zurich Polytechnikum Diplom, in chemical technology.[254]

The curriculum was the following. The first-year or preparatory course was devoted to the study of inorganic chemistry, qualitative analysis (with analytical chemistry lectures), mathematics, mechanics, geology, French or German, and practical and lecture courses in mechanical drawing. During the second or first technological year, students would be required to take advanced inorganic chemistry, quantitative analysis laboratory, technological chemistry, experimental physics or mineralogy, French or German, and drawing. In the third year students would take senior chemistry lectures, organic chemistry lectures, chemical philosophy (Schorlemmer's course), chemical physics laboratory or advanced mineralogy, drawing, and a second course in technological chemistry. The final year was devoted to organic chemistry, the third and fourth courses in technological chemistry, chemical laboratory, and drawing.

The first technological course treated sulphuric acid and alkali manufacture and included visits to chemical works. The second course included study of the products of the destructive distillation of wood, the chemical technology of fuel, and coal-tar products. The third and fourth courses were more heavily organic and stressed the manufacture of coal-tar products, artificial dyestuffs, bleaching, dyeing and calico-printing. The guiding principle in the technical instruction was, as Watson Smith himself put it: "[The teachers] ought to know how best to give the general outlines, and draw general lessons of scientific principle from what they have seen and know, so that the clothing with more minute practical details, proportions and quantities would soon be put on hereafter, when the arena of actual working practice is entered."[255]

The program was immeasurably strengthened in 1885 with the additions of George Herbert Bailey (1852-1924) and Julius B. Cohen (1859-1935) to the staff. Bailey had entered Owens in 1882 already possessing a

London B.Sc. in physical geography. He studied under Roscoe and Arthur Schuster, and later traveled to Heidelberg to work with Bunsen. After receiving the Heidelberg Ph.D. and a London D.Sc., he was appointed demonstrator and assistant lecturer at Owens in 1885. It was upon Bailey's shoulders that much of Roscoe's teaching responsibilities first fell when Roscoe left Manchester in 1886.[256]

Cohen was a former Owens student who upon Roscoe's advice took his Ph.D. with Baeyer at Munich, at a time when his fellow students included Arthur Smithells, W. H. Perkin, Curtius, and Duisberg. Upon returning to Owens as a demonstrator in 1885 Cohen recalled, "The first thing I undertook on my return as demonstrator at Owens College was to introduce the preparation of organic substance as part of the practical course and I believe that my little book on practical organic chemistry which appeared in 1887 was the first of its kind to be published in this country."[257] After Harold Baily Dixon succeeded Roscoe in the chair of chemistry in 1886, the staff was further strengthened by the additions of Gilbert J. Fowler in metallurgy and Arthur Harden as demonstrator in chemistry. Fowler (1868–1953) ultimately left to become superintendent and chemist at the Manchester Corporation sewage works; Harden (1865–1940), a Nobel laureate for 1929 for his biochemical work, was an Owens undergraduate during the early 1880s. He studied under Cohen, and it was Cohen too who suggested his first research work. He returned in 1888 as demonstrator and lecturer but left in 1897 to head the chemistry department of what ultimately became known as the Lister Institute.[258]

The objective form of these important changes came in 1895 with the opening of the Schorlemmer Memorial Laboratory, the first chemical laboratory in Great Britain solely devoted to organic chemistry. It was a small affair, about 1800 square feet, designed for a single professor, two demonstrators, and thirty-six students. But it was up-to-date, an Alfred Waterhouse design based upon the Munich organic labs, and cost about £4800, much of which was specially collected by a committee headed by Ludwig Mond. Within a few years, with the advice and support of Levinstein, a technical organic laboratory was opened under the direction of Schorlemmer's successor, W. H. Perkin, the son of the aniline dye pioneer. Levinstein offered, in addition, an exhibition or scholarship of £50 to enable a promising chemistry graduate to work an additional year in technical organic chemistry.[259]

It seems clear that Owens College had responded positively to the changes wrought by increasing self-consciousness and increasing organization on the part of chemical entrepreneurs. As science-based industry developed new institutions in Manchester for itself, it drew the academic scientists with them, altering the structure and content of the chemical curriculum of the College.

The Manchester vision of a northern, science-oriented university took concrete, if premature, form in October 1873 with the opening of the new Owens campus on Oxford Road. Over 500 citizens attended the elaborate opening ceremonies. The mood was at once exultant and expectant. The Bishop of Manchester reflected it clearly: "You really must not be content — to let Owens College occupy exactly its present position. You aspire, and you have a right to aspire, to take a position equal to the proudest amongst the universities of the land." Humphrey Cole endorsed what was clearly the commanding sentiment; he had "a perfect conviction that before many years were passed Manchester would have a university."[260]

Henry Roscoe took the podium, clearly jubilant. His new chemical laboratory, to the design of which he himself had greatly contributed, was in his view "more complete probably than any in this kingdom." The *Guardian* in a very laudatory leader called the opening an "historic event" which would right a long-standing wrong: "Our reputation falls very far short of our resources. . . . Owens College was founded to help redress the balance." The college's achievement "constitutes it, strictly speaking, a university. Nobody will deny," the *Guardian* went on, "especially after a visit to Dr. Roscoe's magnificent new laboratory, that Owens College is a great school of physical science; but it is much more than this."[261] Owens College would be a university. It could already claim to be an accomplished and interesting new venture in higher education. It had evolved far beyond the cultural ornament for the city which its citizens had originally sought.[262]

6

University Science:
Arthur Schuster and the Organization
of Physics in Manchester

Building a Northern University

The opening of Owens' new campus was, it should be stressed, only a continuation of the battle already begun for a science based, research oriented university. As early as 1868, before the Samuelson Select Committee on Scientific Instruction, Henry Roscoe responded to the question "Is it your opinion that Owens College might become an important northern university?" with the reply that "It is so even now; we are practically the university of the north and we only require putting into an acknowledged position." Even granting the slight exaggeration, much truth remains. Before the Devonshire Commission Principal Greenwood testified that he felt very strongly that "Applied and Experimental Science is entitled to hold a very important status" and reported upon the demands for increased material and human resources that the growth in importance of the sciences had made upon the college.[1]

These increased demands for laboratory space and for faculty in response to the internal development of the special sciences led to increased Manchester interest in governmental support for the college. The Greenwood-Roscoe report of 1868, presented to the Owens College Extension Committee, turns out to have been a foretaste of the future and bears quoting at some length:

The general conclusions to which we have arrived are as follows:

1. That the thorough efficiency of an institution such as Owens College, in its proposed extension, demands a subdivision of each leading subject more complete than is usual in England, and a provision for the scientific departments far more elaborate.

2. That consequently . . . some measure of aid from the national exchequer . . . especially for the schools of science becomes almost a necessity. . . .

214

3. That in order to ensure the lasting efficiency of such institutions as Owens College it is important strictly to maintain their university character and organisation, and this with a view to the interests not only of abstract science, but also of its applications to the arts and manufactures.[2]

The carrying out of these recommendations, especially with respect to laboratory space and to subdivision of the major chairs, was discussed in the previous chapter. The major changes in the physics discipline at Owens which were elaborated during the 1870s and 1880s remain to be noted. Roscoe and Greenwood were determined to make major appointments in physics to replace the departing Jack; their precise character awaited suitable candidates. Roscoe began with the suggestion of splitting the single chair of natural philosophy into two parts: experimental physics and applied mathematics. He approached the director of the Kew Observatory, Balfour Stewart, who agreed to submit an application.[3]

Stewart (1828–87) was a native of Edinburgh who completed his scientific training at the university under James David Forbes. By the time of his application he was already a wide-ranging scientist who had published in astronomy, meteorology, chemistry, mathematics, and physics. For his work on radiant heat he was awarded the Royal Society's prestigious Rumford Medal in 1868.[4] Given his varied interests, it is not surprising that Stewart was reluctant to accept the proposed division of the chair: "I have thought how the chair could be worked," Stewart wrote to Roscoe in June 1870. "I begin by supposing there are to be two chairs. I do not think the division of the subject into applied mathematics and Experimental Physics a good one for Natural Philosophy without experiment is merely a mathematical exercise while experiment without mathematics will neither sufficiently discipline the mind nor sufficiently extend our knowledge in a subject like Physics."[5]

Stewart proposed instead the division of the subject into two parts. The first, the A course, would include mechanics (including statics and dynamics) and the forces and properties of matter; the second, the B course, would include those areas of special interest to Stewart including the "energies of nature and their laws of transmutation" and "Cosmical physics including a sketch of astronomy, meteorology and terrestrial magnetism." Stewart would, of course, on this scheme assume the responsibilities of the B course and another man the A course. Stewart would also assume the duties of director of the proposed physical laboratory about which he had definite ideas:

The Physical Laboratory would not only be used in experimental [illustrations] of certain laws enunciated in the lectures but I think that some observational and also some experimental research ought always to be going on in order that the more advanced students should be brought into contact with nature. Then they ought to be taught the use of the various instruments and set to devise and work out experiments. They ought also to be taught the philosophy of experiment.

1. To pay attention to and evaluate all sources of error giving due weight to each and dismissing those that ought to be disregarded (thus it is a very common mistake to give inordinate importance to some utterly useless refinement)

2. To pay strict attention to [natural] phenomena as . . . something new

3. To reach the legitimate conclusion from an experiment no more and no less.[6]

Stewart's suggestions, or, viewed from another perspective, his conditions, were accepted. The second, junior, professorship was awarded to Thomas H. Core, M.A., after the original choice, James T. Bottomley, decided instead to accept a position at Glasgow. Core (1836–1910), like Stewart, was a graduate of Edinburgh, where he took honors in classics and mathematics. He lectured at the Established Church Training College in Edinburgh and served as vice-principal of the International College, Isleworth, before accepting the position at Owens in 1870. Core's appointment was especially fortunate, for in November 1870 Stewart was involved in a disastrous train wreck leaving him physically disabled and unable to function fully for some time during his long convalescence.[7]

In his last two years (1868–70), Jack had taught four classes: a mechanics class with an enrollment of sixty-two which prepared students for the London matriculation examination; a physics class of thirty-four students which treated heat, light, and sound; a junior mathematical class of fifteen students which treated elementary statics, dynamics, geometrical optics, and hydrostatics; and a senior mathematical class requiring calculus dealing with rigid dynamics, and physical optics. This last class had only one student in it.[8] Stewart and Core reorganized the curriculum into three parts: an experimental course comprising a mechanics class taught by both Core and Stewart, a physics class, similar to Jack's but with an experimental section, taught by Stewart, and an energy class (Stewart); a mathematical course comprising junior and senior sections, the former sometimes taught by Stewart and the latter by Core; and finally a practical physics course under the direction of Stewart. There were 148 students in the physics classes, including 7 in the physical laboratory for 1871–72, a threefold increase since Jack's departure.[9]

The physical laboratory, with its practical course, was an especially interesting innovation. The practical course was designed for three kinds of students: those who "wish for a complete course of Laboratory practice"; those who "wish to confine themselves to those branches of Physics most allied to Chemistry"; and those who "wish to perform experiments of a less refined nature, chiefly with the view of verifying the laws stated in the class-room." The laboratory was fitted out with special attention to optical, electrical, and heat measurements. The 1871–72 session had eight students including two future physicists of distinction, Arthur Schuster and John Henry Poynting. Work was undertaken with the balance (to determine specific gravities). The Wollaston goniometer was used to determine the angles of crystals, and the refractive indices of a glass prism for sodium, lithium, and thallium lines were found. Roscoe supplied a spectroscope for work in spectrum analysis; a thermopile was employed for heat measurements; barometric and electrical resistance measurements were made; and experiments on the elasticity of copper were carried out. Original researches in spectrum analysis were performed by Arthur Schuster, who communicated a paper entitled "On the Spectrum of Nitrogen" to the Royal Society.[10] Schuster joined Francis Kingdon as demonstrator in physics in 1873. In that year the laboratory moved to the Oxford Road campus where three rooms in a basement were provided for it. From time to time rooms were added, but the laboratory quickly outgrew its confines with twenty-seven students by 1880–81 and sixty-two students by 1885–86.[11]

It was not only lack of space but lack of money as well which Stewart believed hampered his efforts. In a letter presented to the Devonshire Commission in March 1871, Stewart maintained that institutions which provide laboratories for original research as well as the professors who carry out the research have claims upon a government anxious to support scientific work. He suggested a body similar to the Grant Committee of the Royal Society be constituted with substantial funds. A laboratory director, having fixed upon a research project, could make direct application for support of his work. Similarly, such a laboratory might be part of an observational network of similar institutions around Britain, and such work would be paid for by the government.[12]

The theme of governmental support was taken up in Stewart's presidential address before section A of the British Association in 1875. Discussing what would later be termed "astrophysics" Stewart noted its interdisciplinary character:

We have here a field which is of importance not merely to one, or even to two, but almost to every conceivable branch of research. Why should we not erect in it a sort of science-exchange into which the physicist, the chemist, and the geologist may each carry the fruits of his research, receiving back in return some suggestion, some principle, or some other scientific commodity that will aid him in his own field? But to establish such a mart must be a national undertaking, and already several nations have acknowledged their obligations in this respect. . . . But what has England as a nation done?[13]

Britain must, as a nation, remove what Stewart termed "the intolerable burden" of lack of funds, lack of access to information, and lack of facilities for research. "Expected to furnish their tale of bricks," he stated almost bitterly, "they have been left to find their own straw."[14]

Stewart could make a forceful case precisely because Owens College itself had been relatively generous with its support. At the beginning of his tenure as professor, the Quay Street building provided Stewart virtually nothing in the way of physics laboratory facilities, save a room for demonstration equipment. In 1871 the trustees approved a request for £200 to erect a magnetic observatory. The basement rooms occupied at Oxford Road after 1873, though modest, represented what may have been England's largest laboratory outside London, and in it practical classes — unusual for England — were held. To underline its support, the college named Stewart Langworthy Professor of Physics in 1874, forestalling a second time the institution of a professorship of applied mathematics.[15]

Stewart's great strength was as laboratory director. Schuster reports: "He was not a good lecturer and had difficulty in keeping order in the lecture room. . . . In the laboratory he was an inspiring teacher, and it would not be an exaggeration to say that he was the godfather of much of our modern science, for both Poynting and J. J. Thomson received their first lessons in physics from him."[16] To the short list of Poynting and Thomson, Schuster should have added, of course, himself.

Thomson has left a brief account of his introduction to physical methods with Stewart in his *Recollections and Reflections*:

We were allowed considerable latitude in the choice of experiments. We set up the apparatus for ourselves and spent as much time as we pleased in investigating any point of interest that turned up in the course of our work. This was much more interesting and more educational than the highly organised systems which are necessary when the classes are large. Balfour Stewart was enthusiastic about research and succeeded in imparting the same spirit to some of his pupils.[17]

Stewart's infectious enthusiasm for research, his distinguished scientific reputation, and his acknowledged position within the physics discipline were weighty counters in the college's drive for university status.

As early as 1868 Roscoe argued before the Extension Committee for his vision of Owens College "as an advanced school for science." But it was not until the spring of 1875, when the extended college was comfortably settled into its new quarters, that Roscoe, Greenwood, and some others drew up and circulated a pamphlet arguing for university status. In November the senate voted to ask the principal to bring the matter up with the council. Roscoe, Greenwood, A. W. Ward, J. E. Morgan of the medical faculty, and T. Barker, the mathematician, were responsible for circulating the pamphlet which became a matter for public discussion in the spring of 1876.[18]

The timing was particularly appropriate. The college had been for some years uncomfortable with the London examinations for matriculation and for degrees. These examinations, often set by examiners unfamiliar or unsympathetic with the Owens curriculum, were thought to be unfairly severe or irrelevant to the candidate's training. The situation was exacerbated in March of 1876 by the announcement that new requirements for

London science degrees were in the offing which would require a major overhaul of the curriculum at Owens.[19] The Owens science faculty, which believed itself second to none in England, had no statutory voice in these changes.

The pamphlet, which was signed by Roscoe, Ward, Morgan, and Greenwood, became a storm center of controversy. It argued that there existed a need for additional teaching universities in England and that Owens College already qualified to fill that need. The document[20] began with a declaration of independence: "Where a college has already attained to a life and a character of its own, it is impossible to accommodate to institutions of an altogether different historical growth without depressing its vitality." The demands of modern society were not being met by the existing universities, it went on to say. In some subjects, such as "the various branches of physical science, of engineering and mechanics, and of medicine," the older universities have "little or no advantage over more modern schools or are placed in a positively disadvantageous relation toward them." Moreover, *social* changes of a momentous sort had occurred and universities must recognize the new situations: "A new stratum of society, and a constantly increasing one has been opened to the influence of university life and it is absolutely necessary for the progress of the country at large, and for the greatness and serenity of its future that means should be found for meeting the growing demand."

Owens College, in a university form, could meet this demand for a modern, science-oriented institution securely fastened to a middle-class base, the document declared. Manchester already met the newer as well as older requirements for a university because of its geographical position (halfway between the English and Scottish foundations) and because of its academic strengths: "Nor can the fact be overlooked that a desire for higher training, especially in science, has made itself manifest in this district." Manchester was the city of Joule and of Dalton—it had a history of distinguished scientific research: "The foundation of a university in Manchester would greatly contribute towards the important object of stimulating original powers and original work."

A special committee was convened, including Roscoe, Greenwood, Balfour Stewart, the Duke of Devonshire, Sir Joseph Whitworth, A. W. Ward, and John Hopkinson, to solicit and publish opinions from numerous dignitaries and educators and the editors of newspapers and journals.[21]

The response was generally favorable, despite some notable demurrers. Robert Lowe, MP. for the University of London, wrote a stinging reply in which he reduced the application for university status to the proposition that the "name of university will be an excellent puff for the institution" and advocated that a *modus vivendi* with London be worked out. The *Saturday Review* insisted that "all work and no play, especially if that work has to be done in a very large and very smoky town, will certainly make Jack a dull, and probably make him a vicious boy." Goldwin Smith, in a letter to the *Times,* advocated instead a set of affiliated colleges, staffed by supernumerary fellows of the colleges of Oxford and Cambridge.[22]

Despite these lurid warnings, the response was on the whole gratifying, and the Owens group proceeded to make application to the government.[23] The most elaborate analysis appeared, as might be expected, in *Nature,* which indicated "that certain branches of knowledge and their applications have developed of late years in a very wonderful manner, so as almost to fix a new epoch in the progress of our race. The present is eminently the scientific age of the world." Oxford and Cambridge have not, unfortunately, recognized the new situation, for they have "unwarrantably neglected the scientific training of their graduates."[24] Owens, as a university, could meet these needs.

On 20 July 1877 a large and prestigious delegation met with the president of the Privy Council to present the college's memorial on the subject. The delegation included clerical and political parties, educators, and representatives of Owens College, including Balfour Stewart and Roscoe. Owens' application was opposed by a firestorm of opposition from other northern communities, including Liverpool, but especially by Yorkshire College, Leeds.[25] Mediation conferences were held between the institutions, and a compromise was reached. The new northern university was to be federal, with Owens as its first and, for a time, only college. Liverpool was admitted in 1884, and Yorkshire was admitted to the university in 1887. The chancellor of the new corporation, entitled the Victoria University, was to be the president of Owens, and the vice-chancellor its principal. In 1880 the new university officially was constituted, and Manchester had its university—albeit federal and without (for a time) the ability to grant medical degrees, but a university nonetheless. *Nature* congratulated the new foundation and assessed its future: "It would be imprudent to ignore the fact that Manchester has special opportunities for becoming a great scientific school, and the eminent teachers who represent its scientific faculty may be confidently trusted to maintain the position which they have secured for their subjects."[26]

The Articulation of Physics at the Victoria University

University status, at least in federated form, gave new impetus in 1881 to Roscoe's original plan for a professorship in applied mathematics as a complement to those already established in physics. Roscoe's choice was his former student, Arthur Schuster, who had worked in Stewart's laboratory, served as demonstrator, and subsequently trained at Heidelberg, Berlin, and the Cavendish laboratory. The post was, however, advertised in the summer of 1881. Three major candidates emerged: Cambridge's J. J. Thomson, who had already published some promising work in the *Philosophical Magazine* but had not yet been awarded his Adams' Prize; Oliver Lodge of University College, who was subsequently to make a career at Liverpool and Birmingham; and Schuster, who was at thirty a youthful fellow of the Royal Society and who had about five years experience at the Cavendish. Schuster already had a substantial record of publication in physics and well-developed interests in what would later be called spectroscopy and astrophysics.

Although Schuster's fame would eventually be eclipsed by his former student, Thomson, Owens College's decision to appoint him was a sound one. Schuster was older, more experienced, and possessed a track record worthy of close examination. Lodge's charge that the post was designed for Schuster, and that his appointment was a foregone conclusion may not be very far from the truth, but it is of little significance.[27] Schuster's was a solid appointment and one which ultimately shed credit upon the college, the university, and the city.

Schuster was the son of Francis Joseph Schuster, a well-to-do Jewish textile merchant of Frankfurt with business connections in Great Britain. After the Seven Weeks' War, when the Rhenish city was annexed by Prussia, the family firm shifted to Manchester, England. Arthur, baptized as a young boy, was educated privately and at the Gymnasium. In 1868 he attended the Geneva Academy until he joined his parents in Manchester in the summer of 1870.

While only sixteen years of age, owing especially to Henry Roscoe's elementary textbook in spectrum analysis, Schuster had developed an interest in physical science. His parents saw at once that he lacked enthusiasm for business and they consulted Roscoe, then professor of chemistry at Owens College, who arranged for Arthur to enroll as a day student in October 1871. He studied physics under Balfour Stewart and was directed in research in spectrum analysis by Roscoe. Within a year he produced his first research paper "On the Spectrum of Nitrogen." Again at Roscoe's suggestion, Schuster enrolled at Heidelberg under Kirchhoff and received his Ph.D. after a less than brilliant examination in 1873.

Schuster served at Owens in 1873 as unpaid demonstrator in the new physics laboratory and afterwards, at the request of Norman Lockyer, joined an eclipse expedition to the coast of Siam. Upon his return to England in 1875 Schuster remained at Owens for a semester and then joined Maxwell as a researcher at the Cavendish Laboratory, where he remained for five years, ultimately joining Lord Rayleigh in an absolute determination of the ohm.

In 1879 Schuster applied for a post at Mason College, Birmingham, and was rejected in favor of his friend J. H. Poynting. Two years later the professorship of applied mathematics was founded at Owens, and Schuster was selected for the chair.

At the beginning of his Owens College career, Schuster took up once again his former interest in what was by now termed "spectroscopy." In 1881 he published an important paper "On Harmonic Ratios in the Spectra of Gases,"[28] in which he refuted G. J. Stoney's explanation of spectral lines using simple harmonic series by demonstrating statistically that the spectra of five chosen elements conform more closely to a random distribution than to Stoney's "law." However, he concluded that "most probably some law hitherto undiscovered exists which in special cases resolves itself into the law of harmonic ratios," and in 1897 Schuster independently discovered the relationship known as the Rydberg-Schuster law, a formula describing regularities in spectral line distribution.

Schuster's interests led him to investigate the spectra produced by the discharge of electricity through gases in otherwise evacuated tubes. The subject of such electrical discharge was imperfectly understood, and he began a series of detailed investigations which led to his Bakerian lectures before the Royal Society in 1884 and 1890. Schuster's findings were of first importance: he showed that an electrical current was passed through gases by ions and that once a gas was "dissociated" (ionized) a small potential will suffice to maintain a current.

Schuster was also one of the first to indicate the path towards determining the ratio e/m for cathode rays using a magnetic field, a method which ultimately led to the discovery of the electron. It was he, too, who first suggested in 1896, shortly after the appearance of Roentgen's researches, that X-rays were small wavelength transverse vibrations of the ether.

Also appointed in 1881 was William W. Haldane Gee (1857-1927). For a decade Haldene Gee served as demonstrator and assistant lecturer in physics, collaborating with Stewart on textbooks of practical physics. Ultimately Haldene Gee left Owens for the post of chief lecturer in physics and electrical engineering at the Municipal Technical School.[29]

In 1885 another major appointment was made in the physics-mathematics area: Horace Lamb (1849-1934). Lamb, like Haldene Gee, was a Lancashire native. After his graduation from the Stockport Grammar School Lamb proceeded to the Owens College and afterwards to Trinity College, Cambridge, where he began his career as a pure and applied mathematician, attaining great fame in the area of hydrodynamics. After almost a decade at the University of Adelaide in Australia he returned to England as professor of pure mathematics at Owens.[30]

By 1885, then, Owens had assembled a stellar array of scientific workers in the physical science area. Stewart, though far from his prime, still enjoyed a major reputation. Schuster, Lamb, and the even better-known Osborne Reynolds all were on the threshold of their most important work. The teaching staff was bolstered by Haldane Gee, Core, R. F. Gwyther, and F. T. Swanwick, the latter two Cambridge graduates. In chemistry Roscoe and Schorlemmer were still a potent combination; upon Roscoe's departure H. B. Dixon and G. H. Bailey served the attempt to maintain Owens' already lustrous reputation. Roscoe's dream of a science oriented university of the north was for the most part realized.

Of this group it was Schuster who was earmarked for future leadership. Upon his official arrival the mathematical physics classes taught by Core and Gwyther were taken over by Schuster, offerings were expanded and upgraded. This permitted Stewart and Core to expand offerings in other areas of physics to meet demand.[31]

Schuster threw himself into his subject with great verve. His Cambridge experience reinforced what was doubtless already present: a conviction of the importance of applied mathematics to the modern scientist. His inaugural address of 1881, while presenting some conventional thoughts on the subject, also reveals in retrospect the budding Manchester educator. For Schuster, as it had been for Roscoe, the important task of an educational institution is the training of scientific judgment:

More important for the ultimate progress of truth than a solitary success is the training of the faculty which enables the scientific man to judge correctly, and to appreciate the results of those who strike out new roads and extend the boundaries of knowledge. It seems to me to be one of the chief objects of an institution like this to bring up men, who by conscientious considerations of scientific speculations may help to give that solidity and elasticity to public opinion which is necessary for the rapid advance of science.[32]

Roscoe must have heard this address with considerable satisfaction. His "protégé" Schuster certainly appreciated the intricacy and success of his program, uniting the importance of the pursuit of pure science with the college's not less noble social goals. Schuster could be relied upon not to neglect Owens College's civic role.

Although Schuster had no direct part in the direction of the Oxford Road laboratory, his interest in spectroscopy placed him strategically between the chemists and the physicists, and several students, notably G. H. Bailey and T. T. Best worked with him in this area. He maintained his own private laboratory, aided by his assistant Arthur Stanton (B.Sc., Edinburgh). He worked closely also with Roscoe; they jointly published articles on spectrum analysis and worked upon a revision of Roscoe's famous lectures on the subject. In this period, too, Schuster pursued his astronomical and astrophysical interests, joining two British eclipse expeditions. By 1885 he had been selected to serve on the Council of the Royal Society and as honorary secretary of the Manchester Literary and Philosophical Society to which he had been elected in 1873.[33]

His work on the discharge of electricity through gases was begun during this period as well and provided the subject for his first Bakerian lecture

224 Science in Victorian Manchester

before the Royal Society in 1884. He was by this time widely recognized as a bright young physicist; despite the fact that his application for the directorship of the Cavendish Laboratory was passed over in 1884 in favor of J. J. Thomson's, on the Manchester scene he was the obvious successor to Balfour Stewart. Upon Stewart's death in 1887, he was named without competition to the Langworthy professorship at the Victoria University. The post was never advertised.

Upon assuming the directorship of the laboratory Schuster relinquished the duties of the applied mathematics professorship to the capable hands of Lamb. The challenge of his new post was substantial. Interest in physics was surging, owing especially to the opportunities opened by the applications of electricity and magnetism. A dynamo house had recently been added to the existing laboratory for exploitation of the large currents which could therein be produced. The number of students in the laboratory, only eight in 1871, had grown to 62 in 1885 and leaped to 105 by 1890.[34]

Schuster's inaugural discourse as Langworthy professor of physics and director of the physical laboratory addressed itself to the future of the Victoria University, Owens College, and the region it was to serve. In it Schuster returned to the history of science which for him was the proper path towards understanding the progress of science. The history of science teaches, he maintained, the existence of a symbiosis between the pure and the applied. While "it is true that powerful impulses are still given by practical men and without these impulses science would soon lose interest and cease to flourish," it is likewise true that "all these practical applications were only rendered possible by a series of theoretical researches undertaken without a view to commercial application." The teaching of the industrial applications of science, therefore, cannot in any meaningful sense be separated from the teaching of science itself. This conclusion reflects the great social truth which has emerged from the Victorian period and, appropriately, at the Victoria University in Manchester:

We have now no longer a separate caste of learned men. The barriers are broken down which divided the [scholar] from the one who gave himself up to commercial and industrial pursuit. . . . No one will regret that university training is brought within the reach of the commercial classes. This result could only have been achieved by a considerable range of studies taught in the universities.

Science was growing not only in the laboratory and the classroom but also in the workshop and the factory; consequently, the "universities should not neglect to provide for a proper teaching of the industrial applications of science because they only could do the higher parts of the technical instruction."[35]

The address was an appropriate keynote for Schuster's tenure as Langworthy professor. It was his aim to fashion a Manchester school of physics appropriate for both the state of the discipline and the university's social context. Schuster's goal was nothing less than a physics *praxis*, articulating the unity of thought and practice. In pursuit of this goal Schuster was required to pay serious attention not only to the frontiers of his discipline but equally to the institutional character of its setting. The next two decades were devoted to the working out of his plan.

Almost at once the offerings of the physical laboratory were expanded. Stewart had supervised a "course of practical physics" supplemented by offerings in chemical physics and "practical electricity" which included some "special electrical and magnetic measurements." Schuster's more ambitious plans included three years of physical laboratory work with the third year emphasizing "an original investigation"; in addition chemical physics was offered in an expanded version to stress spectrum analysis and electrolysis and a wholly revised course of technical physics (ultimately retitled "electrical technology") made use of the new dynamo room with its d-c dynamo and two a-c machines made available for student researches.[36]

New staff were added to reflect a differentiation of function. In 1891 Charles Herbert Lees (1864-1952) was added as senior assistant lecturer. Lees was a graduate of Owens who subsequently studied under Kundt at Strassburg and under Ayrton in electrical technology at South Kensington before returning to Owens in 1891. He took his D.Sc. at Victoria in 1895 and afterwards served Schuster as assistant director of the physical laboratory after 1900.[37]

Lees' special training and interests especially suited him for supervision of student work in electrical technology.[38] In addition, Robert Beattie, an electrical engineer (B.Sc., Durham), was added as demonstrator; Beattie's *forte* was instrumentation, and after 1900 he was named lecturer in electrical engineering. Albert Griffiths, later to head the physics department at Birkbeck College, joined the staff in 1895 as demonstrator and assistant lecturer.

Griffiths, as a student, had joined with Schuster in establishing the physics colloquium in 1894 for "the encouragement of research and the acquisition of knowledge of recent advances in physics and allied subjects."[39] Open to all above the freshman level, the colloquium consisted of original contributions by independent investigators associated with the laboratory and of discussions of recently published papers of significant interest. Mrs. Schuster graciously served as hostess.

Under Schuster's direction, the physical laboratory was an exciting place. Sophisticated researches on the latest and most interesting problems were executed by staff and students; communication at all levels was free and easy. Among the students to pass through since 1881 and who were in some sense Schuster's students were C. T. R. Wilson, Arthur Eddington, Griffiths, R. S. Hutton, G. A. Hemsalech, D. E. Jones, Lees, Walter Makower, J. W. Nicholson, Joseph Petavel, and G. C. Simpson — with the one exception of the Cavendish Laboratory, Britain's most impressive list. When Lees became Victoria's first D.Sc. in physics in 1895, a milestone had been passed. Schuster's Manchester laboratory could rightfully claim a place as a major scientific center.

The old physics laboratory which Schuster inherited from Stewart was outmoded at the start and was rendered intolerable by its successes: by the fall of 1895, 183 students were enrolled as students in it, up from the original eight of 1871–72. In addition to a lecture room and an apparatus room, the laboratory occupied the basement of the west side of the quadrangle. The small dynamo building in the courtyard added some much needed space and equipment, but it was clear to Schuster that his plans for a great Manchester school of physics depended upon a major breakthrough in scientific organization.

For the overcrowding to be relieved capacious new quarters were required. But space was not Schuster's only concern. The increasing complexity of the physics discipline and Schuster's sensitivity to interdisciplinary concerns—already reflected in the new junior appointments to what was in effect a physics "department"—demanded that considerable thought and expertise be brought to bear upon plans for a new building. Schuster devoted his sensitive intelligence, his time, and his fortune to the enterprise.

The new laboratory would reflect Schuster's philosophy of education; it would be the incarnation of the ideas expressed in his inaugural address. A laboratory for him was a "workshop in which eyes and hands unite in producing a combined result." Schuster was prepared to draw upon his own experience, upon those of other successful enterprises like the Cavendish, and upon the needs of industry. As an innovator he was often charged with excessive liberalism, even radicalism; he admitted to having "offended the conservative sensibility which favours old usages."[40] But Schuster, while by conviction a progressive, was by temperament a conservative in the most attractive sense of the term. His philosophy of change was one of *organic evolution*. A clear exposition of his views was presented in his presidential address before section A of the British Association in 1892. Before the assembled workers in mathematical and physical science, Schuster argued that in planning changes "we may learn something from the way in which Nature improves its organisms"—in short, a natural selection to preserve the best and most useful. Unfortunately the usual pattern of reform is quite different: "our attention is drawn to some failure or to something in which we are excelled by other nations, and attempts are made to cure what perhaps had better be left to become rudimentary."

Each nation, owing to a number of circumstances, possesses its own peculiarities, which render it better fitted than its neighbours to do some particular part of the work on which the progress of science depends. . . . I doubt whether the efforts to transplant the research work of the German Universities into this country will prove successful. Does it not seem well to let each country take that share of the work for which the natural growth of its character and of its educational establishments best adapts it?[41]

Schuster was of course not alone in arguing for the existence of national character. The application to scientific institutions was, on the other hand, uncommon. What, however, was peculiarly British in science aside from Britain's strong heritage of mathematical physics? Schuster chose to point to one element which to him was particularly timely. "Is it not true," he asked, "that the one distinctive feature which separates this from all other countries in the world is the prominent part played by the scientific amateur, and is it not also true that our modern system of education tends to destroy the amateur?" By amateur, Schuster meant a practitioner who has had no academic training, one "who learns his science as he wants it and when he wants it" or, employing another term, the *devotee*.[42] He included in this rank men of the stature of Faraday and Joule and would also add, one suspects, his friend, the litigious Henry Wilde (1833–1919).[43]

Schuster was alert to a strength present in the devotee scientist which was valuable and endangered by professionalization. The devotee is often untrammeled by preconceived notions, by dogma. The devotee, especially when reinforced by genius like Faraday or Joule or Darwin, can bring to bear on old problems an entirely fresh perspective. The devotee, Schuster went on, has often supplied originality to the discipline. In the engine of research,

the amateur supplies the steam and Universities supply the cold water; the former, boiling over often with ill-considered and fanciful ideas, does not like the icy bath, and the professional scientist rebels against the latent heat of the condensing steam, but nevertheless the hotter the steam and the colder the water the better works the machine. Sometimes it happens that boiler and cooler are both contained in the same brain, and each country can boast of a few such in each century but most of us have to remain satisfied with forming only an incomplete part of the engine of research.[44]

Formal training in science necessarily works against undisciplined inspiration. But the universities must take care not to insulate themselves from the community from which the devotee springs. A veil of examinations and requirements must not be permitted to keep the devotee apart and out. "The gradual disappearance of the amateur may be a necessary consequence of our increased educational facilities," Schuster lamented, "and we must inquire whether any marked advantages are offered to us in exchange."[45]

The Manchester context, the heritage of civic science, is discernible in this address. Schuster decried the ivory tower. His laboratory and his university could not be permitted to pose aloof from society. The 1892 presidential address contains, therefore, the germ of his plan for physics in a civic university.

But Schuster was also raising a question which even today remains unanswered. Much has been written concerning the benefits of professionalization to the scientific disciplines. Have we considered as well the unavoidable costs?

Creating the New Physical Laboratory

In 1897, when an anonymous donor gave a substantial sum of money (£10,000) towards the erection of a new laboratory, Schuster easily convinced the council to approve the building. Engaging the architect J. W. Beaumont of Manchester, Schuster set about to plan his new building. Beaumont was sent to the continent to survey the most modern of German laboratories before commencing his own design. Schuster and Beaumont worked closely together in order to incorporate the director's ideas into the edifice. The handsome Victorian building was to be of red brick, faced with stone. The rooms were to be lined with white- and brown-glazed bricks; the corridors and stairs were to have blue-glazed tiles. The space for research and teaching was to be ample. The new laboratory was to rank fourth in the world in size, behind Johns Hopkins, Darmstadt, and Strassburg.[46]

The layout of the laboratory was carefully thought out. Schuster expressed his views on the matter at the cornerstone-laying ceremony on 4 October 1898. The most important influence upon Schuster in this regard was his experience in Helmholtz' research and teaching enterprise at Berlin. "I always considered," he later wrote, "Helmholtz' laboratory, such as it was at the time, the ideal of what a teaching laboratory should be." There was an intimacy in the old laboratory which was encouraged by the laboratory plan and which aided in the progress of the students' research. The students worked side by side at tables, and Helmholtz was able to make daily rounds, conversing with each student and advising on a variety of topics of interest to all. The students thus became aware of each other's problems and methods. The shared experience was stimulating and invigorating. After Helmholtz' new laboratory was built, something was lost: "In place of a small number of badly furnished and crowded rooms, there was now a noble building, impressive on the outside and perhaps inside to the casual visitor, but all the soul and scientific spirit of the old place had gone."[47]

This was not mere nostalgia. In the new ordering of things, Helmholtz no longer had the time personally to see each student. Moreover, the common bond between the students was lost; no longer did they share their successes and their problems. They worked "mainly for the purpose of completing a dissertation of sufficient merit for their degree."[48] Each student, working in his private room no longer, had access to the sense of community which was fostered in the old laboratory. Schuster was determined not to make what he considered to be one of Helmholtz' few mistakes. At the cornerstone-laying ceremony Schuster confided to his audience his plans: he insisted on resisting the modern fashion in laboratory design, which provided single rooms for individual research students. In the Manchester laboratory the rooms would be large enough to accommodate at least several students. Their space would be adequate to insure their apparatus and work against disturbance, but he was determined to facilitate that sense of research community. He recalled that in his own student days "we became interested in each other's work, and thus increased our experience and obtained a much broader view of the range of physics. I consider the experience thus gained to be quite invaluable."[49]

Even if specialization was required of the modern physics student, its dominance was not to be unrelieved. In addition the laboratory design was to serve a social purpose as well. In the industrial life of Britain, Schuster maintained, two kinds of men are needed. The first, the discoverers and inventors, must be encouraged at all costs. The new laboratory must encourage the spark of originality and individuality which, as we have seen, Schuster saw endangered:

Where any pronounced originality exists our whole effort should be to foster and develop it. I feel no doubt that the success of our university will entirely depend on the manner in which we allow room for individuality and originality in our courses, while the continued success of our college must depend on the freedom which we claim for individual teaching, even if in special cases the students should be kept out of the university altogether. Far better that a man of original mind should go through life without a degree, than that he should artificially be driven into the broad path of common-place reasoning.[50]

The access of special students must be maintained; similarly, the researchers must not be isolated from the second type of student — those who are destined to fill responsible places in industry as managers and supervisors. Schuster saw the new laboratory as a first step towards removing the artificial barriers between the two types, thus striking at "the distrust with which the greater part of the industrial community still looks upon education." Difficulties will remain, to be sure, especially with regard to courses of instruction, but Schuster looked forward to the help of the various departments of the college and to the "help and advice of the manufacturers in whose workshops our students will have to complete their education."[51]

The first floor was to be devoted to general instruction in practical physics. The ground floor was to be devoted to electrical engineering and electrochemistry. The second floor was to contain research rooms, lecture rooms, and a large theater for 200 persons. The basement was given over entirely to research purposes. On the roof a small astronomical observatory was erected, with a ten-inch refracting telescope, the gift of Sir Thomas Bazley.[52]

Schuster paid special attention to advanced research in electricity and electrical engineering. He was keenly aware of the importance of interdisciplinary work in this area, and of the shared resources supplied for the two subjects. For him, the "association is little short of a necessity." In addition, with special funding, the annex was to contain the John Hopkinson Memorial Wing, named in honor of the late Dr. John Hopkinson, a former Owens student. In it was to be the dynamo house, accumulator room, photometer room, and electrochemical laboratory. In the main building were to be located the standardizing room, electrical and magnetic testing laboratory, alternating current laboratory, and drawing room. For the future he planned a high-voltage transmission laboratory.[53]

Other special laboratories were instituted to reflect the present research interests of the laboratory staff and of Schuster's anticipations of the future. For spectroscopical research a grating room with a large Rowland concave grating was planned, containing the special grating ruled by Professor Henry Rowland of Johns Hopkins, an echelon grating, and apparatus built by Max Wolz of Bonn for measuring spectrum photographs. Other areas specially provided for included meteorology, atmospheric electricity studies, radioactivity, thermal conductivities, low temperature research (with apparatus provided by Henry Roscoe), and high pressure research. The Whitworth Meteorological Observatory in Whitworth Park had been presented to the university in 1892, and a special lectureship had been created for G. C. Simpson in the subject. The electrochemical laboratory for research in electrochemistry and electrometallurgy, under the direction of Schuster's future son-in-law, R. S. Hutton, drew upon the resources of the Hopkinson laboratory's gas engines and dynamos and offered the use of its electric furnaces to related areas such as physical chemistry.[54]

These special facilities illustrate the working out of Schuster's themes. On the one hand, they demonstrate the keen interest in interdisciplinary cooperation which Schuster saw as so essential for physical science research in the twentieth century: the future, as he divined it, would lie in the intelligent exploitation of the crumbling of disciplinary barriers. On the other hand, the actual daily cooperation of physicists with the more applied areas was to fulfill the goals outlined by Schuster since 1889.

The electrical switchboard reveals, in its way, the structure of his thinking. In the standardizing room Schuster planned a main (electrical) distribution board, an auxiliary heavy-current board, and a battery charging board. By inserting various plugs into the main board, any voltage up to

120 could be obtained and distributed to any of the various research or teaching rooms. Using the heavy current board, higher currents than 50 amperes (up to 300 amperes from the dynamo house and 200 amperes from storage cells) could be provided to the lecture halls or to the research and teaching laboratories. If Stewart's laboratory can be likened to a cottage industry, Schuster's can be seen as moving into the era of the factory. The factory system, pioneered in Manchester not far from the Oxford Road site, demonstrated to the world the efficiency and utility of the central power source. Schuster helped bring this lesson into the world of science.

At the cornerstone-laying ceremony Henry Simon, an engineer, represented the city of Manchester's vision, which was in fact coincident with that of Schuster: " 'Greater Manchester' the name of which was synonymous with unfettered progress and trade, represented one of the most glorious centres of industry and enterprise in the whole world. Why then should [the city] in these times of hard competition with other nations, not provide a technical university in every way equal to the importance and standing which Manchester occupied in the world?" Simon went on to express his hopes for combining the efforts of practical men with those of the man of science, for it is after all, natural science which enables us in "the most economical way to bring into subjection the forces of nature and utilise them for our purposes."[55]

The laboratory was opened two years later. After the guest of honor, Lord Rayleigh, praised the effort which city and college had made and expressed high hopes for the future, Schuster reiterated his theme. The Owens College was founded with a dual purpose: to serve the community and to serve science. His laboratory would train young men and women to serve as the general staff for Manchester enterprise as well as to educate researchers to advance the frontiers of knowledge.[56]

In 1906, at the jubilee celebration convened to honor the twenty-fifth anniversary of Schuster's election as professor in the Owens College, Schuster could look back with justifiable pride on his accomplishments. The successful laboratory, well-known throughout the world as a thriving enterprise, matching or exceeding even the most enthusiastic expectations, was very much his own achievement. The curriculum, the research program, and the physical layout of the laboratory all reflected his philosophy. At the celebration Schuster explained to his audience that "we must consider ourselves not so much as individuals but as parts of an organisation or combination. I have had this idea always before me in my efforts as a teacher, and I have tried to work towards it, for instance, in the planning out of these buildings."[57] To claim that Schuster was a noteworthy pioneer in modern scientific organization is surely not to claim too much.

Of the three partners in the federal Victoria University, Owens College was by far the most important. Not only was it the administrative seat for the examining university, but its teaching college was the most prestigious and its research accomplishments towered over the others, Liverpool and Leeds. If the Victoria University was a serious rival to the older universities, especially in the sciences, it was entirely the burden of the Owens College. It was, however, Liverpool which moved first to disrupt the delicately balanced institution. In 1900, backed politically by Joseph Chamberlain and financially by Andrew Carnegie, Birmingham was granted a university charter. Birmingham's success loosed the powerful centrifugal forces within the Victoria University, and Liverpool's demands were the first to be made public.[58] Leeds, the weakest of the constituent colleges, was solidly against disruption; Liverpool, in the main, was strongly for independent status. Manchester took time to decide. A heated debate was begun.

Schuster's laboratory points up the contradictions inherent within the university. On the one hand, the new laboratory was the brilliantly realized result of the close relationship between Owens and the Manchester region: a relationship advocated by Frankland, made concrete by Roscoe and Reynolds, and shaped by Schuster. On the other hand, the administrative ties which bound it were subject to federal construction. In short, Owens' real character, rooted in its history and its aspirations, was in sharp conflict with its actual state. There was, however, much sentiment in Manchester for retaining federal status. The Victoria University was beginning to make itself well-known and prestigious; graduates were fearful of possible depreciation of their degrees. Schuster was one of the first to assume leadership in the independence movement. In a letter to the *Guardian* on "The Multiplication of Universities and the Liverpool Movement," he decried the "timidity" of those who feared the future. Manchester was strong and need not fear rivalry. It was after all intense competition which made the German universities great. Samuel Alexander, the philosopher, eloquently supported the independence position in his "Plea for an Independent University in Manchester" in which he stressed the "growth of that sense of organic connection of a university with its city and district." Alexander had read the history of the college and of the scientific community well. While not parochial, academic work is a form of citizenship. A Manchester University "may well aspire," he wrote, "to represent worthily the intellectual energy of a city which is illustrated by the names of Dalton and Joule."[59]

The forces for independence carried the day. On 15 July 1903 the university was reconstituted as the Victoria University of Manchester, and the Act of Incorporation of the new university, combining both teaching and examining functions in what had been Owens College, became law in June 1904. Schuster was named its first Dean of Science. A man with the keenly honed sense of history like Schuster could appreciate that an important chapter in the history of the scientific community of Manchester was drawing to a close. At the celebrations of the twenty-fifth anniversary of his election to a professorship at Owens in June 1906, Schuster foresaw that "the time cannot be far distant when I shall have to ask some younger man to take the command of this large building [the physical laboratory]."[60] In fact he was shortly to begin the search for his own successor.

In September, Schuster wrote his choice, a young New Zealander, Ernest Rutherford, then working at McGill University in Montreal. Schuster declared that he was interested in giving up the directorship of the laboratory and the professorship, retaining some connection with the university perhaps in mathematical physics. "My work," he wrote, "is drifting away from the experimental," and he wished to turn to the writing of a book long planned on "Cosmical Physics." After the death of his assistant Arthur Stanton, Schuster's experimental work was practically nonexistent; his many other interests required time. He deemed it the proper moment to retire despite his relative youth. Schuster tried to encourage the younger man: "Manchester is not at all a bad place," and there would be "plenty of time for research work and the laboratory is beginning to be known."[61]

Rutherford was in fact keenly interested. He had been approached by Yale and King's College, London. He declined the former, hoping to return to Great Britain, and he refused to consider the latter because of the sacrifice of research facilities which it would involve. At Manchester, however, the situation was brighter. Rutherford was "inclined to consider very favourably the suggestion of becoming a candidate for the position you propose. The fine laboratory you have built up is a great attraction to me as well as the opportunity of more scientific intercourse than occurs here."[62]

When Rutherford was approached for the post of secretary of the Smithsonian Institution in November 1906, matters were already far along; "I should much prefer," Rutherford wrote, "that position at Manchester under the conditions you outlined." Rutherford was eager to return to Europe: "We are too near the scientific periphery here for America as yet does not count very seriously."[63]

Why Rutherford? Schuster was convinced that his successor must be a young vigorous experimentalist who would dominate the exciting new areas opened up by the discovery of X-rays and radioactivity. Rutherford's prodigious activity at McGill in these areas and his classic Bakerian lecture of 1903 on the radioactive changes in the radium, thorium, and actinium families equipped him with a substantial reputation, and, to Schuster, with the requisite credentials. Moreover, their views on what the future held coincided. As Rutherford wrote to Schuster in early 1907:

I quite agree with you about the choice between experimental and mathematical physics. The latter as a rule know a certain amount of theoretical physics but as far as I have seen are almost entirely destitute of experimental knowledge or instinct. Even the best of them have a tendency to treat physics as purely a matter of equations. I think this is shown by the poverty of the theoretical communications on the problems which face experimentalists today.[64]

Rutherford easily met Schuster's criteria; Schuster indicated that "I am so strongly attached to the place that I could not bear to leave my position except to someone who will keep up its reputation and increase it. There is no-one to whom I would leave it with greater freedom from anxiety than yourself." Rutherford arrived in Manchester on the sixth of June 1907.[65]

The subsequent history of physics has further justified the intelligence of Schuster's choice. Rutherford went on to build upon Schuster's excellent beginnings what was perhaps in its time the most important school of physics in the world. The list of co-workers in the laboratory during Rutherford's Manchester years was nothing less than glittering: Chadwick, Moseley, Geiger, Marsden, Bohr, Fajans, Boltwood, Andrade, Hevesy, Nuttall, C. G. Darwin, A. S. Russell, H. R. Robinson, and still others, whose work was of fundamental importance. Manchester had evolved into a world center for scientific research.

Preparation for the Future

Clearly discernible during the course of the nineteenth century is a large change in our understanding of the nature of science. At the beginning of the century science was generally viewed as a collection of God's laws to be discovered by gifted individuals and applied to man's needs. This conception of natural knowledge was transformed into science as *method*—widely conceived as expertise to be applied to all areas of human endeavor. During the reign of the former, the production of scientific knowledge and of scientists was a happy accident—or, at best, the beneficence of the Almighty. After the change, scientific institutions were seen as *producers* not only of scientific knowledge but of scientists as well—science as enterprise. The gentleman-amateurs of Thomas Henry's day welcomed science as a mark of sharpening class differences; science was an avocation at once gentlemanly and utilitarian which could set the now-wealthy manufacturer or merchant apart from his erstwhile drinking companions. For the aggressive merchants and manufacturers of the 1820s and 1830s, science as useful knowledge was a justification of the position of their class, for it was the middle classes who were the custodians, dispensers and "diffusers" of such knowledge. With the coming of the civic scientists in the 1840s and 1850s, we see the beginnings in the city of a new world. The new professionals' ideology of science as method justifies their own aspirations and positions as civic experts but, extended, justifies also the claims of the professional-managerial segments. They were the custodians of expertise.

The academic science of Owens College refined and extended this new view. The college was the institution which not only was a repository of expertise but could also produce managers and professionals through its teaching role, and could produce new science itself through the research of its professors. Academic science demonstrates to society that science and the scientist are not gifts from God or good fortune but are reproducible items. University science is but a more sophisticated version. New ideas in organization are developed which make research and teaching under the old college seem as quaint as the factory system rendered cottage industry.

This emergence of Manchester as a world science center of brilliance and productivity was the result of evolution through adaptation. The contours of the scientific community — its interests, concerns and approaches to what it came to see as critical problems — were shaped by its civic environment; its character in addition responded to the development of the disciplines represented within it. As the sciences increased in specialization and complexity during the course of the nineteenth century, they placed new stresses upon the community, affecting the composition of its membership and the nature of its leading institutions. The scientific frontier reacted upon Manchester just as Manchester acted upon it.

The scientific institutions of the city were able to flourish by fitting their nature and purposes to the values of the elite groups within the city and region who were called upon to support them. The practitioners of science developed a view of science which was at once civic and cosmopolitan. The philosopher Samuel Alexander noted a "growth of that sense of organic connection of a university with its city and district." This "civic or communal sentiment" was only the end point of a long history of the intimate relationship between science and society in Manchester; it continued to reflect its origins. "Academic work, the life of the teacher and the learner and the investigator is not something remote from daily interests but [is] a form of citizenship."[66]

The "Idea" of science incarnate in its institutions and in the practice of its investigators evolved, as we have seen, through successive stages from the gentlemanly ideal of Thomas Henry and Thomas Percival, through the devotee, the early professional civic science of Playfair and the Liebig students, to the subtly changing academic science of Frankland, Roscoe, and Schuster. Each successive stage of Manchester science both confronted and drew upon the prior. Each stage was, however, never far from a sense of urban context: an "ivory tower" was never erected. As science itself became more and more relevant to the central needs of society — for the proper functioning of the city and the health of its industries — external pressures created, extinguished and shaped institutions and called into being or pushed to the fore new types of scientific practitioners.

Similarly the ever increasing complexity of the disciplines and the ever more refined specialization of scientists limited or caused the extinction of the gentleman amateur and of the devotee and brought to the fore an elite group of frontier researchers. As Schuster took pains to point out, there was a price to pay, but that price was unavoidable, as even Schuster, with some regret, came to see. Natural philosophy evolved slowly into the dual vision of the twentieth century: science in its social role as *expertise* and *enterprise,* and science in its cultural form as empirical knowledge. In Manchester, the two were clearly recognized as sides of a single coin.

The result was a scientific community of remarkable intrinsic vitality and social viability. Manchester, insofar as science was concerned, appeared well prepared for the new century. The twentieth century for both good and ill, would be an "age of science" and science-based technology. It was the period, really the first, in which science would move to the center of the economy and the politics of advanced industrial nations. Despite the decline of the cotton economy of the city during the first half of the century, Manchester's place as a center for exciting and important scientific work remained sturdy. The evolution of science in Manchester in many ways was an exemplar for the rest of Great Britain.

Nature's editorial, "The New Century," in January 1901 remarked upon the new role for science in relation to society in the coming era:

The enormous and unprecedented progress in science during the last century has brought about a perfectly new state of things, in which the "struggle for existence" which Darwin studied in relation to organic forms is now seen, for the first time, to apply to organised communities, not when at war with each other but when engaged in peaceful commercial strife. . . . The scientific spirit must be applied as generally in England as elsewhere. The increasing complexity of industrial and national life requires a closer adjustment of means to ends, and this can only be attained by those who have had education on a scientific basis, and have therefore acquired the scientific habit.[67]

Nature's vision of a scientific society was beginning to diffuse slowly but surely from its already convinced readership to other segments of the ruling elite. The future prime minister, Arthur Balfour, glimpsed the future before the Cambridge Summer Meeting in August 1900, when he cautioned his audience to be aware that

If we were allowed a vision of the embryonic forces which are predestined most potently to affect the future of mankind, we should have to look for them, not in the Legislature, nor in the Press, nor on the platform, nor in the schemes of practical statesmen, nor in the dreams of political theorists, but in the laboratories of scientific students whose names are but little in the mouths of men, who cannot themselves forecast the results of their own labours, and whose theories could scarce be understood by those whom they chiefly benefit.[68]

R. B. Haldane put the matter even more strongly in his 1905 pamphlet, *The Executive Brain of the British Empire*. For Haldane, "the surest and best way to secure national efficiency is to educate our manufacturers and merchants liberally along scientific lines, and to enlist the cooperation of distinguished men of science in the work of national administration." Haldane urged the formation of a *corps scientifique* as an advisory body for the state; this *corps* would include the "exceptional talent" which the state increasingly requires.[69]

The Manchester scientific community, along with the rest of the industrialized world, stepped into a new world, increasingly technocratic and bureaucratic, but one which its nineteenth-century forebears would have recognized. On one issue — the symbiotic relationship between science and modern society — its history provided apt preparation for the future.

Notes

Abbreviations

McG	*Manchester Guardian*
MCRL	Manchester Central Reference Library
MGS Trans.	Manchester Geological Society *Transactions*
MLPS Mem.	Manchester Literary and Philosophical Society *Memoirs*
MLPS Proc.	Manchester Literary and Philosophical Society *Proceedings*
MTMEP	"Minutes of the Trustees' Meetings for Educational Purposes," Owens College
PP	*Parliamentary Papers*
UCL	University College, London
UMIST	University of Manchester Institute of Science and Technology

Preface

1. R. Bendix, "Science and the Purposes of Knowledge," *Social Research* 42 (1975):345.

2. L. Strachey, *Eminent Victorians* (London, 1929), p. vii.

3. C. Levi-Strauss, *The Savage Mind* (Chicago, 1966), p. 15.

Chapter 1

1. John Heilbron, *H. G. J. Moseley: The Life and Letters of an English Physicist* (Berkeley, 1974), p. 47.

2. W. H. Chaloner, "The Birth of Modern Manchester," in *Manchester and Its Region* (Manchester, 1962), p. 134; B. R. Mitchell and P. Deane, *Abstract of British Historical Statistics* (Cambridge, 1971), pp. 24–26; Leon Marshall, *The Development of Public Opinion in Manchester* (Syracuse, N.Y., 1946), p. 30.

3. Léon Faucher, *Manchester in 1844* (London, 1844), p. 16.

4. "Cotton manufacture was the most important in the kingdom in value of product, capital invested and numbers employed" (David Landes, "Technological Change and Industrial Development in Western Europe," in *Cambridge Economic History of Europe,* vol. 6, pt. 1 [Cambridge, 1965], pp. 274–275).

5. Marshall, p. 12. See also Stephen Marcus, *Engels, Manchester and the Working Class* (New York, 1975), pp. 1–27.

6. Chaloner, "Birth," p. 133.

7. Ibid., p. 134: A. E. Musson and E. Robinson, *Science and Technology in the Industrial Revolution* (Toronto, 1969), pp. 427–58.

8. Faucher, p. 17.

9. Ibid., p. 21.

10. Faucher's anonymous translator added: "With respect to science, the whole phenomenon of Manchester society is but a continual series of investigations into, and practical application of, scientific knowledge" (Ibid., pp. 21–22).

11. Hepworth Dixon, "Manchester," *People's Journal* 3 (1847): 246.

12. *MLPS Mem.* 1 (1785): vi

13. Ibid., p. vii.

14. Ibid., p. xi.

15. Ibid., p. xiii.

16. Francis Nicholson, "The Literary and Philosophical Society 1781–1851," *MLPS Mem.* 68 (1924): 104.

17. On Henry, see W. V. Farrar, Kathleen Farrar, and E. L. Scott, "The Henrys of Manchester. Part I: Thomas Henry (1734–1816)," *Ambix* 20 (1973): 183–208.

18. *MPLS Mem.* 1 (1785): 9.

19. Ibid., p. 11.

20. Ibid., pp. 12–13.

21. Ibid., p. 14.

22. Ibid., p. 15.

23. Ibid., p. 25.

24. Ibid., p. 27.

25. *MLPS Mem.,* 2d ser. 2 (1813): 233–34.

26. *MLPS Mem.,* 2d ser. 3 (1819): 208.

27. Ibid., p. 234.

28. *MLPS Mem.* 2 (1785): 21, 22, 23.

29. A. Thackray, "Science and Technology in the Industrial Revolution," *History of Science* 9 (1970): 84.

30. A. E. Musson and E. Robinson, "Science and Industry in the Late Eighteenth Century," *Econ. Hist. Rev.* 13 (1960): 222–44.

31. See Frank Greenaway, *John Dalton and the Atom* (Ithaca, N. Y., 1966), pp. 86–87; cf. Thackray, pp. 81–82.

32. E. Patterson, *John Dalton and the Atomic Theory* (Garden City, N.Y., 1970), p. 74. See also A. Thackray, *John Dalton* (Cambridge, Mass., 1972), p. 49.

33. Greenaway, p. 192.

34. See, for example, that of Lyon Playfair, in T. Wemyss Reid, *Memoirs and Correspondence of Lyon Playfair* (London, 1899), p. 57.

35. *MLPS Mem.* 9 (1883): 289.

36. *Jour. Stat. Soc.* 4 (1841): 265.

37. Joseph Thompson, *The Owens College: its Foundation and Growth* (Manchester, 1886), pp. 264–67; W. C. Williamson, *Reminiscences of a Yorkshire Naturalist* (London, 1896), pp. 60 ff.

38. *McG,* 1 Dec. 1847, p. 6.

39. S. D. Cleveland, *The Royal Manchester Institution* (Manchester, 1931), pp. 5–6.

40. *McG,* 4 Oct. 1823, p. 3.

41. *McG,* 29 Nov. 1823, p. 2.

42. Cleveland, p. 9.

43. *Manchester Iris* 2 (20 September 1823): 302 (hereafter cited as *Iris).*

44. *Iris* 2 (27 Sept. 1823): 1.

45. *Iris* 2 (4 Oct. 1823): 1.

46. *Iris* 2 (11 Oct. 1823): 323–26.

47. On the character of the Royal Institution see also R. F. Bud, "The Royal Manchester Institution," in D. Cardwell, ed., *Artisan to Graduate* (Manchester, 1974), pp. 119–33.

48. Mabel Tylecote, *The Mechanics' Institutes of Lancashire and Yorkshire before 1851* (Manchester, 1957), pp. 129–30. This section depends largely on the work of Mrs. Tylecote.

49. *McG,* 10 April 1824, p. 1.

50. Ibid.

51. Quoted in Tylecote, *Mechanics' Institutes,* p. 151.

52. *McG,* 8 May 1824, p. 1.

53. Ibid.

54. B. Heywood, *Addresses Delivered at the Manchester Mechanics' Institution* (London, 1843), p. 3.

55. Ibid., p. 7.

56. Ibid., p. 14.

57. Ibid., p. 33.

58. Tylecote, p. 135.

59. On Detrosier, see G. Williams, *Rowland Detrosier* (York, 1965), pp. 3–36.

60. Papers of the Banksian Society, Manchester Central Reference Library (MCRL); James Cash, *Where There's a Will There's a Way, or Science in the Cottage* (London, 1873), pp. 59, 61.

61. Cash, p. 63; Papers of the Banksian Society, MCRL.

62. Williams, *Detrosier,* p. 16. The most complete account of the New Mechanics' Institution is R. G. Kirby's "An Early Experiment in Workers' Self-Education," in Cardwell, *Artisan,* pp. 87–89.

63. R. Detrosier, "Address Delivered to the Members of the New Mechanics' Institution 25 March, 1831" (London, n.d.), p. 7.

64. R. Detrosier, *Lecture on the Utility of Political Unions* (London, 1832), p. 24.

65. R. Detrosier, *Memoir,* prefixed to the second edition of *An Address on the Necessity of Moral and Political Instruction to the Working Classes* (London, n.d.), p. 10.

66. Heywood, pp. 35–36.

67. Tylecote, pp. 137–38.

68. *Iris* 2 (29 Nov. 1823): 1.

69. Quoted in Tylecote, *Mechanics' Institutes,* p. 37.

70. *McG,* 4 March 1840, p. 3.

71. *McG,* 19 Jan. 1838, p. 3.

72. Ibid.

73. *MGS Trans.* 1 (1841): 32 (hereafter cited as *MGS Trans).*

74. Ibid., pp. 33–34.

75. Ibid., p. 35.

76. Ibid., pp. 61–62.

77. *MLPS Mem.* 1 (1785): vii.

78. Ibid., p. vi.

79. *McG,* 5 Feb. 1840, p. 3.

80. Heywood, pp. 29–34.

81. *McG,* 5 Feb. 1840, p. 3.

82. See for example the views of Z. in *The Manchester Iris* 2 (16 Sept. 1823): 302.

83. *British Association Report,* 1842, pp. xxxi-xxxvi.

84. *MLPS Mem. and Proc.* 42 (1898): lx-lxi.

85. "Objects and Rules of the Association," *British Association Report,* 1842, p.v.

86. *McG,* 18 June 1842, p. 3.

87. *British Association Report,* 1842, pp. xxxi-xxxii.

Chapter 2

1. Max Weber, "Science as a Vocation," *From Max Weber: Essays in Sociology,* ed. Gerth and C. W. Mills (London, 1948), pp. 129-56.

2. Edward Shils, "The Profession of Science," *Advancement of Science* 24 (1968): 469.

3. Donald Cardwell, *The Organisation of Science in England* (London, 1972), pp. 59-62.

4. Oxford English Dictionary, "Scientific."

5. W. Whewell, *Philosophy of the Inductive Sciences* (London, 1840), 1: cxiii. See also S. Ross, "Scientist: the Story of a Word," *Ann. Sci.* 18 (1962): 65-85.

6. Mabel Tylecote, *The Mechanics' Institutes of Lancashire and Yorkshire before 1851* (Manchester, 1957), p. 149.

7. B. Heywood, *Addresses Delivered at the Manchester Mechanics' Institution* (London, 1843), p. 22; Tylecote, pp. 147-48.

8. *McG,* 27 March 1839, p. 3.

9. Ibid.

10. For Fairbairn's views see *McG* 13 April 1850, p. 8.

11. *McG,* 8 May 1839, p. 3; *McG,* 30 March 1839, p. 1.

12. See chapter 1.

13. See letters to the editor of the *Manchester Guardian,* 8 May 1839 and 20 April 1839.

14. *McG,* 16 April 1839, p. 1.

15. *McG,* 22 February 1840, p. 1.

16. *McG,* 30 May 1840, p. 1.

17. A. Armytage, *Social History of Engineering* (London, 1961), p. 146.

18. *Catalogue of the Royal Adelaide Gallery* (London, 1835), p. 15.

19. *Bradshaw's Manchester Journal* 1 (1841): 9–11, 134–35.

20. W. W. Haldane Gee, *Electrician* 35 (1895): 132.

21. Osborne Reynolds, *Memoir of James Prescott Joule* (Manchester, 1892), p. 30.

22. J. Joule, *MLPS Mem.*, 2d ser. 14 (1857): 83.

23. Ibid., pp. 78–79.

24. *McG,* 29 March 1843, p. 6.

25. Joule, *MLPS Mem.*, 2d ser. 14 (1857): 77.

26. *McG,* 11 Jan. 1843, p. 1.

27. Joule, *MLPS Mem.*, 2d ser. 14 (1857): 83.

28. Ibid., p. 82.

29. T. Wemyss Reid, *Memoirs and Correspondence of Lyon Playfair* (London, 1899), p. 58.

30. Elizabeth Patterson, *John Dalton and the Atomic Theory* (Garden City, 1970), pp. 277–83.

31. W. C. Williamson, *Reminiscences of a Yorkshire Naturalist* (London, 1896), p. 76.

32. Ibid., p. 81; *MLPS Mem.* 10 (1852): 203 ff.

33. Williamson, p. 81.

34. Ibid., pp. 76–77.

35. Wemyss Reid, p. 57.

36. *Slater's Manchester, Salford and Suburban Directory* (Manchester, 1832), p. xxi.

37. Robert Rawson, "Memoir of the Late Eaton Hodgkinson," *MLPS Mem.*, 3d ser. 2 (1865): 151.

38. Rawson reports that he read Lagrange, Laplace, Euler, and Bernoulli (Ibid., p. 152).

39. Ibid., p. 162; *MLPS Mem.* 4 (1824): 225–89.

40. *Proc. Roy. Soc.* 12 (1863): xii; W. Pole, *Life of William Fairbairn* (London, 1877), p. 181.

41. F. Nicholson, "The Literary and Philosophical Society 1781–1851," *MLPS Mem.* 68 (1924): 133–34.

42. Pole, p. 181.

43. *MLPS Mem. and Proc.* 14 (1874–75): 132.

44. Pole, p. 103.

45. Ibid., p. 114.

46. A. Musson and E. Robinson, *Science and Technology in the Industrial Revolution* (Toronto, 1969), p. 481.

47. For Fairbairn's life, see J. Burnley's article "Sir William Fairbairn" in the *Dictionary of National Biography; Fortunes Made in Business* (London, 1884–87), II, passim.

48. Pole, p. 157.

49. Pole, p. 156.

50. Ibid.

51. *Proc. Man. Sci. and Mech. Soc.* 1 (1872): 1, 4.

52. *Proc. Man. Sci. and Mech. Soc.* 2 (1873):1.

53. J. F. Bateman, *History and Description of the Manchester Waterworks* (London and Manchester, 1884).

54. *MLPS Mem.* 75 (1930-31):110.

55. Ibid.

56. Reynolds, *Joule,* p. 30. On Joule's early work, see G. Jones, "Joule's Early Researches," *Centaurus* 13 (1968):198-219.

57. James Joule, *MLPS Mem.* 75 (1930-31):112.

58. James Joule, *Scientific Papers,* I, 66-81 (hereafter cited as *SP*); see also Arthur Schuster, *Biographical Fragments* (London, 1932), pp. 201-2.

59. Joule, *SP,* I:59-60.

60. *Phil. Mag.* 19 (1841):260-77.

61. Joule, *SP,* I:78.

62. Joule, *SP,* I:123.

63. Schuster, *Biographical Fragments,* p. 201.

64. *Brit. Assn. Rpt.,* 1842, p. 31.

65. L. Rosenfeld, "Joule's Scientific Outlook," *Bull. Brit. Soc. Hist. Sci.* 1 (1952):173.

66. Joule, *SP,* II:1-2.

67. Joule, *SP,* I:120.

68. Joule, *SP,* I:156, 158.

69. Joule, *SP,* II:215.

70. Ibid., pp. 211-15.

71. Joule-Playfair Correspondence, MCRL.

72. *Brit. Assn. Rpt.,* 1845, p. 31.

73. Reynolds, *Joule,* p. 105.

74. Joule, *MLPS Mem.* 75 (1930-31):113.

75. Wemyss Reid, p. 74.

76. Joule-Playfair Correspondence, MCRL; emphasis supplied. The bride-to-be was Amelia, daughter of John Grimes, comptroller of customs, Liverpool (*Nature* 26 (1882):619).

77. Roscoe, *Life and Experiences of Sir Henry Enfield Roscoe* (London, 1906), p. 120.

78. Joule, *SP,* II:215.

79. J. T. Bottomley, *Nature* 26 (1882):619.

80. Reynolds, *Joule,* pp. 117-18.

81. Bottomley, *Nature* 26(1882): 619.

82. Robert Angus Smith, "A Centenary of Science in Manchester," *MLPS Mem.* 9 (1883):399.

83. Warren Hagstrom, *The Scientific Community* (New York, 1965), p. 9.

84. James Joule, Manuscripts, UMIST; unless otherwise indicated, the following quotations are from this manuscript.

85. Rosenfeld, "Joule's Scientific Outlook," p. 176.

86. Hagstrom, p. 52.

87. Joule, *MLPS Mem.* 75 (1930-31):110.

88. Reynolds, *Joule,* p. 105.

89. Schuster, *Biographical Fragments,* p. 205.

90. Reynolds, *Joule,* p. 168.

91. J. Joule, *Nature* 26 (1882): 293–94.

92. *MLPS Proc.* 21 (1882): 143.

93. James Binney, *The Centenary of a Nineteenth Century Geologist* (Taunton, 1912), p. 5.

94. Ibid., p. 8.

95. Ibid., p. 18.

96. Smith, "Centenary of Science," p. 447.

97. Binney, *MGS Trans.* 5 (1866), p. 158.

98. On the popularity of geology in Great Britain, see Charles Gillispie, *Genesis and Geology* (New York, 1959), esp. chap. 8.

99. J. S. Flett, *The First Hundred Years of the Geological Survey of Great Britain* (London, 1937), p. 14.

100. Flett, p. 31; E. Bailey, *Geological Survey of Great Britain* (London, 1852), p. 27.

101. H. B. Woodward, *History of Geology* (London, 1911), p. 97.

102. A. C. Ramsay, *Record of the School of Mines* (London, 1852), I: 90.

103. Ibid., p. 92.

104. Ibid.

105. Smith, "Centenary," p. 415.

106. *Royal Society Catalogue of Scientific Papers* 1 (1867): 372–73.

107. J. Binney, *Centenary,* p. 29.

108. Joule, "Binney," p. 294.

109. *Roy. Soc. Phil. Trans.* 145 (1855): 149–56.

110. Joule, "Binney," p. 294.

111. Smith, "Centenary," p. 462.

112. *MLPS Mem. and Proc.* 21 (1881–82): 143.

113. See, for example, his son's accounts of slights at the hands of Williamson: J. Binney, "Centenary," pp. 50–51; see also *MLPS Proc.* 14 (1875): 50.

114. *MGS Trans.* 8 (1868–69): 10.

115. *MGS Trans.* 5 (1866): 178.

116. Ibid., pp. 178–79.

117. *Health Journal* 5 (1887): 17.

118. *Phil. Mag.* 8 (1836): 571–73.

119. Smith, "Centenary," pp. 267–76.

120. *McG,* 31 March 1838, p. 1.

121. City of Manchester Gas Department, *One Hundred and Forty-three Years of Gas in Manchester* (Manchester, 1949), pp. 7–10.

122. W. V. Farrar, "Richard Laming and the Coal-Gas Industry, with His Views on the Structure of Matter," *Ann. Sci.* 25 (1969): 246.

123. Henry Roscoe, *Proc. Roy. Inst.* 11 (1886): 453.

124. Georg Lunge, *Coal, Tar and Ammonia* (London, 5th ed., 1916), I: 225, 82.

125. John Leigh, *MLPS Mem.,* 2d ser. 9 (1851): 299, 301–4.

126. *McG,* 9 October 1844, p. 6.

127. *McG,* 14 October 1844, p. 6.

128. *McG,* 9 October 1844, p. 6.

129. Smith, "Centenary," pp. 291–98.

130. *McG,* 5 May 1849, p. 8.

131. John Leigh and N. Gardiner, *History of Cholera in Manchester in 1849* (Manchester and London, 1850), p. 3; unless otherwise noted, all quotations in the discussion of the cholera study are from this source.

132. *McG,* 5 Sept. 1867, p. 3.

133. *McG,* 9 Jan. 1868, p. 3.

134. *The Free Lance* 3 (1868): 76.

135. *McG,* 5 March 1868, p. 6.

136. *McG,* 2 May 1872, p. 6.

137. *McG,* 7 May 1868, p. 6.

138. *McG,* 5 Feb. 1869, p. 6.

139. *MLPS Proc.* 20 (1881): 69.

140. *Monthly Notices of the Roy. Ast. Soc.* 48 (1888): 157–60.

141. *MLPS Proc.* 15 (1875): 166.

142. *Roy. Ast. Soc. Mon. Not.* 9 (1848–49): 37–38.

143. J. Bottomley, *MLPS Proc.* 27 (1887): 31–32.

144. *Roy. Ast. Soc. Mon. Not.* 48 (1888): 160.

145. *MLPS Mem.* 4 (1871): 128, 147; Balfour Stewart, *Nature* 36 (1887): 585.

146. *McG,* 15 Dec. 1887, p. 6.

147. Stewart, *Nature* 36 (1887): 505.

148. *McG,* 5 Nov. 1845, p. 4.

149. Baden Powell, *Quart. Jour. Ed.* 4 (1832): 197–98.

150. Everett Mendelsohn, in Karl Hill, ed., *The Management of Science* (Boston, 1964), pp. 3–48.

151. Leigh, of course, came from a distinguished old Cheshire family but apparently had to make his own way in Manchester; only Sturgeon came from truly humble origins.

152. Edward Binney, "On Some of the Objects of the Manchester Geological Society," *MGS Trans.* 5 (1866): 180.

153. *MGS Trans.* 8 (1868–69): 10.

154. Henry Roscoe, *Life and Experiences* (London, 1906), p. 121.

155. Hagstrom, p. 9.

156. *McG,* 22 Dec. 1881, p. 5.

157. The sociologists S. Box and J. Ford have argued that experiences of social marginality and "identity crisis" are solved for some by turning to science. See S. Box and J. Ford, *Sociology* 1 (1967): 225–38.

158. F. Nicholson, *MLPS Mem.* 18 (1924), pp. 137, 134.

159. *Manchester Faces and Places* 2 (1890): 153–56; *McG,* 21 Nov. 1904, p. 12.

160. *Proc. Manchester Field Naturalists' Society* 1 (1860): 3.

161. Ibid., p. 4.

162. Ibid., p. 12.

163. *Manchester Faces and Places* 2 (1890): 156.

164. L. Grindon, *Joseph Sidebotham, The Lesson of a Lovely Life: a Memoir* (Manchester, 1886), p. 15.

165. Letter, 8 April 1863, MCRL, MSC 606.42/3.

166. Letter, 18 April 1863, MCRL, MSC 606.42/3.

167. Letter, MCRL, MSC 606.42/4.

168. Letter, 18 April 1863, MCRL, MSC 606.42/5.

169. Letter, 20 April 1863, MCRL, MSC 606.42/6.

170. Circular, 20 April 1863, MCRL, MSC, 606.42/7.

171. Smith, "Centenary," p. 322.

172. *MLPS Proc.* 1 (1857–60): 91–92; *Proc.* 3 (1862–63), pp. 77–81.

Chapter 3

1. Robert Angus Smith, *Life and Works of Thomas Graham* (Glasgow, 1884), p. 65.

2. Ibid., p. 66.

3. T. Wemyss Reid, *Memoirs and Correspondence of Lyon Playfair* (London, 1899), pp. 36–37.

4. Ibid., p. 40.

5. In this year Graham became an editor of Liebig's prestigious journal and had already begun to send students to Germany to prepare for the Ph.D. degree. Graham doubtless met Liebig at the 1837 Liverpool meeting of the British Association.

6. Liebig's *Annalen der Chemie u. Pharmacie* 37 (1841): 152–64.

7. *Brit. Assn. Rpt.* (1840): 76–77.

8. Wemyss Reid, p. 43; see also Buckland's letter to Peel in Wemyss Reid, p. 78.

9. Ibid., p. 44.

10. *Manchester Faces and Places* 9 (1897): 4; Wemyss Reid, p. 53.

11. For Watt, Davy, and Beddoes, see Harold Hartley, *Humphry Davy* (London, 1966), p. 18. Mr. T. R. Underwood, a proprietor of the Royal Institution of Great Britain, was one of the first to urge Davy's appointment to the R.I. He discussed the matter with Count Rumford, who, according to Underwood, had received full powers to make the appointment. Underwood introduced Rumford to James Thomson, who convinced the count to make the appointment (John Paris, *Life of Sir Humphry Davy, Bart.* [London, 1831], 1: 115).

12. H. Davy, *Works* (London, 1839), 2: 22.

13. Ibid., pp. 303–6.

14. *Nicholson's Journal* 23 (1809): 174–82; *Philosophical Magazine* 5 (1834): 355–65; J. Thomson, *Notes on the Present State of Calico Printing in Belgium* (Clitheroe, 1841).

15. E. Baines, *History of the Cotton Manufacture in Great Britain* (London, 1835), pp. 277–79.

16. It was Thomson who provided the materials for Baines's sketch of the history of chemistry as applied to calico printing (Baines, p. 279).

17. A. Musson and E. Robinson, p. 350; *Manchester Faces and Places* 9 (1897): 4.

18. Baines, *Cotton,* p. 278.

19. See *Brit. Assn. Rpts.* (1842): 32-33, and *Chem. Soc. Mem.* 3 (1845-48): 348-69.

20. Mercer to Playfair, 30 June 1846, Mercer-Playfair Correspondence, MCRL.

21. E. A. Parnell, *Life and Labours of John Mercer* (London, 1886), pp. 103-4; R. A. Smith, "Centenary," p. 348.

22. *Annals of Electricity* 7 (1841): 421-27.

23. *McG,* 16 March 1842, p. 3.

24. *Brit. Assn. Rpts.* (1842): 42-54.

25. On Heywood, see *MLPS Mem. and Proc.* 43 (1898): xi.

26. *McG,* 29 March 1843, p. 5.

27. Wemyss Reid, pp. 56-57.

28. Ibid., p. 79.

29. On Dalton and Playfair see Wemyss Reid, p. 57 and *passim.*

30. Wemyss Reid, pp. 60-63; *Brit. Assn. Rpts.,* 1845, pp. 142-86.

31. *McG,* 3 July 1844, p. 6.

32. *McG,* 21 Aug. 1844, p. 7.

33. *McG,* 5 Oct. 1844, p. 9.

34. *McG,* 19 Oct. 1844, p. 7.

35. Ibid.

36. *McG,* 26 Oct. 1844, p. 7

37. *McG,* 18 June 1845, p. 4.

38. Smith, "Centenary," p. 348, n. 1. On Young and Binney, see John Butt, "Technical Change and the Growth of the British Shale Oil Industry (1680-1870)," *Eco. Hist. Rev.* 17 (1965): 511-21.

39. The best account of Smith's life and work is A. Gibson and W. V. Farrar, "Robert Angus Smith and 'Sanitary Science,' " *Notes and Records of the Royal Society* 28 (1974): 241-62.

40. John published a paper "On the Origin of Colour and the Theory of Light," *MLPS Mem.,* 3d ser. 1 (1862): 1-96. He served for many years as the master of the Perth Academy.

41. J. Liebig, "On the Azotised Nutritive Principle of Plants," *Annals of Electricity* 10 (1843): 483-84.

42. *Nature* 30 (1884): 104.

43. On Smith's early life, in addition to the article by Gibson and Farrar, see P. J. Hartog in the *Dictionary of National Biography* and Edward Schunck's "Memoir of Robert Angus Smith," *MLPS Mem.,* 3d ser. 10 (1887): 90-102.

44. *Ann. der Chem. u. Pharm.* 39 (1841): 1-25.

45. *Phil. Mag.* 20 (1842): 495-500.

46. Ibid., p. 495.

47. Ibid., pp. 498-99.

48. *McG,* 12 June 1844, p. 4, and 23 Dec. 1848, p. 9.

49. *MLPS Mem. and Proc.* 48 (1903-04): 18.

50. His father was, at least, an annual subscriber to the British Association for its Manchester meeting in 1842.

51. *MLPS Mem. and Proc.* 42 (1897-98): li-lii.

52. On Allan, see *Jour. Chem. Soc.* 20 (1867): 386–87; on Gilbert see *Proc. Roy. Soc.* 75 (1905): 236–42; *Dictionary of National Biography,* 2d supp.; *Jour. Chem. Soc.* 81 (1902): 625–28.

53. I am relying for this information on Crace-Calvert's own account in a letter 13 March 1849, at the library of University College, London. I am grateful to Dr. G. Roberts for calling this letter to my attention.

54. *Journal de Pharmacie* 2 (1842): 388–94.

55. See Crace-Calvert, *Dictionary of National Biography.*

56. *McG,* 26 Dec. 1849, p. 6; ibid., 17 Feb. 1849, p. 7; ibid., 24 Feb. 1849, p. 8; ibid., 7 Feb. 1849, p. 9.

57. *McG,* 30 Oct. 1847, p. 4.

58. Ibid.

59. *McG,* 2 July 1845, p. 1.

60. *McG,* 7 July 1849, p. 1; ibid., 1 Sept. 1849, p. 1.

61. *McG,* 17 June 1873, p. 5.

62. E. M. Mitchell, "The English and Scottish Cotton Industry," *Scottish Historical Review* 22 (1925), pp. 101–114; W. Marwick, "The Cotton Industry and the Industrial Revolution in Scotland," *Scottish Historical Review* 21 (1924), p. 207; Henry Hamilton, *The Industrial Revolution in Scotland* (London, 1966), pp. 148–149.

63. *McG,* 30 Oct. 1847, p. 4.

64. See Jack Morrell, "The Chemist Breeders," *Ambix* 19 (1972): 1–46.

65. *Proc. Roy. Soc.* 24 (1876): xxxi–xxxii.

66. A. W. Hofmann, "The Life-Work of Liebig in Experimental and Philosophic Chemistry," in *Lectures delivered before the Chemical Society, Faraday Lectures 1869–1928* (London, 1928), p. 48.

67. Hofmann, p. 51.

68. Quoted in J. Partington, *History of Chemistry* (London, 1964) 4: 296–97. When Dumas asked him why he withdrew from pure chemistry he replied, "With the theory of substitution as a foundation, the edifice of chemical science may be built up by workmen. Masters are no longer needed."

69. Partington, IV, p. 297.

70. Liebig to Wohler, 23 Nov. 1837, in *Aus Justus Liebigs und Friedrich Wohlers Briefwechsel in den jarhren 1829–1873* (Braunschweig, 1888), 1: 113.

71. *Brit. Assn. Rpt.,* 1837, p. 41.

72. Liebig claimed to be the first to opine that ammonia is the source of nitrogen in plants. See *Chem. News* 7 (1863): 268.

73. *Chem. News* 27 (1873): 206.

74. E. A. G. Robins, "The Changing Structure of the British Economy," *Advancement of Science* 11 (1954–55): 184.

75. Hofmann, p. 52.

76. *Proc. Roy. Soc.* 24 (1876): xxxv.

77. Liebig to Peel, 14 March 1843, letter 433, Imperial College Library, London; See also Wemyss Reid, pp. 69–70.

78. *Chem. News.* 14 (1866): 289.

79. J. L. Dumas, "Liebig et son empreinte sur l'agronomie moderne," *Rev. d'Hist. des sciences* 18 (1965): 2.

80. *Chem. News* 17 (1868): 19. An account of the formation of the company appears in E. K. Muspratt, *My Life and Work* (London, 1917), pp. 161–62.

81. *Chem. News* 15 (1867), p. 264; ibid., 18 (1868), p. 213.

82. Hofmann, p. 50.

83. *Chem. News* 1 (1859): 39.

84. James Joule, "On the Utilization of the Sewage of London and Other Large Towns," *MLPS Proc.* 1 (1857–58): 57.

85. Quoted in Partington, 4: 300.

86. Liebig, *Familiar Letters on Chemistry,* 3d ed. (London, 1851), p. 5.

87. Wemyss Reid, p. 46.

88. Ibid., pp. 46–47.

89. Liebig, *Letters,* p. 20.

90. Beatrice and Sidney Webb, *English Local Government* (London, 1922), 4: pp. 400, 401.

91. Asa Briggs, *Victorian Cities* (New York, 1965), p. 92.

92. Friedrich Engels, *The Condition of the Working Class in England* (1845), trans. W. D. Henderson and W. H. Chaloner (Oxford, 1958), p. 50.

93. Alexis de Tocqueville, quoted in W. H. Thomson, *History of Manchester to 1852* (Altrincham, 1967), p. 333.

94. Ibid., p. 357.

95. Or, one had to flee them. L. H. Hayes, in his *Reminiscences of Manchester . . . from 1840* (Manchester, 1905), recalls: "In Manchester about the year 1840 the middle classes began to realise that town life was not very desirable, and families began migrating and settling in the various suburbs" (p. 151).

96. M. W. Flinn, *"Public Health Reform in Britain* (New York, 1968), p. 13.

97. Great Britain, Poor Law Commissioners, *Report on the Sanitary Condition of the Labouring Population of Great Britain* (1842), ed. M. W. Flinn (Edinburgh, 1965), p. 223. (Hereafter cited as "Chadwick Report").

98. Ibid.

99. Engels, p. 60.

100. T. Southwood Smith, *Treatise on Fever* (London, 1830), pp. 348–49.

101. Chadwick Report, p. 63.

102. John Burnett, "History of Food Adulteration in Great Britain in the Nineteenth Century," *Bull. Inst. Hist. Res.* (London) 32 (1959): 105.

103. J. Burnett, *Plenty and Want* (London, 1966), p. 191. See especially Chapters 5 and 10.

104. F. Filby, *History of Food Adulteration and Analysis* (London, 1934), chap. 8.

105. On this issue see Ernest Stieb, *Drug Adulteration* (Madison, Wis., 1966), especially parts 1 and 2.

106. *McG,* 28 May 1842, p. 1.

107. Ibid., p. 2. The anti-smoke contingent could count on the *Guardian's* editor Jeremiah Garnett who was a leading participant.

108. *McG,* 11 June 1842, p. 3; *Brit. Assn. Rpts.,* 1842, p. 107; *McG,* 7 June 1843, p. 3.

109. E. K. Welsch, "Victorian Cities and Pollution Problems," *Univ. of Wisconsin Library News* 5, no. 7 (1970): 1.

110. *McG,* 23 Aug. 1843, p. 5. See also Ann Beck, "Some Aspects of the History of Anti-Pollution Legislation in England, 1819–1854," *Jour. Hist. Med.* 14 (1959): 475–89.

111. *PP,* vol. 7 (1843), "Report of the Select Committee to Inquire into the Smoke Nuisance," p. 14.

112. Ibid., pp. 55, 42, 13–14.

113. Ibid., pp. 105, 177, 178.

114. *Quarterly Review* 71 (1843): 420–21.

115. *PP,* vol. 43 (1846), "Report Upon the Means of Obviating the Evils Arising from Smoke," p. 9.

116. *Quarterly Review* 71 (1843): 418–20, 421.

117. Quoted in D. Roberts, *Victorian Origins of the British Welfare State* (New Haven, Conn., 1960), p. 183.

118. *McG,* 28 May 1845, p. 9; ibid., 31 May 1845, p. 9.

119. *McG,* 13 August 1845, p. 6.

120. *McG,* 9 October 1844, p. 6.

121. *McG,* 6 Nov. 1844, p. 5.

122. Ibid.; 13 Nov. 1844, p. 8.

123. *McG,* 2 Nov. 1844, p. 5.

124. 20 Nov. 1844, p. 5.

Chapter 4

1. Wemyss Reid, p. 65.

2. Playfair to Chadwick, 16 April 1842, Chadwick Papers, UCL.

3. Playfair to Chadwick, 15 Jan. 1844, Chadwick Papers UCL.

4. *Lancet,* 1842–43, pt. 1, p. 130.

5. *Lancet,* 1843–44, pt. 1, pp. 171, 175.

6. *Lancet,* 1845, pt. 1, p. 104.

7. *Brit. Assn. Rpt.,* 1840, p. 73.

8. *Annals of Electricity* 7 (1841): 425, 427.

9. See for example the later statement of C. Miller, "Remarks on the Aid which Pathology and Therapeutics derive from Chemistry," *Lancet,* 1849, no. 2, p. 121.

10. Lyon Playfair, *Second Report of the Health of Towns Commissioners* (London, 1845), p. 347; the quotations in the following discussion are from this source.

11. "*Report Upon Means of Obviating the Evils Arising from Smoke,*" *Parl. Papers* 43 (1846): 3.

12. *MLPS Mem,* 2d ser. 8 (1848): 446.

13. Ibid., p. 448.

14. B. Kershaw, *Modern Methods of Sewage Purification* (London, 1911), p. 4.

15. Lyon Playfair, *Report of the Metropolitan Sanitary Commission* (London, 1848), p. 7.

16. Henry Austin, *Report on the Means of Deodorizing and Utilizing the Sewage of Towns* (London, 1857), pp. 25, 95.

17. *Chem. News* 14 (1866): 34.

18. *Trans. Nat. Assn. Soc. Sci.*, 1858, p. 448.

19. *Brit. Assn. Rpt.*, 1861, p. 127.

20. Ibid., p. 128.

21. *Chem. News* 4 (1861): p. 157.

22. *Chem. News* 4 (1861): pp. 201–202.

23. See the *Dictionary of National Biography, q.v.* "Joseph Lister." Carbolic acid had a long prior history as a disinfectant. See especially M. J. Lemaire, "On the Use of Phenic Acid for Disinfecting Purposes," *Chem. News* 4 (1861): 60, and the report on Lemaire's work in *Chem. News* 8 (1863): 89–90, concerning the action of carbolic acid on "viruses," ferments and miasmatic poisons. See also F. Crace-Calvert, "On Some Applications of Carbolic Acid," *Chem. News* 5 (1862): 19, where the use by Manchester physicians and surgeons is outlined. For a review of practical disinfection, see Dr. Letheby, "On the Practice of Disinfection," *Chem. News* 14 (1866): 267–68.

24. See, for example, his lectures before the French Society for the Encouragement of National Industry, which he addressed at the invitation of Dumas (*Chem. News* 16 (1867): 296, 310, 320).

25. William Thomson became chief assistant to Crace-Calvert in 1869 and partner in 1873 (*MLPS Mem. and Proc.* 68 (1923–24): 149).

26. *McG*, 1 Nov. 1848, p. 7; ibid., 23 Dec. 1848, p. 7.

27. *McG*, 6 Dec. 1848, p. 7; ibid., 9 Dec. 1848, p. 9.

28. *Report of the Select Committee on Adulteration of Food* (London, 1855), p. 131.

29. *Manchester and Salford Sanitary Assn.*, First Report (1853); Second Report (1854).

30. *Chem. News* 9 (1869): 105.

31. *Phil. Mag.* 30 (1847): 478–82.

32. Ibid., p. 478.

33. Ibid., p. 481.

34. *Brit. Assn. Rpt.*, 1848, p. 16.

35. Ibid., pp. 16, 30.

36. *Report of the Commission to Inquire into the Condition of Mines* (London, 1864), Appendix B, p. 165.

37. Report of the Metropolitan Sanitary Commission (London, 1848), p. 34.

38. *Brit. Assn. Report*, 1851, pp. 67, 76, 52.

39. *PP*, vol. 44 (1878), "Report of Comm. on Noxious Vapours," pp. 93–94, 118–19, 101.

40. *Chem. News* 6 (1862): 208; see also Ann Beck, "Some Aspects of the History of Anti-Pollution Legislation in England 1819–1854," *Jour. Hist. Med.* 14 (1959): 484.

41. *PP*, "Rpt. of Comm. on Noxious Vapours," p. iii.

42. Ibid., p. 146.

43. On the work of the committee, see R. M. MacLeod, "The Alkali Acts Administration 1863–1884: the Emergence of Civil Scientist," *Victorian Studies* 9 (1965): 85–112.

44. L. F. Haber, *The Chemical Industries during the Nineteenth Century* (Oxford, 1958), p. 209.

45. F. S. Taylor, *A History of Industrial Chemistry* (London, 1957), pp. 184–185; D. W. F. Hardie and J. D. Pratt, *History of the Modern British Chemical Industry* (Oxford, 1966), p. 28. A good account of the process is given in Haber, pp. 252–54.

46. Taylor, p. 184.

47. Hardie and Pratt, p. 31.

48. *Chem. News* 6 (1862): 202.

49. MacLeod, "Alkali Acts," pp. 90–91.

50. *PP*, vol. 20 (1865), "Report of the Alkali Inspector," p. 61.

51. MacLeod, "Alkali Acts," p. 93.

52. *PP*, vol. 17 (1866), "Second Report . . . ," p. 8.

53. *PP*, vol. 18 (1868), "Fourth Report . . . ," p. 4.

54. *PP*, vol. 16 (1867), "Third Report . . . ," p. 1.

55. Ibid., pp. 48, 52, 53.

56. MacLeod, "Alkali Acts," p. 94; *PP*, vol. 14 (1869), "Fifth Report . . . ," pp. 3, 5, 34.

57. *PP*, vol. 19 (1873), "Ninth Report . . . ," pp. 3–4.

58. *PP*, vol. 16(1872), "Eighth Report . . . ," p. 4.

59. *PP*, vol. 19 (1873), "Ninth Report . . . ," p. 35; ibid., vol. 25 (1874), "Tenth Report . . . ," p. 395.

60. MacLeod, "Alkali Acts," p. 96.

61. *PP*, vol. 16 (1875), "Eleventh Report . . . ," p. 4.

62. Haber, p. 152, and Taylor, pp. 186–187.

63. Taylor, p. 186. See also *PP*, vol. 44 (1878), "Rpt. of Comm. on Noxious Vapours," p. 8.

64. Hardie and Pratt, p. 33.

65. Hurter, who developed the Deacon-Hurter chlorine process, was a pupil of R. W. Bunsen. Hardie and Pratt, p. 32.

66. *PP*, vol. 16 (1876), "Interim Rpt. of the Alkali Inspector," p. 12.

67. O. MacDonagh, "The Nineteenth-Century Revolution in Government: Reappraisal," *Hist. Jour.* 1 (1957): 52–67. MacDonagh's model has been attacked by Henry Parris, "The Nineteenth-Century Revolution in Government, a Reappraisal Reappraised," *Hist. Jour.* 3 (1960): 17–37, and by J. Hart, "Nineteenth-Century Social Reform: a Tory Interpretation of History," *Past and Present,* no. 31 (1965): 39–61. See also V. Cromwell, *Vic. Stud.* 9 (1966): 245–55, and E. Midwinter, *Past and Present,* no. 34 (1966): 130–33, and especially Roy MacLeod, "Social Policy and the Floating Population," *Past and Present,* no. 35 (1966): 101–32.

68. MacDonagh, "Revolution," pp. 58–60, 61.

69. *PP*, vol. 44 (1878), "Rpt. of Comm. on Noxious Vapours," pp. 1–2, 11, 37. For the later history of the legislative results, and especially for Smith's role in drafting the Alkali Act of 1881, see MacLeod, "Alkali Acts," pp. 104–10.

70. Hart, "Social Reform," p. 38.

71. National Association for the Promotion of Social Sciences *Trans.* (1857), p. 518, 520, 524.

72. Justus Liebig, *Familiar Letters on Chemistry,* (London, 1851), p. 20.

73. NAPSS *Trans.* (1857), p. 522.

74. *Chem. News* 40 (1879): 304.

75. NAPSS *Trans.* (1857), pp. 524–25.

76. Smith to Chadwick, 19 Sept. 1861, Chadwick Papers, UCL.

77. R. A. Smith, *Air and Rain* (London, 1872), dedication.

78. L. Playfair, *Record of the School of Mines* 1 (1852): 24, 25, 26, 27, 44, 46.

79. E. Baines, *History of the Cotton Manufacture in Great Britain* (London, 1835), p. 285.

80. Love and Barton, *Manchester as It Is* (Manchester, 1839), pp. 220–21.

81. L. Faucher, *Manchester in 1844* (London, 1844), p. 21, 22–22n7.

82. A. E. Musson and E. Robinson, *Science and Technology in the Industrial Revolution* (Manchester, 1969), chaps. 7–9.

83. Musson and Robinson, pp. 235–39.

84. *MLPS Mem.* 1 (1785): 26–27; *MLPS Mem.* 3 (1790): 343–408.

85. Musson and Robinson, p. 243.

86. *MLPS Mem.* 5 pt. 1 (1798): 298–313.

87. Quoted in Musson and Robinson, p. 315. On the introduction of chlorine bleaching see ibid., chap. 8.

88. See Musson and Robinson, chapter 9, for the chemical developments in dyeing in the Manchester region.

89. A. Clow, "Industrial Background" in D. C. Cardwell, *John Dalton and the Progress of Science* (Manchester, 1968), p. 134.

90. A. Clow and N. Clow. *The Chemical Revolution* (London, 1952), p. 253; *City of Manchester Gas Department, 143 Years of Gas in Manchester* (Manchester, 1949), p. 49.

91. Clow, "Industrial Background," p. 137. On Gossage see J. Fenwick Allen, *Some Founders of the Chemical Industry,* 2d ed. (London, 1907), pp. 1–36.

92. *Rpt. Select Committee on Scientific Instruction* (London, 1868), pp. 297, 300.

93. *PP,* vol. 20 (1865), "Report of the Alkali Inspector," p. 26.

94. Clow "Industrial Background," p. 136.

95. *Jour. Soc. Chem. Ind.* 8 (1889): 528.

96. *MLPS Mem.,* 4th ser 3 (1890): 293.

97. Ibid., pp. 295–96.

98. On Martius, see *Ber. der Deutsch. Chem. Gesell.* 53 (1920): 72–75; on Caro see *Jour. Soc. Chem. Ind.* 29 (1910): 1143, and S. Miall, *History of the British Chemical Industry* (London, 1931), p. 80; E. F. Ehrhardt, "Reminiscences of Dr. Caro," *Jour. Soc. Chem.* 43 (1924): 561–65, which includes Caro's views of Britain's chemical decline; John Beer, "Heinrich Caro" in the *Dictionary of Scientific Biography;* and especially E. Bernsthen, *Ber. der Deutsch. Chem. Gesell.* 45 (1912): 1987–2042.

99. Bernsthen, p. 1991.

100. *Jour. Soc. Chem. Ind.* 8 (1889): 529.

101. See below, chapter 5, for Ivan Levinstein's part and the subsequent history of artificial dyestuffs in Manchester.

102. *MLPS Proc.* 4 (1865): 149–52.

103. *Jour. Soc. Chem. Ind.* 8 (1889): 530.

104. Hart wrote, in 1860, the article on that subject for Sheridan Muspratt's *Dictionary of Applied Chemistry.*

105. Haber, p. 160.

106. *McG,* 6 July 1883, p. 5.

107. J. F. Allen, *Some Founders,* pp. 26–47.

108. T. E. Thorpe, *Dictionary of Applied Chemistry,* 3 vols. (London, 1890–93), 1: 80.

109. Ibid., p. 80, and *Chem. News* 12 (1865): 222, 235.

110. Thorpe, p. 80.

111. *McG,* 24 Aug. 1857, pp. 3.

112. Ibid., p. 4.

113. *McG,* 25 Aug. 1857, p. 4.

114. See the obituary in *McG,* 6 July 1883, p. 5; see also the obituary in *Jour. Soc. Chem. Ind.* 2 (1883): 321.

115. Allen, *Some Founders,* 274, 275; see also *McG,* 9 Dec. 1854, p. 7.

116. Allen, *Some Founders,* p. 276.

117. *Chem. News* 14 (1866): 207.

118. *McG,* 23 Feb. 1853, p. 7; *McG,* 8 Nov. 1854, p. 9.

119. *McG,* 24 Dec. 1850, p. 8.

120. *Brit. Assn. Rpt.,* 1854, p. 65.

121. F. Crace-Calvert, *Dyeing and Calico Printing,* 2d ed. (London, 1876), p. 450.

122. *Jour. (Roy.) Soc. of Arts* 15 (1867): 730; *Chem. News* 1 (1862): 20.

123. *Jour. Soc. Arts* 15 (1867): 730.

124. Ibid., pp. 732–33.

125. *Chem. News* 3 (1861): 126.

126. Poggendorf's *Handworterbuch* determines the degree as a Giessen doctorate.

127. E. E. Fournier-D'Albe, *Life of Sir William Crookes* (London, 1923), pp. 86–87.

128. Ibid., pp. 88–90.

129. *Brit. Assn. Rpt.,* 1887, p. 624.

130. *MLPS Proc.* 41 (1897): xlvii.

131. Tylecote, pp. 132–33.

132. Ibid., p. 160.

133. *Sci. Misc.,* no. 1 (1840): 1; ibid., no. 6 (1840): 48.

134. *McG,* 7 April 1852, p. 5.

135. *Sci. Misc.,* no. 4 (1840): 28–29.

136. *McG,* 31 Jan. 1852, p. 7.

137. Harold Silver, *The Concept of Popular Education* (London, 1965), pp. 217–20. See also the chapter, "How Useful Is Thy Dwelling Place," in C. C. Gillispie, *Genesis and Geology* (New York, 1959).

138. P. W. Musgrave, "The Definition of Technical Education (1860–1910)," Vocational Aspect, May 1964, pp. 105–11; idem, "Constant Factors in the Demand for Technical Education," *Brit. Jour. Educ. Stud.* 14 (1966): 173–87; and idem, *Technical Change, the Labour Force and Education* (Oxford, 1967), chaps. 2–4.

139. K. B. Smellie, *A History of Local Government,* 4th ed. (London, 1969), p. 166.

140. *McG,* 29 Aug. 1867, p. 5.

141. *Report of the Select Committee on Scientific Instruction* (London, 1868), p. 268.

Chapter 5

1. *McG,* 16 Jan. 1836, p. 3.

2. Ibid.

3. Quoted in E. Fiddes, *Chapters in the History of Owens College and of Manchester University 1851–1914* (Manchester, 1937), p. 21.

4. Ibid., pp. 21–24; Joseph Thompson, *The Owens College, its Foundation and Growth* (Manchester, 1886), pp. 20–22; Fiddes, p. 23.

5. J. Thompson, pp. 25–26.

6. *McG,* 16 Nov. 1836, p. 3.

7. J. Thompson, p. 19.

8. Ibid., p. 29.

9. *McG,* 4 Nov. 1837, p. 3.

10. Owens' will is quoted in B. W. Clapp, *John Owens, Manchester Merchant* (Manchester, 1965), p. 173.

11. J. Thompson, p. 51.

12. J. Thompson, chapter 3, has an extended account of the first trustees.

13. MTMEP, 30 January 1949, vol. 3, pp. 9–10, MCRL.

14. MTMEP, 3: 20–21, MCRL.

15. *Sketches from the Life of Edward Frankland* (London, 1902), p. 120.

16. On Stenhouse, see *Jour. Chem. Soc.* 39 (1881): 185–88; on Penny see *Jour. Chem. Soc.* 23 (1870): 301–06.

17. Quoted in *Life of Frankland,* pp. 122–23; see also *McG,* 4 Jan. 1851, p. 6.

18. *McG,* 4 Jan. 1851, p. 6.

19. *Life of Frankland,* p. 123.

20. MTMEP, 3: 21, 23–25, MCRL.

21. *McG,* 18 Jan. 1851, p. 6; W. C. Williamson, *Reminiscences of a Yorkshire Naturalist* (London, 1896), chaps. 4 and 8.

22. *McG,* 15 Mar. 1851, p. 8.

23. Ibid.

24. The list included a J. A. Smith, perhaps a misprint for R. A. Smith, *McG,* 22 March 1851, p. 8.

25. *McG,* 22 March 1851, p. 8; see also *Introductory Lectures on the Opening of Owens College* (Manchester, 1852), p. 96.

26. *McG,* 22 March 1851, p. 8.

27. Ibid., and *Introductory Lectures,* p. 99.

28. Ibid., pp. 99–100.

29. *McG,* 22 March 1851, p. 8.

30. *McG,* 13 August 1851, p. 5.

31. John Burnett, *A History of the Cost of Living* (Harmondsworth, Middlesex, 1969), pp. 202–03.

32. John Burnett, *Plenty and Want* (London, 1966), p. 92, and *Cost of Living*, p. 253. See also F. Musgrove, "Middle Class Education and Employment in the 19th Century," *Economic History Review*, 2d ser. 12 (1959): 99, who takes the range £200 to £1000 as the middle middle-class range after midcentury.

33. *Life of Frankland*, p. 135.

34. *McG*, 7 July 1852, p. 9.

35. *Illustrated London News* 8 (1846): 348.

36. *McG*, 7 July 1852, p. 9.

37. *McG*, 29 Jan. 1853, p. 6.

38. *McG*, 2 Feb. 1853, p. 7.

39. J. Thompson, pp. 146–48.

40. *McG*, 2 July 1853, p. 6.

41. MTMEP, 6 Nov. 1854, p. 41, MCRL.

42. *McG*, 30 June 1855, p. 7.

43. Owens College Examination Papers, July 1851, University of Manchester Library; ibid., July 1852.

44. Owens College, "Annual Report," 1 July 1859, University of Manchester Library.

45. *Proc. Roy. Soc.* 84 (1910): xxx; *Nature* 82 (1909): 101; and *Dictionary of National Biography*.

46. *Jour. Soc. Chem. Ind.* 34 (1915): 1230. It may have been Gerland who introduced Henry Roscoe to the subject of vanadium chemistry.

47. *Nature*, 4 Nov. 1886, pp. 8–10.

48. *McG*, 24 Dec. 1851, p. 6. See also Edward Frankland, *Experimental Researches in Pure, Applied and Physical Chemistry* (London, 1877), pp. 480–536; *MLPS Mem.*, 2d ser. 10 (1852): 71.

49. *McG*, 21 Oct. 1854, p. 8.

50. MS M33/1/1/1., MCRL.

51. Frankland to Bunsen, 3 March 1856, Bunsen Papers, Heidelberg University.

52. *McG*, 1 July 1854, p. 6.

53. *McG*, 5 July 1854, p. 6.

54. *McG*, 25 Oct. 1854, p. 7.

55. MTMEP, 3: 40–41, MCRL.

56. Owens College, Miscellaneous Materials, "Principal's Report," 20 May 1856, MCRL.

57. *McG*, 30 June 1855, p. 7.

58. *McG*, 3 July 1856, p. 3.

59. Owens College, Miscellaneous Materials, MCRL.

60. MTMEP, 3:53, 55.

61. *McG*, 9 July 1858, p. 2; ibid., 12 July 1858, p. 4; ibid., 14 July 1858, p. 3; 15 July 1858, p. 2; ibid., 22 July, p. 2.

62. MTMEP, 11 Sept. 1857, p. 85, MCRL.

63. *McG*, 1 Sept. 1857, p. 4.

64. MTMEP, 11 Sept. 1857, p. 85, MCRL.

65. For a brief sketch of the life of Roscoe, see R. Kargon, "Henry Enfield Roscoe," *Dictionary of Scientific Biography.*

66. *McG,* 15 Oct. 1857, p. 3.

67. See, for example, Fiddes, p. 48; H. B. Charlton, *Portrait of a University* (Manchester, 1951), pp. 54 ff; Thompson, p. 212.

68. Owens College, Misc. Mat., MCRL; *McG,* 5 July 1858, p. 3.

69. Owens College, Misc. Mat., MCRL; *McG,* 29 June 1861, p. 5; ibid., 4 July 1863, p. 4.

70. Fiddes, p. 57; J. K. Wright, "Owens in the Sixties," *Record of the Owens College Jubilee* (Manchester, 1902), pp. 46–49. See also Henry Brierly, "Further Reminiscences of Owens College Forty Years Ago," *Owens College Jubilee,* p. 50, and the early volumes of the *Old Owensian Journal.*

71. Charlton, p. 54.

72. J. R. T. Hughes, "Problems of Industrial Change," in *1859: Entering an Age of Crisis,* ed. P. Appleman, W. Madden, and M. Wolff (Bloomington, Ind., 1959), pp. 131–32.

73. H. Hale Bellot, *University College London 1826–1906* (London, 1929), pp. 249–303.

74. Fiddes, pp. 59–60.

75. John Roach, *Public Examinations in England 1850–1900* (Cambridge, 1971), p. 26.

76. J. Thompson, p. 211; Roach, pp. 89–90.

77. University of London, *Report of the Committee [of the Senate] Appointed to Consider the Propriety of Establishing a Degree or Degrees in Science* (London, 1858), p. iii. Signatories of the memoriam included Hofmann, Tyndall, Graham, Lyell, Huxley, Wheatstone, Williamson and Grove—a stellar array of British scientists representing a number of disciplines.

78. D. S. L. Cardwell, *Organisation of Science in England* (London, 1972), pp. 92–93.

79. University of London, *Rpt. of Comm. [on] Degrees in Science,* p. v.

80. Cardwell, p. 94, *University College Calendar 1859–1860,* p. 363.

81. *Lancet,* no. 4 (Jan. 1859): 564; ibid., no. 29 (Oct. 1859): 442.

82. *All the Year Round,* July 16, 1859, p. 283.

83. These figures are obtainable from the *University College Calendar 1860–61 through 1871–72,* and University College London, *Proceedings at the Annual General Meeting of the Members of the College. Report of the Council and Financial Statements, 1858–1865,* UCL.

84. *U. C. Calendar,* 1870–71, p. ix, UCL.

85. *Lancet,* no. 24 (Sept. 1859): 303–04.

86. *U. C. Calendar,* 1861–62, pp. 363, 379; *U. C. Calendars,* 1864–1872.

87. The University College Calendars contain descriptions of these examinations; see also Geoffrey Millerson, *The Qualifying Examinations* (London, 1964), pp. 12–24.

88. J. G. Fitch, "Examination Schemes and their Incidental Effects on Public Education," NAPSS *Trans.,* 1858, p. 220.

89. John Roach, pp. 195, 199, 208.

90. *PP,* vol. 24 (1860), *"Fifth Report of the Civil Service Commissioners,"*

p. 310. See also *PP*, vol. 55 (1876), *"Selection and Training of Candidates of the Indian Civil Service,"* p. 277.

91. *PP*, vol. 30 (1856), "First Report . . . Commissioners," pp. xviii; *Fifth RCSC*, xxiv (1860), p. 310.

92. MTMEP, 3:84, MCRL.

93. *McG*, 2 Oct. 1860, p. 3; MTMEP, 3:99, MCRL; Owens College, Miscellaneous Materials, MCRL.

94. MTMEP, 3:85, MCRL.

95. Ibid., 3:89–90.

96. Ibid., 3:96.

97. *Proc. Roy. Soc.* 99 (1921): viii–ix.

98. *McG*, 29 June 1861, p. 5; ibid., 25 June 1864, p. 7.

99. H. E. Roscoe, "Bunsen Memorial Lecture," *Memorial Lectures Delivered Before the Chemical Society 1893–1900* (London, 1901), p. 545; idem, "Robert Wilhelm Bunsen," *Nature* 23 (1881): 597.

100. Roscoe Papers, University of Manchester; Michael Sanderson, *The Universities and British Industry 1850–1970* (London, 1972), p. 82–84; *The Health Journal* 4 (1886–87): 66.

101. Thompson, pp. 229–36.

102. *Brit. Assn. Rpt.*, 1861, pp. 108–28.

103. H. E. Roscoe, *Life and Experiences of Sir Henry Enfield Roscoe* (London, 1906), pp. 124–26; D. Thompson, "Henry Enfield Roscoe and Technical Education" (M.A. Thesis, University of Manchester), p. 84.

104. Roscoe, *Life*, p. 126; D. Thompson, "Roscoe," pp. 91–98; *Science Lectures of the People Delivered in Manchester 1866–67 and 1870–71.* (Manchester, 1871), preface to the first series and preface to the second series.

105. Fairbairn to Roscoe, 22 May 1861, Chemical Society Archives, London.

106. Roscoe, *Life*, p. 103. See also H. E. Roscoe, *Record of Work Done in the Chemical Department of the Owens College 1857–1887* (Manchester, 1887), p. 2.

107. *Report of the Select Committee on Scientific Instruction* (London, 1868), p. 277.

108. Ibid., p. 283; *Report of the Royal Commission on Scientific Instruction and the Advancement of Science* (London, 1872), 1: 497–99.

109. Roscoe, *Record*, p. 3.

110. *Rpt. Sel. Comm. on Sci. Inst.* (1868), p. 278.

111. Ibid., p. 280.

112. *Rpt. Roy. Comm. Sci. Ins. and Adv. of Sci.* (1872), p. 499.

113. Ibid., p. 499; *McG*, 24 June 1865, p. 6.

114. Roscoe, *Record*, p. 5.

115. Ibid., p. 7.

116. Ibid., p. 12.

117. Ibid., pp. 23–26. See also H. E. Roscoe, "Science Education in Germany," *Nature* 1 (1869): 159, where he writes of Bunsen in these terms.

118. *Brit. Assn. Rpt.*, 1884, p. 668.

119. *Rpt. Roy. Comm. Sci. Ins. and Adv. of Sci.* (1872), 1: 468.

120. Roscoe, *Record*, pp. 24–25.

121. *Rpt. Sel. Comm. on Sci. Ins.* (1868), p. 284.

122. Ibid., p. 289.

123. *Economist,* 8 Feb. 1868, p. 153.

124. On the redefinition of technical education after 1860, see "The Definition of Technical Education, 1860-1910," *Vocational Aspect,* May, 1964, pp. 105-11, and P. W. Musgrave, *Technical Change, the Labour Force and Education* (Oxford, 1967), chaps. 2 through 4, *passim.*

125. See, for example, Henry Brierly, "Owens College Forty Years Ago," and "Further Reminiscences," *O.C. Jubilee,* p. 50.

126. S. Wilkinson, "Some College Friendships," *O. C. Jubilee,* p. 51. Both Swanwick and Poynting later attended Trinity College, Cambridge, and returned to teach at Owens College.

127. *Proc. Roy. Soc.* 99 (1921): vi.

128. *Nature* 113 (1924): 540-41.

129. See Musson and Robinson, chapters 3, 12, and 15, *passim,* and pp. 439, 455, 475.

130. R. Lewis and A. Maude, *Professional People* (London, 1952), p. 28; Millerson, *Qualifying Associations,* pp. 68, 129.

131. *Manchester Association of Engineers 1856-1956* (Manchester, 1956), sections 3 and 4; Thomas Ashbury, *The Jubilee of the Manchester Association of Engineers* (Manchester, 1905), pp. 8-12; A. C. Doan, *Some Episodes in the Manchester Association of Engineers* (Manchester, 1938), p. 16.

132. Manchester Institution of Engineers, *Proceedings,* pt. 1 (1867): 1; *Manchester Scientific and Mechanical Society Reports* (1873), prospectus.

133. MTMEP, 3: 118, 114, MCRL.

134. J. Thompson, p. 295.

135. *McG.* 29 Nov. 1867, p. 1; Thompson, p. 295.

136. Ibid., p. 296.

137. See R. H. Kargon, "Osborne Reynolds," in the *Dictionary of Scientific Biography;* Jack Allen, "The Life and Work of Osborne Reynolds," *Osborne Reynolds and Engineering Science Today,* ed. J. D. Jackson and D. M. McDowell (Manchester, 1970), pp. 1-82.

138. Quoted in Allen, "Life and Work," pp. 2-3.

139. Ibid., pp. 15-16.

140. Lyon Playfair, "The Progress of Applied Science and Its Effect Upon Trade," *Contemporary Review* 53 (1888): 368.

141. J. J. Thomson, *Recollections and Reflections* (London, 1936), p. 15.

142. MTMEP, 4: 69, MCRL.

143. *MLPS Mem.* 4 (1871): 279-86; *Brit. Assn. Rpts.,* 1870, pp. 222-24. The *Royal Society Catalogue of Scientific Papers* contains a fuller list of Reynolds' published work.

144. *Man. Sci. and Mech. Soc. Rpts.* 1871, pp. 4, 5.

145. *Man. Sci. and Mech. Soc. Rpts.,* 1874, p. 1.

146. *Man. Sci. and Mech. Soc. Rpts.,* 1877, p. 4.

147. Ashbury, *Jubilee of the Manchester Association of Engineers,* p. 19.

148. *MLPS Proc.* 11 (1872): 171; ibid., 13 (1874): 165.

149. *McG,* 25 June 1870, p. 5; ibid., 22 June 1872, p. 7.

150. Allen, "Life and Works," p. 6.

151. *Second Report of the Royal Commissioners on Technical Instruction* (1884), 1: 439, 440.

152. Ibid.

153. MTMEP, 4: 88, MCRL.

154. *Brit. Assn. Rpt.,* 1887, p. 859.

155. Ibid., p. 860.

156. Ibid.; J. Thompson, pp. 561–63.

157. MTMEP, 4 April 1884, 10: 107, MCRL.

158. J. Thompson, pp. 566–67.

159. On Reynolds' work and its implications, see Allen, "Life and Work," pp. 67–80.

160. Thompson, p. 245.

161. MTMEP, 4: 16, 17, MCRL; and Thompson, pp. 247, 248.

162. Owens College, *Substance of a Report by the Principal and Professors to a Committee of the Trustees Appointed to Consider the Subject of Obtaining New College Buildings and Connected Subjects* (Manchester, 1865).

163. J. Thompson, pp. 312–13.

164. *McG,* 2 Feb. 1867, p. 1.

165. Ibid., p. 4.

166. Ibid., p. 5.

167. *McG,* 5 Feb. 1867, p. 5.

168. *Spectator,* Feb. 1867, quoted in *McG,* 11 Feb. 1867, p. 4.

169. *McG,* 22 Jan. 1898, p. 7; *MLPS Proc.* 42 (1897–1898): xliv–xlvi.

170. Roscoe, *Life,* p. 111.

171. J. Thompson, pp. 341, 365; see also ibid., chapter 12, *passim,* for the abortive efforts to gain government aid.

172. The story of the Extension Act is told in J. Thompson, chapter 14.

173. On Gladstone see *MLPS Proc.* 15 (1875–76): 165–66. Gladstone (1816–1875) was an engineer and astronomer who was an active member of the Lit. & Phil. and a strong supporter of the Owens College.

174. A list of the members of these bodies is given in Thompson, pp. 622–25. See also Fiddes, pp. 65–69, and A. Angellier, "Étude sur Owens College," *Bull. de la Société pour l'étude des questions d'enseignement supérieur* (1880): 367–480.

175. In 1856 the trustees had rejected a proposed merger, MTMEP, 3: 45–47.

176. MTMEP, 4: 141, MCRL.

177. J. Thompson, chap. 17, *passim;* E. M. Brockbank, *The Foundation of Provincial Medical Education in England* (Manchester, 1936), pp. 104–5.

178. Brockbank, pp. 108–10.

179. MTMEP, 7: 48, MCRL.

180. Williamson, *Reminiscences,* p. 142.

181. MTMEP, 6:123–25 and 10:56, MCRL; Williamson, *Reminiscences,* p. 153.

182. MTMEP, 6, p. 19; see also *Proc. Edinburgh Roy. Soc.* 46 (1925–26): 352–53. For the growth of the Owens College physics department, see below, chapter 6.

183. MTMEP, 6, p. 102.

184. On Dittmar, who became a professor at the Anderson's College, Glasgow, see *Jour. Soc. Chem. Ind.* 11 (1892): 116–47; A. Crum Brown, *Nature* 14 (1892): 493.

185. *Proc. Roy. Soc.* 52 (1893); vii; [A. Schuster], memorial notice, *McG,* 28 June 1892, p. 6. Schuster's authorship is noted by F. Engels and H. Roscoe, 30 June 1892, in Marx-Engels *Werke,* b. 38 (Berlin, 1968), pp. 379–80. See also J. B. Cohen, "The Development of Organic Chemistry in Great Britain," *Old Owensian Journal* 8 (1930): 46.

186. *Jour. Soc. Chem. Ind.* 11 (1892): 594.

187. Engels wrote a eulogy for *Vorwarts,* no. 153 (3 July 1892): 1.

188. See the letter from Engels to Roscoe, 28 May 1892, in Marx-Engels, *Werke,* b. 38, p. 353.

189. MTMEP, 10:118, MCRL.

190. *Minutes Sel. Comm. on Sci. Instr.,* 1868, p. 278.

191. *Brit. Assn. Rpt.,* 1842, p. 39; Walter Gardner, *The British Coal Tar Industry* (London, 1915), pp. 110–11.

192. F. A. Mason, "The Influence of Research on the Development of the Coal-Tar Dye Industry," *Science Progress* 1 (1915): 239.

193. Mason, p. 241. Twelve gallons of gas tar would yield 1.1 pounds of aniline (Gardner, p. 114).

194. Mason, p. 242.

195. Haber, pp. 82–83.

196. Mason, p. 247.

197. Ibid., p. 245.

198. Ibid., p. 244.

199. See Haber, p. 83, and J. J. Beer, "Coal Tar Dye and the Origins of the Modern Industrial Research Laboratory," *Isis* 49 (1958): 126–27.

200. Mason, p. 249.

201. Mason, pp. 249–50; R. Rose, "Growth of the Dyestuffs Industry," *Jour. Chem. Educ.* 3 (1926): 997.

202. O. Witt, "Wechselwirkungen zwischen der chemischen Forschung und der chemischen Technik," in *Die Kultur der Gegenwart,* Teil 3, abt. 3, Bd. 2 (Berlin, 1913): 520–21.

203. Mason, p. 252.

204. *Royal Commission on Scientific Instruction and the Advancement of Science,* 1872, 1: 371.

205. Quoted in I. Levinstein in *Jour. Soc. Chem. Ind.* 5 (1886): 351.

206. J. J. Beer, "The Emergence of the German Dye Industry," *Illinois Studies in the Social Sciences* 44 (1959): 64–66.

207. *Brit. Assn. Rpt.,* 1901, p. 587.

208. The German-British trade rivalry in the 1870s and the 1880s, and the debate surrounding it, has been explored elsewhere. See R. J. S. Hoffman's *Great Britain and the German Trade Rivalry 1875–1914.* (Philadelphia, 1933), pp. 28–101; D. H. Aldcroft, ed. *The Development of British Industry and Foreign Competition (1875–1914)* (London, 1968), pp. 18–21.

209. *PP,* vol. 23 (1867), appendix, pp 438, 437.

210. Ibid., p. 443.

211. Ibid., pp. 445–46.

212. Ibid., p. 463.

213. *Edinburgh Review* 127 (1868): 435.

214. Gardner, p. 57; David Landes, *The Unbound Prometheus* (Cambridge, 1969), p. 275.

215. *McG,* 28 Aug. 1867, p. 4.

216. *McG,* 29 Aug. 1867, p. 4.

217. *Nature* 1 (1869): 239.

218. *Chem. News* 25 (1872): 309. See also W. V. Farrar, "The Society for the Promotion of Scientific Industry 1872–1876," *Ann. Sci.* 29 (1974): pp. 81–86, upon which I have relied for some of the following.

219. *McG,* 16 Oct. 1872, p. 3.

220. *Engineer* 37 (1874): 64–65.

221. *Jour. Soc. Prom. Sci. Ind.* 1 (1874): 19, 21.

222. Ibid., p. 21.

223. Ibid., p. 191.

224. Ibid., p. 185; *Jour. Soc. Prom. Sci. Ind.* 2 (1875): 1; ibid., 1 (1874–75): 193.

225. Despite apparent opposition from firm owners, the chemists met regularly, first at St. Helens and Widnes and later at Liverpool (*Jour. Soc. Chem. Ind.* 34 [1915]: 749). Its moving force was George E. Davis (1850–1907), a chemist at the Gerard's Bridge Chemical Works and later at the Runcorn Soap and Alkali Company. In 1880 he settled in Manchester as a consulting and analytical chemist (*Jour. Soc. Chem. Ind.* 26 [1907]: 598). Davis recalled that "in 1879 some members who had refused to join us at the start wanted to come in and as they were not allowed endeavoured to form a society of their own in Widnes," under the name South Lancashire Chemical Society (*Jour. Soc. Chem. Ind.* 34 [1915]: 749). John Hargreaves, an early member, recalls the situation differently: after attending a meeting of the Tyne Society he attempted to form a similar group in Widnes (*Jour. Soc. Chem. Ind. [Review]* 4 [1921]: 86R–87R).

226. *Jour. Soc. Chem. Ind., Jubilee Number,* 1931, p. 10.

227. Ibid., p. 19.

228. *Jour. Soc. Chem. Ind.* 1 (1882): 250, 252.

229. *Jour. Soc. Chem. Ind.* 2 (1883): 213; see also Society of Chemical Industry (Manchester Section), Minutes of General Meetings, 3 vols. (1884–1904), unpaginated, Manchester Central Reference Library (hereafter cited as S.C.I. (Manchester) Minutes.

230. S.C.I. (Manchester) Minutes, October 21, 1884.

231. On Levinstein see *Jour. Soc. Chem. Ind.* 35 (1916): 458, and *McG,* 16 Mar. 1916, p. 7. See also W. J. Reader, *Imperial Chemical Industries, a History* (London, 1970), 1: 111, 261–64. Reader describes Levinstein as a "German Jew who . . . timelessly deplored the state of British scientific education while underpaying his own chemists" (p. 111), and his business dealings as "semi-piratical" (p. 261).

232. *Jour. Soc. Chem. Ind.* 16 (1897): 600. See also *Jour. Soc. Chem. Ind.* 21 (1902): 893 ff, and ibid., 22 (1903): 843 ff.

233. *McG,* 16 Mar. 1916, p. 7.

234. *Jour. Soc. Chem. Ind.* 16 (1897): 600.

235. *McG,* 16 Mar. 1916, p. 7.

236. Harry Grimshaw, S.C.I. (Manchester) Minutes, 3 Dec. 1869.

237. S.C.I. (Manchester) Minutes, 24 Feb. 1885; *Jour. Soc. Chem. Ind.* 11 (1892): 876.

238. *Jour. Soc. Chem. Ind.* 5 (1886): 157, 158.

239. S.C.I. (Manchester) Minutes, 3 Dec. 1889.

240. S.C.I. (Manchester) Minutes, 8 April 1892; ibid., 3 March 1893; ibid., 19 June 1891.

241. *Jour. Soc. Chem. Ind.* 10 (1891): 895.

242. S.C.I. (Manchester) Minutes, 19 June 1891.

243. Ibid.; Minutes, 7 Dec. 1894. On Dreyfus, see *Jour. Soc. Dyers and Colourists* 52 (1936): 140–41.

244. *McG,* 12 June 1886, p. 9; *Jour. Soc. Chem. Ind.* 5 (1886): 637.

245. Royal Jubilee Exhibition, *Official Catalogue* (Manchester, 1887), p. 20.

246. *Jour. Soc. Chem. Ind.* 6 (1887): 624, 626–27, 628–43.

247. *McG,* 30 Nov. 1887, p. 8; J. H. Reynolds, "The Origin and Development of Technical Education in the City of Manchester," *Old Owensian Journal* 2 (1924): 8–12; *Nature* 37 (1887): 111.

248. Fiddes, *Chapters,* p. 161; *Jour. Soc. Chem. Ind.* 5 (1886): 637.

249. *Jour. Soc. Chem. Ind.* 11 (1892): 875; ibid., 8 (1889), p. 963.

250. Chamber of Commerce, Chemical and Allied Trades Committee, Minutes of the Meetings 1890–1900, pp. 1–4, 64–65, 97–102, MCRL.

251. *Calendar of the Owens College, 1871–72,* p. 50; ibid., *1875–1876,* p. 50; Victoria University of Manchester, *Register of Graduates up to July 1, 1908,* 3d ed. (Manchester, 1908), pp. 350–51.

252. J. B. Cohen, "The Development of Organic Chemistry in Great Britain," *Old Owensian Journal* 8 (1930): 46.

253. On Smith, see P. P. Bedson's Obituary Notice, *Jour. Chem. Soc.* 117 (1920): 1637–38.

254. *Jour. Soc. Chem. Ind.* 4 (1885): 87–88.

255. Ibid., p. 89.

256. H. B. Dixon, "George Herbert Bailey," *Jour. Chem. Soc.* 125 (1924): 2677–79.

257. H. S. Roper, "Julius Behrend Cohen," *Jour. Chem. Soc.,* 1935, p. 1332; Cohen, "Development," p. 46.

258. I. Smedley-Maclean, "Arthur Harden," *Biochemical Journal* 35 (1941): 1071. See also A. Ihde, "Arthur Harden," in the *Dictionary of Scientific Biography;* F. G. Hopkins and C. J. Martin, "Arthur Harden," *Obit. Not. Fell. Roy. Soc.* 4 (1942): 3–14.

259. *Jour. Soc. Chem. Ind.* 14 (1895): 525, 554, and *Nature* 52 (1895): 63; P. J. Hartog, *The Owens College Manchester* (Manchester, 1900), pp. 67–68.

260. *McG,* 8 Oct. 1873, pp. 6, 7.

261. Ibid., pp. 6, 5.

262. See also the Duke of Devonshire's address at the opening ceremonies in Owens College, *Essays and Addresses by Professors and Lecturers* (London, 1874), pp. ix–x.

Chapter 6

1. *Report of the Royal Commission on Science Instruction* (1872), 1: 476, 479–80.

2. Ibid., p. 508.

3. Stewart to Roscoe, 20 May 1870, British Chemical Society Archives, Burlington House, London.

4. On Stewart's life, see P. J. Hartog, "Balfour Stewart," in the *Dictionary of National Biography;* Arthur Schuster, "Memoir of the Late Balfour Stewart, LL.D., F.R.S.," *MLPS Mem.*, 4th ser. 1 (1888): 253–72; unsigned obituary (also by Schuster) in *McG*, 20 December (1887), p. 5.

5. Stewart to Roscoe, 2 June 1870, B.C.S. Archives.

6. Ibid.

7. *Manchester Faces and Places* 4 (1893–94): 242; *McG*, 11 July 1910, p. 6; Hartog, "Stewart," p. 1162.

8. *Calendar of Owens College, 1869–1870*, pp. 37–38, 67.

9. *Calendar of Owens College, 1872–1873*, pp. 37–39, 74.

10. Ibid., pp. 38, 64–65.

11. *The Physical Laboratories of the University of Manchester* (Manchester, 1906), p. 1.

12. *Rpt. Roy. Comm. on Sci. Inst.*, (1872) 1: 511.

13. *Brit. Assn. Rpt.*, 1875, p. 8.

14. Ibid., p. 9.

15. MTMEP, 7: 47; J. J. Thomson, *Recollections and Reflections* (London, 1936), p. 19; MTMEP, 6: 104.

16. Arthur Schuster, *Biographical Fragments* (London, 1932), pp. 206–07.

17. J. J. Thomson, *Recollections*, pp. 19–20.

18. MTMEP, 25 Nov. 1868, 6: 66; ibid., 13 Nov. 1875, 6: 121; ibid., 11 Mar. 1876, 6: 121–23, MCRL.

19. J. Thompson, p. 517; MTMEP, 11 Mar. 1876, 6: 123, MCRL.

20. Reprinted in the *McG*, 29 June 1876.

21. A list is provided by J. Thompson, pp. 525–28.

22. *Fortnightly Review*, n.s. 219 (1877): 166, 171. There is an effective rejoinder by E. Freeman in *MacMillan's* 35 (1877): 407–16; *Sat. Rev.* 42 (1876): 203; quoted in *Sat. Rev.* 44 (1877): 636–37.

23. On the entire controversy see E. A. Fiddes, "The University Movement," in *Historical Essays in Honour of James Tait*, ed. J. G. Edwards *et al.* (Manchester, 1933), pp. 97–109.

24. *Nature* 14 (1876): 225, 226.

25. James Heywood, "The Owens College, Manchester and a Northern University," *Jour. Stat. Soc.* 41 (1876): 536–47.

26. See E. Fiddes, *Chapters in the History of Owens College and of Manchester University 1851–1914* (Manchester, 1937), pp. 69–106; *Nature* 20 (1879): 22; ibid., 22 (1880): 261.

27. Oliver Lodge, *Autobiography* (London, 1931), pp. 151–52.

28. *Proc. Roy. Soc.* 31 (1881): 337–47.

29. *MLPS Mem. and Proc.* 52 (1927): v–vi.

30. A. E. H. Love, "Sir Horace Lamb," *Obit. Notices of Fellows of the Roy. Soc.* 1 (1932–35): 375–91.

31. *Calendar of the Owens College, 1880–1881,* pp. 65–69.

32. *Nature* 25 (1882): 400–401.

33. *Physical Laboratories,* pp. 65, 69. *Jour. Chem. Soc.* 41 (1882): 283–87; *MLPS Mem.* 7 (1882): 80–83.

34. *Physical Laboratories,* p. 1.

35. Arthur Schuster, *University Teaching in Its Relation to the Industrial Applications of Science* (Manchester, 1889), pp. 7, 8; *McG,* 2 Oct. 1889, p. 7.

36. *Calendar of the Owens College, 1887–1888,* pp. 64–65; ibid., *1888–1889,* pp. 64–65; *Victoria University Calendar 1889–1890,* pp. 68–71.

37. William Wilson, "Charles Herbert Lees," *Obit. Not. Fell. Roy. Soc.* 8 (1952–53): 523–28. Lees was elected F.R.S. in 1906, the year he left Manchester for what became Queen Mary College, London.

38. Lees' research interests were in the areas of electrolysis and the measurement of thermal conductivities.

39. *Physical Laboratories,* p. 22.

40. *Nature* 66 (1902): 617. M. Sanderson, *The Universities and British Industry* (London, 1972), pp. 92–93. Chaim Weizman, *Trial ad Error* (New York, 1949), pp. 116–17; Arthur Schuster, *Jubilee Celebration* [Manchester, 1906], p. 13.

41. *Brit. Assn. Rpt.,* 1892, p. 628.

42. Ibid.

43. Wilde, an inventor of the dynamo, was an important Manchester benefactor. See A. Schuster, *Biographical Fragments* (London, 1932), pp. 248–57; W. W. Haldane Gee, "Henry Wilde," *MLPS Mem.* 63 (1920), pp. 1–16.

44. *Brit. Assn. Rpt.,* 1892, p. 629.

45. Ibid.

46. *Physical Laboratories,* p. 2; "The New Physical Laboratories of the Owens College, Manchester," *Nature* 58 (1898): 621.

47. A. Schuster, *Progress of Physics During 33 Years (1875–1908)* (Cambridge, 1911), p. 17; "New Physical Laboratories," p. 622.

48. Schuster, *Progress,* p. 18.

49. "New Physical Laboratories," p. 622.

50. Ibid.

51. Ibid.

52. *Physical Laboratories,* pp. 2–3.

53. Ibid., pp. 13, 16–17.

54. Ibid., pp. 25–38; R. S. Hutton, *Recollections of a Technologist* (London, 1964), p. 39.

55. *McG,* 5 Oct. 1898, p. 10.

56. *McG,* 30 June 1900, p. 11.

57. Schuster, *Jubilee Celebration,* p. 12.

58. Fiddes, *Chapters,* pp. 95–96.

59. *McG,* 7 June 1902, p. 6; 9 June 1902, p. 9; 10 July 1901, p. 12; 11 June 1902, p. 12. See also the debate between Schuster and Arthur Smithells of Leeds, *Nature* 66 (1902): 252–54, 319–20.

60. Schuster, *Jubilee Celebration,* p. 14.

61. A. Schuster to E. Rutherford, 7 Sept. 1906, Rutherford Papers, Cambridge.

62. Rutherford to Schuster, 26 Sept. 1906, Schuster Papers, Royal Society of London. See also J. B. Birks, *Rutherford at Manchester* (New York, 1963), p. 47. Rutherford agreed with the suggestion that Schuster might be able to stay on in the area of mathematical physics.

63. Rutherford to Schuster, 18 Nov. 1906, and 3 Jan. 1907, Schuster Papers, Royal Society.

64. Rutherford to Schuster, 27 Jan. 1907, Schuster Papers, Royal Society.

65. Schuster to Rutherford, 7 Oct. 1906, Rutherford Papers, Cambridge; Birks, pp. 52, 65.

66. S. Alexander, "A Plea for an Independent University of Manchester," in *McG,* 11 June 1902, p. 12.

67. *Nature* 63 (1901): 222–23.

68. *Nature* 62 (1900): 358.

69. Quoted in *Nature* 72 (1905): 185.

Bibliographical
Note

For this study I have drawn rather extensively upon manuscript collections in Manchester, Cambridge, and London, and have consulted briefly materials drawn from other sources. The Manchester Central Reference Library possesses several important collections, including Owens College, "Minutes of the Trustees' Meetings for Educational Purposes," of which I have used volumes three through eleven; Joule-Playfair Manuscripts, 1844-1850 (copyright Manchester Literary and Philosophical Society); Mercer-Playfair Letters; Society of Chemical Industry (Manchester section), "Minutes of General Meetings (1844-1921)" and "Minutes of Committee Meetings, 1886-1932"; Manchester Chamber of Commerce, Chemical and Allied Trades Committee, "Minutes of the Meetings, 1890-1900"; Owens College, "Miscellaneous Materials"; John Graham, "History of Calico Printing"; and Banksian Society Papers.

At the University of Manchester are small collections of the papers of Henry Roscoe and Carl Schorlemmer. The University of Manchester Institute of Science and Technology (UMIST) holds an interesting collection of James Joule manuscripts and the archives of the Manchester Mechanics' Institution. The John Rylands Library, Manchester, also has unpublished Roscoe materials.

In London, at the Imperial College Library, there is a substantial collection of Lyon Playfair papers. At University College, I consulted with profit the Edwin Chadwick Papers. The Chemical Society Archives, Burlington House, possesses an important and useful collection of Roscoe correspondence. The Royal Society of London has Arthur Schuster correspondence of which I made considerable use. Mrs. R. S. Hutton and the late Professor R. S. Hutton permitted me to consult the papers of Mrs. Hutton's father, Arthur Schuster, which remain in the family's possession.

At the Cambridge University I found the papers of J. J. Thomson, Lord Kelvin, and Ernest Rutherford useful for correspondence with or about Manchester figures.

I have consulted materials drawn from the Bunsen Papers, Heidelberg University, which contains important Bunsen-Roscoe correspondence.

Many of the scarcer pamphlets and printed ephemera were consulted at the British Museum, London, the University of Manchester Library, and the Hutzler Collection, the Johns Hopkins University, Baltimore.

The literature on Manchester and its social, political, and economic background is extensive. The following sources were of special interest to me. My study was stimulated in part by remarks on working-class culture in E. P. Thompson, *The Making of the English Working Class* (New York, 1963), and by Friedrich Engels, *The Condition of the Working Class in England* [1845], trans. W. D. Henderson and W. H. Chaloner (Oxford, 1958). I made use of the translation of the Engels work in K. Marx and F. Engels, *On Britain,* 2d ed. (Moscow, 1962), pp. 1–336. On Manchester: Asa Briggs, *Victorian Cities* (New York, 1965); W. H. Brindley, ed. *The Soul of Manchester* (Manchester, 1929); W. H. Chaloner, "The Birth of Modern Manchester," in British Association, *Manchester and its Region* (Manchester, 1962); H. J. Dyos, ed., *The Study of Urban History* (New York, 1968); H. J. Dyos and M. Wolff, *The Victorian City: Images and Realities,* 2 vols. (London, 1974); Léon Faucher, *Manchester in 1844* (London, 1844); N. J. Frangopulo, ed., *Rich Inheritance* (Manchester, 1962); T. W. Freeman, H. B. Rodgers, and R. H. Kinvig, *Lancashire, Cheshire and the Isle of Man* (London, 1966); V. A. C. Gatrell, "The Commercial Middle Class in Manchester c. 1820–1857" (Ph.D. diss., Cambridge University, 1971); B. Love, *Manchester as It Is* (Manchester, 1839); Stephen Marcus, *Engels, Manchester and the Working Class* (New York, 1975); Leon Marshall, *The Development of Public Opinion in Manchester* (Syracuse, N. Y., 1946); E. W. Martin, *Where London Ends* (London, 1958); A. Redford, "The Emergence of Manchester," *History* 24 (1939): 32–49; A. Redford and I. Russell, *The History of Local Government in Manchester,* 3 vols. (London, 1939–1941); W. H. Shercliff, *Manchester: a Short History of its Development* (Manchester, 1965); Arthur Silver, *Manchester Men and Indian Cotton* (Manchester, 1966); W. H. Thomson, *History of Manchester to 1852* (Altrincham, 1967); James Wheeler, *Manchester: Its Political, Social and Commercial History, Ancient and Modern* (London, 1836).

On the economic background to this study of special use were: Derek Aldcroft and Harry Richardson, *The British Economy 1870–1939* (London, 1969); Thomas Ellison, *The Cotton Trade of Great Britain* [1886] (London, 1968); John Foster, *Class Struggle and the Industiral Revolution* (London, 1974); Henry Hamilton, *The Industrial Revolution in Scotland* (London, 1966); Jane Jacobs, *The Economy of Cities* (New York, 1969); David Landes, "Technological Change and Industrial Development in Western Europe," in *Cambridge Economic History of Europe,* vol. 6, pt. 1 (Cambridge, 1965); David Landes, *The Unbound Prometheus* (Cambridge, 1969); E. N. Mitchell, "The English and Scottish Cotton Industry." *Scottish Historical Review* 22 (1925): 101–14; P. L. Payne, *British Entrepreneurship in the Nineteenth Century* (London, 1974); Sidney Pollard, *The Genesis of Modern Management* (Cambridge, Mass., 1965); H. B. Rodgers, "The Lancashire Cotton Industry in 1840," in A. Baker, J. Hamshere, and J. Langton, eds., *Geographical Interpretations of Historical Sources* (Newton Abbot, 1970), pp. 337–56; and Andrew Ure, *The Philosophy of Manufactures* 3d ed. (London, 1861).

The scholarship on Victorian science is likewise rich. I wish to draw special attention to the following which should provide the reader with a rapid and reliable introduction. Indispensable is D. S. L. Cardwell's *Organisation of Science in England* (London, 1972). Also valuable: W. H. G. Armytage, *A Social History of*

Engineering (London, 1961); Charles Gillispie, *Genesis and Geology* (New York, 1959); J. B. Morrell, "The Chemist Breeders," *Ambix* 19 (1972): 1–46; A. E. Musson and E. Robinson, *Science and Technology in the Industrial Revolution* (Toronto, 1969); M. Sanderson, *The Universities and British Industry* (London, 1972); and Arnold Thackray, "Science and Technology in the Industrial Revolution," *History of Science* 9 (1970): 76–89.

Science in Manchester is the subject of an interesting and controversial article, written from a viewpoint very different from my own by Arnold Thackray, "Natural Knowledge in Cultural Context: The Manchester Model," *American Historical Review* 79 (1974): 672–709. A recent collection of essays relating to the history of UMIST, D. S. L. Cardwell, ed., *Artisan to Graduate* (Manchester, 1974), contains considerable new material on scientific institutions in Victorian Manchester. Still useful is Robert Angus Smith, "A Centenary of Science at Manchester," Manchester Literary and Philosophical Society. *Memoirs* 9 (1883): 1–475. On the Royal Manchester Institution, there is S. D. Cleveland, *The Royal Manchester Institution* (Manchester, 1871). On the Literary and Philosophical Society, Francis Nicholson, "The Literary and Philosophical Society 1781-1851," Manchester Literary and Philosophical Society. *Memoirs* 68 (1924): 97–148, still provides useful information. On the Statistical Society, see Thomas Ashton, *Economic and Social Investigations in Manchester 1833–1933* (London 1934), and David Elesh, "The Manchester Statistical Society," *Journal of the History of the Behavioral Sciences* 8 (1972), pt. 1, pp. 280–301 and pt. 2, pp. 407–17. On medicine in Manchester, see E. M. Brockbank, *The Foundation of Provincial Medical Education in England* (Manchester, 1936); Charles Newman, *The Evolution of Medical Education in the Nineteenth Century* (Oxford, 1957); and S. W. F. Holloway, "Medical Education in England 1830-1858, a Sociological Analysis," *History* 49 (1964): 299–324.

On mechanics' institutions and technical education: Michael Argles, *South Kensington to Robbins* (London, 1964); Elaine Storella, " 'O What a World of Profit and Delight': the Society for the Diffusion of Useful Knowledge" (Ph.D. diss. Brandeis, 1969); James Hole, *Essay on the History and Management of Literary, Scientific and Mechanics' Institutions* (London, 1853); Edward Royle, "Mechanics Institutions and the Working Class," *History Journal* 14 (1971): 305–21; Mabel Tylecote, *The Mechanics' Institutes of Lancashire and Yorkshire before 1851* (Manchester, 1957); G. Williams, *Rowland Detrosier* (York, 1965); S. F. Cotgrove, *Technical Education and Social Change* (London, 1958); Thomas Kelly, *George Birkbeck* (Liverpool, 1957); R. B. Hope, "Education and Social Change in Manchester 1780-1851" (M.Ed. thesis, Manchester, 1955).

On Owens College and the Victoria University of Manchester: Joseph Thompson, *The Owens College: its Foundation and Growth* (Manchester, 1886): H. B. Charlton, *Portrait of a University 1851-1951* (Manchester, 1951); P. J. Hartog, *The Owens College Manchester* (Manchester, 1900); E. Fiddes, *Chapters in the History of Owens College and of Manchester University, 1851-1914* (Manchester, 1937); W. H. G. Armytage, *The Civic Universities* (London, 1955); A. J. C. Magian, *An Outline of the History of Owens College* (Manchester, 1931); James Heywood, "The Owens College, Manchester and a Northern University," *Journal Statistical Society* 41 (1878): 536–47; A. Angellier, "Etudes sur Owens College,"

Bulletin, Société pour l'Étude des Questions d'Enseignement Supérieur 1 (1880): 367-480; George Haines, *Essays on German Influence upon English Education and Science 1850-1919* [New London, Conn.], 1969; *Record of the Jubilee Celebrations at Owens College, Manchester* (Manchester, 1902); R. W. Bailey, "The Contributions of Manchester Researches to Mechanical Science," Institution of Mechanical Engineers *Proceedings*, June, 1929, 613-83; *The Physical Laboratories of the University of Manchester* (Manchester, 1906); J. B. Birks, *Rutherford at Manchester* (New York, 1963). George Gissing's novel *Born in Exile* (London, 1894), has some biting scenes of Owens College.

On the chemical industry in Manchester and its background, the following are very useful: C. M. Mellor and D. S. L. Cardwell, "Dyes and Dyeing, 1775-1860," *British Journal History of Science* 1 (1963): 265-79; S. Miall, *History of the British Chemical Industry* (London, 1952); A. Clow, "Industrial Background," in D. S. Cardwell, *John Dalton and the Progress of Science* (Manchester, 1968); D. W. F. Hardie and J. D. Pratt, *History of the Modern British Chemical Industry* (Oxford, 1966); F. S. Taylor, *A History of Industrial Chemistry* (London, 1957); L. F. Haber, *The Chemical Industry During the Nineteenth Century* (Oxford, 1958); A. S. Irvine, *A History of the Alkali Division, formerly Brunner, Mond and Co. Ltd.* (n.p., 1958); *Perkin Centenary London: 100 Years of Synthetic Dyestuffs* (New York, 1958); Donald McCloskey, ed., *Essays on a Mature Economy: Britain after 1840* (London, 1971), especially P. Lindert and K. Trace, "Yardsticks for Victorian Entrepreneurs," pp. 239-74; R. B. Pilcher, *The Profession of Chemistry* (New York, n.d.); J. Fenwick Allen, *Some Founders of the Chemical Industry* (London and Manchester, 1907); Ross Hoffman, *Great Britain and the German Trade Rivalry 1875-1914* (Philadelphia, 1933); Walter Gardiner, ed., *The British Coal Tar Industry* (London 1915); W. A. Campell, *The Chemical Industry* (London, 1971); T. C. Barker, R. Dickenson, and D. W. F. Hardie, "The Origins of the Synthetic Alkali Industry in Britain," *Economica* 23 (1956): 158-171; F. M. Rowe, *The Development of the Chemistry of Commercial Synthetic Dyes (1859-1938)* (London, 1938); H. W. Richardson, "The Development of the British Dyestuffs Industry Before 1939," *Scottish Journal of Political Economy* 9 (1962): 110-29; Paul Hohenberg, *Chemicals in Western Europe 1850-1914* (Chicago, 1967).

Some of the major scientific figures in Manchester (Dalton, Joule, Frankland, Playfair, Smith, Reynolds, Roscoe, Schorlemmer, Schuster, Lamb, Stewart) are included in the recent *Dictionary of Scientific Biography*, Charles Gillispie, ed. (New York, 1970-). Each is worthy of fuller study than can be given to any individual by my study of the Manchester community. Of special note are the following: Elizabeth Patterson, *John Dalton and the Atomic Theory* (Garden City, N.Y., 1970); D. S. L. Cardwell, ed., *John Dalton and the Progress of Science* (Manchester, 1968); A. L. Smyth, *John Dalton 1766-1844, a Bibliography of Works by and about Him* (Manchester, 1966); Arnold Thackray, *John Dalton* (Cambridge, Mass., 1972); Frank Greenaway, *John Dalton and the Atom* (Ithaca, N.Y., 1966); Alfred H. Gibson, *Osborne Reynolds and his Work in Hydraulics and Hydrodynamics* (London, 1946); J. D. Jackson and D. M. McDowell, eds., *Osborne Reynolds and Engineering Science Today* (Manchester, 1970); T. Edward Thorpe, *The Right Honourable Sir Henry Enfield Roscoe* (London, 1916);

D. Thompson, "The Influence of Sir H. E. Roscoe on the Development of Scien-
and Technical Education During the Second Half of the Nineteenth Century"
(M.Ed. thesis, Leeds, 1958); M. N. W[est] and A. J. C[olenso] eds., *Sketches
from the Life of Edward Frankland* (London, 1902); W. Pole, *Life of Sir William
Fairbairn, Bart.* (London, 1877); Osborne Reynolds, *Memoir of James Prescott
Joule* (Manchester, 1892), published as Manchester Literary and Philosophical
Society *Memoirs and Proceedings* 6 (1892): 1-191; L. Rosenfeld, "Joule's Scien-
tific Outlook," *Bulletin of British Society for the History of Science* 1 (1952):
169-76; G. Jones, "Joule's Early Researches," *Centaurus* 13 (1968): 198-219;
Arthur Schuster, *Biographical Fragments* (London, 1932); Arthur Schuster,
"Memoir of the Late Balfour Stewart, LL.D., F. R. S.," Manchester Literary and
Philosophical Society *Memoirs* 1 (1880): 253-72; W. V. Farrar, K. Farrar, and E.
L. Scott, "The Henrys of Manchester," *Royal Inst. Of Chem. Reviews* 4 (1971):
35-47; idem., "The Henrys of Manchester," pt. 1: *Ambix* 20 (1973): 183-208;
pt. 2: *Ambix* 21 (1974): 179-207; pt. 3: *Ambix* 23 (1976): 27-52; Roy Mac-
Leod, "The Alkali Acts Administration 1863-1884: the Emergence of the Civil
Scientist," *Victorian Studies* 9 (1965): 85-112; A. Gibson and W. V. Farrar,
"Robert Angus Smith, F. R. S. and 'Sanitary Science'," Royal Society of London,
Notes and Records 28 (1974): 241-62.

I have used a number of books relating to the sociology of the professions.
Among those not cited above which are of special utility are the following: W. J.
Reader, *Professional Men* (New York, 1966); Kenneth Prandy, *Professional Em-
ployees* (London, 1965); J. A. Jackson, *The Professions and Professionalization*
(Cambridge, 1970); John Rayner, *The Middle Class* (London, 1969).

Index

Fleming, Thomas, 67
Fleming, William, 25, 32, 44
Fletcher, Samuel, 154, 155, 160
Flinn, Michael, 111
Forbes, James D., 215
Former, John, 137
Foster, John Frederick, 155, 160
Fowler, Gilbert J., 212
Fox and Brothers, 89
France, 198, 199
Frankfurt, 221
Frankland, Edward, 35, 126, 140, 144,
 156, 158-67, 170, 174, 175, 187, 200,
 201, 232, 235
Frankland, Percy, 200
Free Lance, 73
Fremy, E., 156
Fyfe, A., 68

Galloway's (Engineers), 183
Gamgee, Arthur, 194
Gardiner, N., 70, 71
Gardner, J., 106
Gay-Lussac, J., 101, 124
Gee, William W. Haldane, 222
Geiger, Hans, 234
General Council of Medical Education
 and Registration, 172
Geological Society. *See* Manchester Geo-
 logical Society
Geological Society of London, 30-32, 61,
 62; *Journal,* 63
Geological Survey, 62, 88, 120
Gerardin, A., 97, 100
Gerland, B. Wilhelm, 163
Germany, 47, 48, 94, 95, 197-202
Giessen, University of, 87, 89, 95, 96, 97,
 100-103, 118, 133, 151, 156, 160,
 161, 195
Gilbert, J. H., 87, 97, 100, 101, 103
Gladstone, Murray, 75, 193
Glasgow, University of, 86, 100, 156,
 182, 188
Glasgow Herald, 182
Gobelins, 97
Gossage, William, 126, 127, 137
Göttingen, University of, 15
Gough, John, 12, 28
Graebe, Carl, 199, 202
Graham, John, 94, 100
Graham, Sir John, 95
Graham, Thomas, 53, 55, 59, 87, 89, 91,

93, 94, 97, 100, 103, 124, 151, 156,
 163, 167
Grainger, R. D., 114, 119
Granville, Lord, 201, 202
Greenford Green Works, 197, 199, 202
Greenough, G. B., 62
Greenwood, J. G., 156, 166, 169, 173, 188,
 190, 192, 214, 215, 218, 219
Greg, R. H., 21, 31, 46, 154
Greg, W. R., 31
Gregory, O. G., 39
Gregory, William, 92, 105, 106, 118, 156
Griess, Peter, 198, 204, 209
Griffin, R., xi
Griffiths, Albert, 225
Grimshaw, Harry, 179, 205, 207, 209
Grindon, Leo, 80-85, 147, 194
Grote, A., 169
Grove, W. R., 35
Guardian. See Manchester Guardian
Guinon, Marnas and Bonnet, 142-43
Guthrie, Frederic, 163
Guy's Hospital (London), 67, 170
Gwyther, R. F., 222

Hagstrom, W., 58, 79
Haldane, R. B., 237
Hall, John, 27
Harden, Arthur, 212
Hardie, Edward, 14
Hargreaves, John (chemist), 204
Hargreaves, John (inventor), 21
Harrison, Ralph, 7
Hart, Jennifer, 132
Hart, Peter, 139, 141, 205
Hart, William, 137, 138
Hartog, Marcus M., 195
Harvey, A. W., 87
Hassall, A. H., 111, 122
Hastings-Birkbeck Bill (1885), 206
Hawkshaw, John, 31, 37, 38
Health of Towns Commission, 91, 124
Heelis, T., 85
Heeren, F., 96
Heidelberg, 92, 162, 163, 167, 192, 195,
 210, 212, 221
Helmholtz, H., 228, 229
Hemsalech, G. A., 225
Henry, Thomas, 7-11, 13, 15, 18, 27-29,
 31, 34, 35, 42, 44, 65, 76, 93, 136, 235
Henry, William, 10, 13, 15, 28, 31, 34, 35,
 42, 65, 68, 69, 159

The Johns Hopkins University Press

This book was set in Compugraphic Baskerville text and display type by Naecker Bros. Associates, Inc., from a design by Susan Bishop. It was printed on 50-lb. Publishers Eggshell Wove paper and bound in Holliston Arrestox cloth by Universal Lithographers, Inc.